SIGNALS AND SYSTEMS FOR SPEECH AND HEARING

Signals and Systems for Speech and Hearing

Second Edition

by

Stuart Rosen
UCL Speech, Hearing and Phonetic Sciences, London, UK

and

Peter Howell
UCL Cognitive, Perceptual and Brain Sciences, London, UK

BRILL

LEIDEN • BOSTON
2013

Originally published by Emerald Group Publishing Limited, Bingley, UK, in 2011
under ISBN 978-1-84855-226-5.

Library of Congress Cataloging-in-Publication Data

Rosen, Stuart.
Signals and systems for speech and hearing / by Stuart Rosen, UCL Speech, Hearing and Phonetic Sciences, London, UK, and Peter Howell, UCL Cognitive, Perceptual and Brain Sciences, London, UK. -- Second edition.
pages cm
Includes bibliographical references and index.
ISBN 978-90-04-25243-1 (pbk. : alk. paper) 1. Hearing. 2. Speech. 3. Auditory pathways. 4. Speech processing systems. I. Howell, Peter, 1947- II. Title.

QP461.R665 2013
612.8'5--dc23

2013009991

ISBN 978-90-04-25243-1 (paperback)

This book is printed on acid-free paper.

For Jan and Zach and Sam
xoxo $\rightarrow \infty$

For Julie and David and Eleanor

Contents

Preface

It is now almost 20 years since this book first appeared. In the intervening years, we have often toyed with the idea of revising the original text, especially in the light of the dramatic changes in the instrumentation now available to even the most casual user of computers. The widespread use of cheap digital computers and electronics mean that more or less anyone can perform extremely sophisticated analyses and syntheses of sound. Given these developments, it is perhaps more surprising how little of the book *required* revision. After all, Fourier's theorem is still as valid as it was when first proposed at the beginning of the 19th century!

What has changed the field immeasurably is the nearly complete replacement of analogue techniques by digital ones in the recording, manipulation, storage and transmission of signals. This has been reflected in changes throughout the book in the kind of instrumentation described. Furthermore, two chapters have been heavily revised. Chapter 11, dealing with spectrograms, has been much extended and describes the two different ways spectrograms can be constructed—through filter banks and time windowing—and the relationship between them. Chapter 14, dealing explicitly with digital signals and systems, has been expanded to give concrete examples of digital systems and digital signal processing, including the notion of infinite impulse response (IIR) and finite impulse response (FIR) filters.

Other changes we have made are in response to our own thinking and extensive teaching experience. One of us (PH) has used the book as the basis for a course on speech production aimed at undergraduate psychologists for 20 years. The other (SR) has taught a one-term (10 week) course based on it for 7 years, mostly to trainee audiologists. SR's experience has led to a substantially revised Chapter 12, which now focuses on the notion of the auditory periphery as a set of systems, showing how its functions are analogous to those used in making a spectrogram. Also, a new section has been added to Chapter 10 concerning windowing and spectral splatter, as these are topics of great practical importance.

Keeping up with modern developments, there is now a web site for the book, in which we plan to include additional material, and deal with questions from puzzled readers, who we hope are not too numerous! This can be found at: http://www.emeraldinsight.com/promo/signals.htm.

As usual, many people assisted us in this task. Special thanks must go to Katharine Mair, who read and made extensive useful suggestions about much of the more heavily revised material; Sam Evans, who recorded most of the speech sounds used to make the spectrograms in Chapter 13; and Kim Foster, from Emerald, who first took us on as our editor, and ensured this revision was seen through.

Finally, we would like to thank our families for their support and understanding during a period in which this revision weighed heavily on both of us, certainly causing additional stress in already very busy lives.

Stuart Rosen and Peter Howell

Preface to the Original Edition

Many people working in the speech and hearing sciences come from the arts or humanities, which makes it difficult for them to master the essential technical underpinnings of the area—in essence, linear systems analysis. Even the two of us, with much mathematical and scientific training (in addition to backgrounds in experimental psychology), found it difficult to come to grips with many of the technical concepts central to the understanding of speech and hearing.

Perhaps the main barrier we've found to acquiring knowledge in this area is that available textbooks nearly always assume an audience consisting of students of engineering. Such students would already have a reasonable amount of training in electrical network theory and differential equations, and, of course, these textbooks take that for granted. It is relatively difficult, even with sufficient mathematical expertise, to get much out of these books because of this assumed background. For readers without a knowledge of calculus, it is well-nigh impossible.

The problem of patchy understanding of basic technical concepts is acknowledged, at least implicitly, in most books dealing with speech and hearing. Nearly all begin with an introductory chapter purporting to cover the topics explored in this book—from relatively straightforward ones like the nature of sine waves and the calculation of decibels to the much more complex ideas of Fourier analysis and synthesis. Our experience has been that only those who already know this material could hope to appreciate its significance, and they do not need to read such a chapter. It seemed to us that only an entire book could cover in a sensible and thorough way the main concepts required. Furthermore, although we realized that there are some students who could handle a highly mathematical text, there are many who would have no mathematical training beyond algebra and trigonometry. Therefore, we decided to try to tell the entire story in words and pictures, with only minimal mathematics. In fact, for the primary audience this book is aimed at (trainee speech and language therapists, audiologists, phoneticians and psychologists) the need, for example, to explicitly calculate a Fourier transform will rarely or never occur. On the other hand, such people do need a good understanding of what a Fourier transform does, and how it applies to speech production, the peripheral auditory system, the operation of hearing aids and so on.

What we have attempted here is to provide the reader with a thorough introduction to the concepts of signals and systems analysis that play a role in the speech and hearing sciences. Few equations are used, and we have tried to maintain an informal, friendly and informative style throughout. Readers who like their technical material straight will, we hope, forgive us our little jokes. Because much of the story is told through figures, we have gone to great lengths to provide figures that show what the text says they do! Many is the time we have looked at other textbooks to find sinusoids that are clearly not sinusoidal, representations of speech signals that are physically impossible, or what can only be called 'impressions' of particular waveforms. We hope the reader will come away with a strong visual understanding of the concepts involved, and so have tried to provide as clear and truthful graphs as we could. An appendix at the back of the book gives details about the construction of the figures where it seems appropriate, in addition to pointers to the literature relevant to topics covered.

This book can be used at many levels, from the student who hasn't heard of a spectrum before, to the experienced worker who has only a fuzzy understanding of the notion of an impulse response. We have tried to keep the underlying conceptual structure of signals and systems analysis explicit, in the hope that even some readers with advanced technical training might find clarification of the basic principles.

Similarly, the book can be used in a number of ways. For those with some appropriate background, topics may be delved into at random. For most beginning students, Chapters 1–8 should provide a basic set of concepts for further work in the field, with Chapters 9 and 10 a desirable addition. Chapters 11–13 are more applied than the previous chapters, in that relatively few new concepts are introduced. Here, the previously developed ideas are put through their paces, with much basic hearing and speech science being developed on the way. Finally, Chapter 14, an introduction to digital signals and systems, may be treated as somewhat of an appendix. Digital techniques can never wholly replace more traditional analogue ones in the speech and hearing sciences, most importantly because we can only produce and listen to analogue signals. On the other hand, digitally based equipment is becoming more and more important in all fields related to acoustics, and therefore there is often the need for at least a rudimentary understanding of the principles involved. Most of the basic concepts in the digital area (e.g. spectra, impulse responses and filtering) are, at least at the level of this book, simple extensions of the ideas developed for analogue signals and systems. Therefore, only the most basic aspects of digitization (including quantization, sampling and aliasing) are discussed.

Although convention requires authors of books to be listed in a certain order, this book has been a joint and equal effort from start to finish. We hope, however, that our separate contributions, reworked in many versions, have fused into a cohesive, uniform whole.

With a book that has had so long a gestation, there are many people who have lent assistance along the way, only a few of whom can be mentioned here. First thanks must go to Cathy Weir, who read the entire manuscript at an early stage, and made many useful suggestions for improvement. Bridget Allen did the same with a nearly final version. Mark Huckvale gladly helped with problems related to a purpose-designed computer-based system for drawing figures. Helen Jefferson-Brown provided important assistance at many points along the way. We also thank the technical staff of our respective departments at University College London for varied help in setting up equipment, in sorting out problems and in drawing figures, in particular, Steven Nevard, Mahen Goonewardane and David Cushing at the Department of Phonetics and Linguistics, and Jonathan Hunwick and Tim Aspden at the Department of Psychology. Andrew Carrick at Academic Press, the only one of our editors to have survived this book's long gestation, managed to cajole us into finishing. We thank him for a consistent, yet never overwhelming, pressure forward. Finally, we thank our families. Never before has it been so clear to us the sacrifice this customary phase reflected. Thanks to Jan Eaton for insisting on giving up family Sundays so this book could be finished; to Julie Howell for similar sacrifices; and to Zachary and Samuel Eaton-Rosen, and David and Eleanor Howell, for time lost with their respective Dads.

Stuart Rosen and Peter Howell

simple extensions of the ideas developed for [illegible] systems [illegible] of the [illegible] aspects of [illegible] (inclusion [illegible] and [illegible] [illegible]).

[illegible]

With a book that has had so long [illegible], there are many people who have helped [illegible] [illegible] to Cathy Wan [illegible] [illegible] Steven [illegible] [illegible] Department of [illegible] Department of Psychology [illegible] Carrick at [illegible] Press, [illegible] [illegible]

[illegible]

CHAPTER 1

Introduction

Flipping through this book for the first time, you may be wondering what it could possibly have to do with speech and hearing, with all the graphs and equations and something about 'signals' and 'systems' on the cover. 'Why should I be studying such a subject?', you may well ask. These feelings are probably particularly prevalent in the minds of just those this book is aimed at—students learning to be audiologists, speech and language therapists, psychologists and phoneticians. Signals and systems sound too much like topics in electrical engineering, which, indeed, they are. Nevertheless, the ideas we'll discuss are of growing importance to many disciplines, providing general methods applicable to problems other than those normally thought of as being in the domain of the engineer.

Let's first get a rough idea about what signals and systems are by considering an example. You'll be aware that the same MP3 file sounds different when played on different hi-fi stereo 'systems'. The designer of hi-fi equipment needs to know how differences in the physical properties of the 'systems' involved (for example, an amplifier or loudspeaker) are related to how they sound, to help him or her design better-sounding hi-fi 'systems'. Generally, speaking, the hi-fi designer is interested in the accurate reproduction of 'signals'. For the moment, it's sufficient to understand what's meant by a signal in only the most general way. When talking about hi-fi, the signal could be the sound coming out of a loudspeaker, electromagnetic waves transmitted from a radio station, or the information contained in an MP3 file that allows you to reproduce a piece of music. The hi-fi designer wants to know what a particular piece of equipment (a 'system') will do to the signals so as to assess the fidelity of the final result.

We always make a distinction between *input* and *output* signals. Suppose our hi-fi designer is working on a new version of an MP3 player. For the designer, the input signal will be the music downloaded from the internet in digital form. The output signal will be the sound coming from the headphones. In this case, the *system* will be nothing other than the entire MP3 player.

Our designer (and buyer) needs a simple and objective way of understanding what happens to the input signal from the encoded file during its transmission through the equipment. In this way, she or he can specify precisely what output signal will result. Different systems can be compared (with proper regard for economic factors) and present systems can be improved. Signals and systems analysis does just this by characterizing in a compact way what a system will do to *any* input signal.

Of course, this is only one example. Here, the digital signal stored in the MP3 file is transformed into sound by electrical equipment. But the techniques we'll develop can be used for a very general class of systems which operate on mechanical, acoustic or even optical signals (although the last will not concern us). For example, DJs still often use vinyl records in their mixing. Here, a turntable produces sound by converting the grooves on a record to an electrical signal which can then be amplified and played from a loudspeaker. The engineer interested in the accuracy with which the needle on the turntable can track the grooves in the record will also use techniques of signals and systems analysis, but here a mechanical, not an electrical, signal is involved.

From the examples that have been considered so far, it might be thought that signals and systems analysis is only important in showing how accurately a signal can be reproduced. This is not so. It is often desirable to deliberately alter a signal in some specific way. A bugle, for example, is designed to alter the sound of the player's buzzing lips to achieve a clarity and brightness of tone.

These types of applications alone suffice to make a study of systems and signals of enough interest, because workers in the speech and hearing sciences depend on the proper functioning of much equipment and need to be aware of its limitations. A good example is the research currently being done in speech perception using speech synthesizers. It's difficult, if not impossible, to do speech synthesis well without some knowledge of signals and systems. But the appropriate analysis of signals and systems allows much more, being applicable not only to the artificial systems used by researchers but also to the very 'signals' and 'systems' they are interested in.

Workers interested in hearing want to know what happens to a sound (the input signal) as it passes through the peripheral auditory system. Others are interested in the signals humans produce. The production of speech can be modelled as an input signal (either the buzz of the larynx vibrations or the hiss of turbulent noise when two articulators come close together) modified by the vocal tract system to produce the output. The shape of the vocal tract determines what speech sound is produced in something like the way that a bugle changes the noise of vibrating lips.

It isn't only aspects of normal functioning that can be approached in this way. The simplest type of hearing aid operates as a crude amplifier. Whether a particular device will be of any use to a person suffering from a hearing loss depends on the degree of hearing loss and the performance of the hearing aid. Crucial aspects of these questions can be examined using signals and systems analysis. For example, it's possible to define the way in which the complex sounds of speech are transformed by a hearing aid from relatively few simple measurements.

Let us not forget people who are impaired in their ability to *produce* speech. When a person has the larynx surgically removed (an operation known as laryngectomy), the re-routing of the breathing passages from the mouth to a hole in the neck means that there is no longer any sound source to make speech. So, although the rest of the vocal tract can function pretty much as normal, nothing can be heard. Some patients learn to use esophageal speech, a sort of controlled burping that substitutes (although imperfectly) for the normal action of the vocal cords. Others never learn this technique and resort to a hand-held vibrator (known as an 'artificial larynx') on the throat. Most only vibrate at one frequency, so the speech sounds monotone, like that of a robot. Also, the quality of speech generated in this way is said to be very 'buzzy'. Signals and systems analysis can help to explain not only the differences between esophageal and normal speech, but also why 'artificial larynxes' make speech sound monotone and buzzy, and can help in the attempt to make them produce more natural speech.

Thus, signals and systems analysis is important in two ways. It helps us to choose appropriate equipment or techniques for our work, and it offers a way to better understand the signals and systems under study.

In the following chapters, the general approach involved in the analysis of signals and systems will be developed. Examples will be drawn mainly from the peripheral auditory processing of sounds and the mechanisms of speech production. From time to time, we'll also consider the equipment used in the study of speech and hearing. This is not, however, a book about the speech and hearing sciences *per se*, and our coverage of topics will be highly selective. It may therefore seem at times that the material covered is rather remote from what you are really interested in. All we can say is 'stick with it'. By the time you get to Chapters 11–13, you'll have enough background to tackle, in a sophisticated way, some very important issues in the speech and hearing sciences. Those chapters will serve as a culmination and will show in just how many ways the ideas that have been developed can be put to use.

Exercises

1. Think about making an audio recording on to your computer using a microphone. What is the input signal to the microphone, and what is its output?

2. If that recording is now played over a loudspeaker, what is the output signal?

3. Analyze a telephone conversation in terms of the signals and systems involved.

4. Write a short essay describing the relevance of signals and systems to some aspect of speech and/or hearing. Try to use examples not already given in this chapter.

CHAPTER 2

Signals in the Real World

The techniques of analysis that we're going to develop through this book are of a very general nature, and hence they require fairly abstract descriptions. It would be easy, in the face of all this abstraction, to lose sight of the fact that our ultimate aim is to further our understanding of things happening in the 'real world'. So, although we'll talk in a general way about 'signals', don't forget that those 'signals' normally refer to some real physical event or chain of events.

As our interests are primarily to do with speech and hearing, you might think that we need only concern ourselves with the class of signals known as *sound*. As you already realize from Chapter 1, though, this isn't so. Sound is always produced by moving objects (like vocal folds and loudspeaker cones), and often transmitted by them (like the vibrating middle ear bones), so we also need to describe how objects move in time (*mechanical* signals). Furthermore, much equipment we use transduces acoustic signals into *electrical* ones, making them easier to record, manipulate and reproduce. In this chapter, we want to show how these three different types of signals can be represented in a common form, making all amenable to the techniques of analysis described later on.

The movement of a tuning fork

Whenever sound occurs, some object must move or vibrate to produce it. This can be illustrated with a simple instrument—the tuning fork. One way to make a tuning fork emit sound is to 'twang' the prongs by squeezing them together and then releasing them, as seen at the top of the next page.

Let's consider the movement of a single prong in detail. Squeezing the tuning fork causes the prong to be displaced inward from its resting position where its springiness makes it prone to return to its initial position. Thus, when the prong is released, it begins to move outward. This movement continues as the prong passes through its resting position, since it is

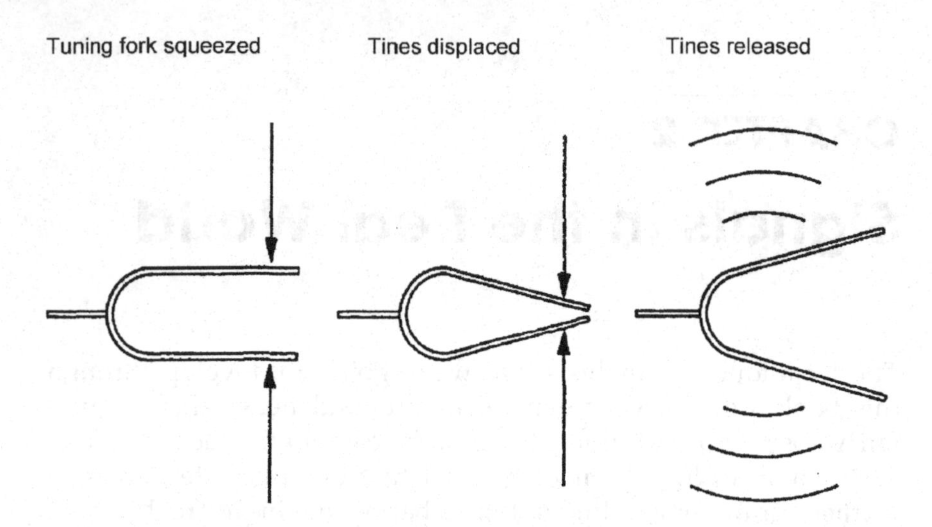

moving at some speed. As the prong moves further and further out from its resting position, the now inward-directed force due to its springiness gets stronger and stronger. This 'restoring force' makes the prong slow down and eventually come to a halt. The position that it stops in isn't the resting position though; it is somewhat outward:

The inward force due to the prong's springiness keeps acting, however, causing the prong now to move *inward* towards its resting position. Just as when the prong was moving outward, however, it passes right through its resting position, and comes to a halt at a position somewhat inward:

The outward restoring force asserts itself once more, and the cycle begins again. If there were no frictional forces operating, this back-and-forth motion would continue indefinitely. Both prongs of the fork would move in exactly the same way, of course, which is why it is only necessary to look at one. Thus, the complete cycle of prong displacement would look like this:

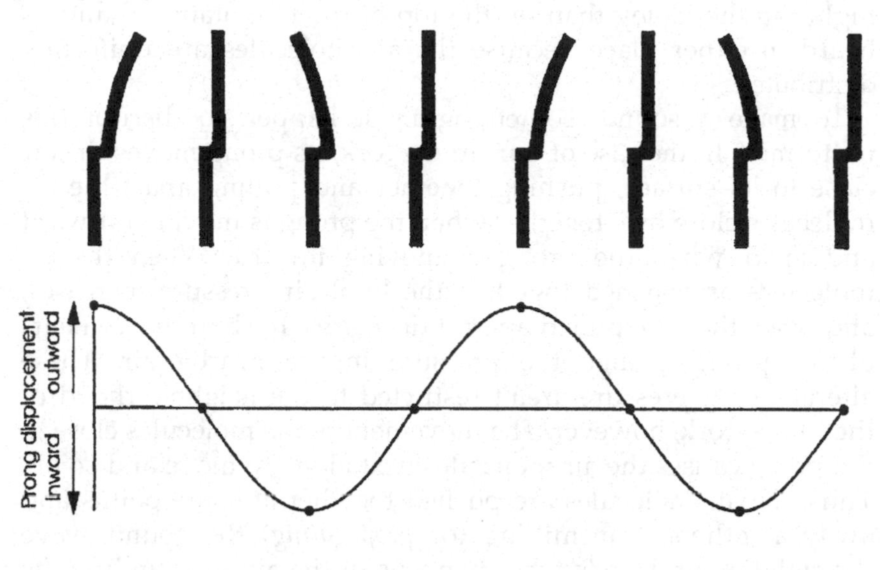

This is one example of a signal—a *mechanical* one, as we're dealing with the motion of an object. Note the axes upon which it is drawn. The horizontal axis, or x-axis, represents time, measured in seconds, or convenient fractions thereof. The vertical axis, or y-axis, represents the position of the prong relative to its resting state, and is measured in metres (or feet) or convenient fractions thereof. Sometimes the prong is to the left of its resting position, and sometimes to the right. We have used the positive y-axis to represent the position of the prong when it is to the right of its resting state, and the negative y-axis to represent its position to the left of the resting state.

What is sound?

So far, all we know about the tuning fork is that it moves in a certain way. This movement is a bit too fast for us to see, so all we experience is a blur. The more important subjective feature of a tuning fork, though, is that it makes a *sound*. How are these movements of the fork converted into sound and transmitted to our ears, causing us to hear something?

Sound is generated when the normally uniformly distributed air molecules are pushed together and apart by a vibrating object. It is the varying distribution of the air molecules in space and time that causes sound to be heard, not the overall density of the molecules *per se*. In other words, if the air molecules are uniformly distributed, no matter what their density, no sound is heard. Consider standing in a quiet valley or on the top of a mountain on a still day. Although the density of the air molecules is higher in the valley than on the top of the mountain, nothing is heard in either place because the air molecules are uniformly distributed.

To make a sound, something must happen to disrupt this uniformity. In the case of our tuning fork, its prong moves the air close to its surface, pushing together and pulling apart the air molecules close by—together when the prong is moving outward and apart when the prong is moving inward. When the air molecules are pushed together, the local air pressure increases, and when they are pulled apart it decreases. In short, movements of the prong change the pressure in the nearby air. These alterations in pressure aren't restricted to the neighbourhood of the tuning fork, however. The movement of the molecules close to the prong causes the air molecules next along to move and so on. Thus, the air molecules are pushed together at some points and away at others, transmitting (or *propagating*) the sound wave through the air. Here's what happens in the air surrounding the tuning fork at a particular instant some time after 'twanging':

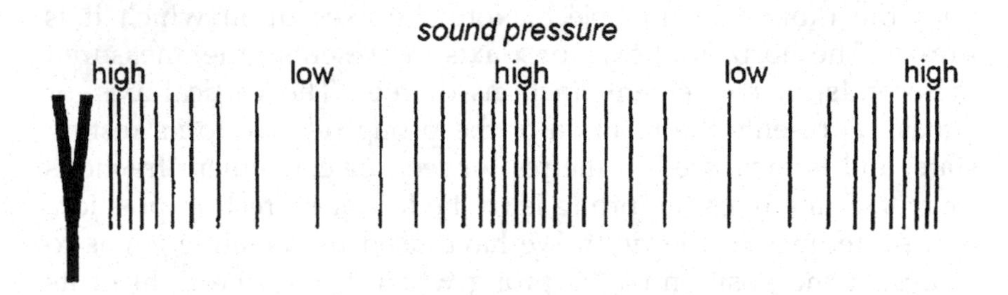

Regions of high pressure are indicated by lines that are close together, while regions of low pressure are indicated by lines spaced fairly widely apart. Note the area of high pressure right at the prong which occurs because we have taken this 'snapshot' at a moment when the prong is positioned outward from its resting state.

In this way the sound wave can be transmitted a considerable distance. Although the sound wave may travel far, the air molecules themselves don't travel all this way. They do move somewhat from their resting position but the wave itself is passed on to the molecules like the wave passing down a whip when it is cracked, or a 'Slinky' toy when it is stretched out and jiggled. Just as the thong of the whip doesn't move away from its handle, nor the rings of the 'Slinky' travel down the room, the air molecules stay in roughly the same place, passing on to their neighbours energy acquired from the tuning fork.

But just as a wave in a whip requires a thong, and a 'Slinky' requires a metal spiral, so the passage of a sound wave requires a *medium of transmission*. Usually, the medium is air, and since it cannot be seen we are normally unaware of it. It has to be there, though. Sound from a vibrating object cannot be heard through a vacuum, as there are no molecules to pass along the changes in pressure.

The alternate increases and decreases in air pressure happen all the while the prong vibrates back and forth. If the air pressure was measured at a point some fixed distance away, it would show a pattern similar to the movement of the tuning fork prong:

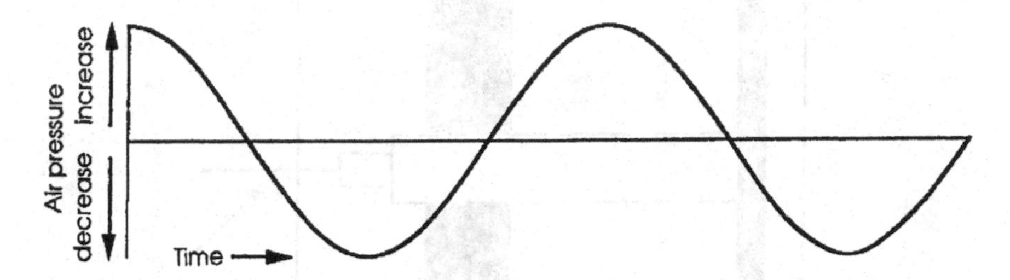

This picture is just like the one we showed for the movement of the tuning fork. It too is a signal, but, since it deals with pressure changes, it is an *acoustic* signal. The *x*-axis still represents time, but now the *y*-axis represents pressure, relative to the normal pressure of the air. Thus, if the air pressure is higher than it would be without the sound, it has a positive value. When the pressure is lower than normal, it is negative. Pressure used to be measured (and sometimes still is) in pounds per square inch, a measure of *force* divided by a measure of *area*. Nowadays, all acoustic work has been 'Europeanized', and so we use the metric unit known as the *Pascal* (Pa). This is equivalent to one Newton per square metre, again a measure of force divided by a measure of area.

Converting pressure changes into a more convenient form

Acoustic signals are clearly important to us, as it is pressure changes in the air that we hear as sound. Unfortunately, they are a very inconvenient kind of signal because they cannot be stored or manipulated very easily. It would be much better if this acoustic signal could be converted into an electrical one, at the same time preserving all the essential information. This is exactly what a *microphone* does, a device designed to convert the air pressure changes into variations in voltage. Generally speaking, any device that converts one form of energy into another (e.g. pressure changes to electrical ones, as here) is known as a *transducer*.

There are several different sorts of microphones that are used for this task, but the one common factor is that they first change the acoustic signal into a mechanical one. To make our discussions concrete, we will discuss one particular type which is popular for research—a *condenser microphone*. Inside a condenser microphone are two plates, one of which is permanently fixed to the housing of the microphone, while the other is movable:

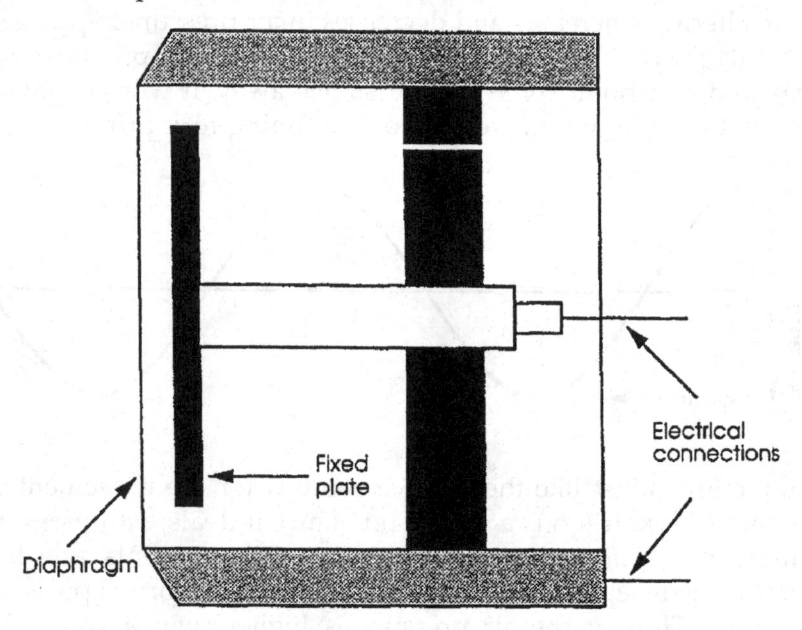

When it is held in a sound field, variations in air pressure at the face of the microphone cause the movable thin plate (the *diaphragm*) to move because the air pressure between the two plates is held more or less constant. When the local air pressure is above atmospheric, the diaphragm moves towards the fixed plate, and when the local air pressure is below atmospheric it moves away from the fixed plate. So far, so good. We've converted one mechanical signal (the movement of the tuning fork) into another (the movement of the microphone diaphragm). This doesn't seem to have gained us much!

In fact, the crucial thing about a microphone is that we can convert this diaphragm movement into an electrical signal. Both plates in the microphone have electrical connections which result in a circuit element known as a capacitor or *condenser* (hence the name). This circuit works in such a way as to produce a variation in voltage that is directly proportional to the variations in air pressure at the face of the microphone. If we measured the electrical output of our microphone to the mellifluous sound of our tuning fork, then, we would find this:

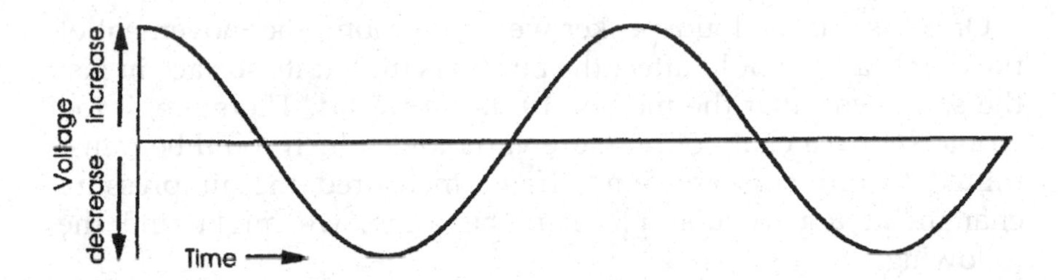

The *x*-axis, as always when we discuss waveforms, represents time. The *y*-axis, though, is specified in *volts*, as we're dealing with an electrical signal. The crucial thing about this electrical signal is that it looks just like the variations in air pressure from the tuning fork. Therefore, it has the same information. The advantage of the electrical signal over the acoustic one is that voltages can be relayed to a measuring or recording device (like a computer), allowing easier storage and manipulation.

Generally speaking, other types of microphone have a similar relationship between sound pressure variation and voltage, although they differ in the way this conversion is performed. The essential point is that the variation in voltage at the output of the microphone is an accurate representation of the variation in air pressure. As you might expect, the better the microphone, the closer the voltage waveform matches the pressure waveform.

Getting back to acoustic signals

Although electrical signals are convenient from many points of view, they cannot be listened to directly. They need to be converted back into air pressure variations. This is usually done with a *loudspeaker*, a device that works something like a microphone in reverse. Variations in electrical current fed to a loudspeaker cause equivalent movements of its diaphragm. Therefore, if we looked at the position of the loudspeaker diaphragm relative to its resting point when it was fed with the electrical signal representing the sound of our tuning fork, we'd get a curve that looked like this:

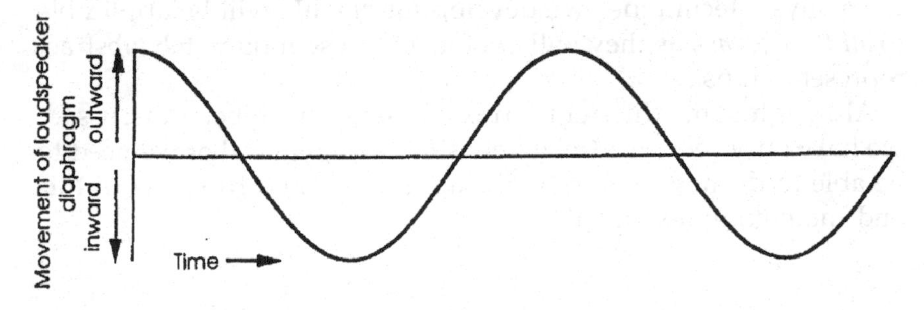

Of course, if the loudspeaker were in a room, the movement of the diaphragm would affect the air molecules at its surface in just the same way that the moving tuning fork did. Therefore, it too would set up a chain of pressure variations which could be transmitted to our ears as sound. If we measured the air pressure changes at a particular place in the room, we might find the following:

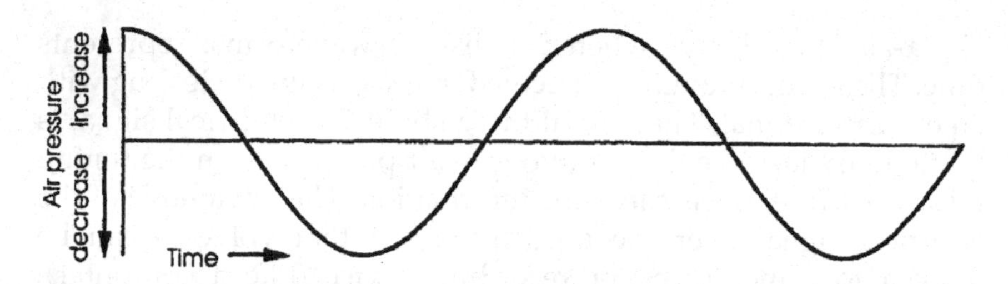

In this way, we could reproduce the sound of the tuning fork. Considering this example casually, it may not be readily apparent what the use of such a system could be. All we've done is gone from one set of air pressure variations (caused by the tuning fork) to another (caused by the loudspeaker). However, if we could greatly *amplify* the extent of the air pressure variations, we could make the sound of the tuning fork audible to an entire lecture theatre (or stadium!). This is the basis, then, for a public address system, which picks up sounds with a microphone, and re-creates them in greatly amplified form.

Summary

In this chapter, we have seen examples of all the different kinds of signals we will need to deal with—acoustic, mechanical and electrical. Although each has its own set of characteristics (e.g. in the way they are transmitted through a medium), all can be described in a common form—as a graph or waveform. Therefore, the analysis techniques we develop for signals will be applicable to *all three forms*, as they will operate on these moderately abstract representations.

Although some different forms of energy have been discussed, we have only seen one (relatively simple) signal. What we need to be able to do now is to describe accurately the properties of this and more complex signals.

Exercises

1. Describe what is vibrating and causing sound in the following cases:

   (a) a person whistling.
   (b) a clarinet.
   (c) a siren.
   (d) a flute.
   (e) a drum.
   (f) a violin.
   (g) a xylophone.

2. Consider an expensive hi-fi stereo set which consists of a CD player, a preamplifier, amplifier and loudspeakers. Which of these systems are transducers, and which are not?

3. What are the input and output signals for each system in Exercise 2, and in what form is their energy (acoustic, mechanical or electrical)?

4. Describe how a microphone other than a condenser microphone (e.g. a carbon microphone, or a moving-coil microphone) transforms pressure variations into an electrical signal (requires outside reading).

5. Give examples of systems that perform the following transformations:

   (a) electrical to mechanical.
   (b) mechanical to electrical.
   (c) mechanical to acoustic.
   (d) acoustic to mechanical.

6. Try to name at least two routes for the sound of your own voice to reach your ears. How might this contribute to the fact that your voice sounds different from a recording than it does when you speak directly?

CHAPTER 3

Introduction to Signals

Now that we've shown how various types of signals can be represented in a common form (as a graph), let's examine in more detail one particular signal—the sinusoidally varying sound produced by a tuning fork:

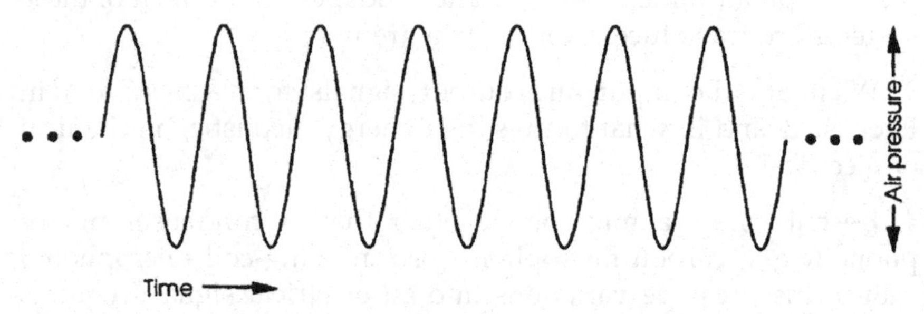

The y-axis represents air pressure and the x-axis time. You'll note that the air pressure goes up, then comes down, and then goes up again, back to where it started. It then repeats itself exactly and continues doing so. Signals that continually trace the same path in this way are called *periodic* and, conversely, those that do not repeat exactly are called *aperiodic*. We've shown just a short stretch of the periodic signal that results from the vibrations of the prongs of a tuning fork, but truly periodic signals go on repeating indefinitely. In the figure above, this is indicated by the three circles on each side of the waveform. Sinusoids are only one, rather simple, type of periodic signal, albeit very important ones. They are important not only because they are simple but, as we'll see in a later chapter, because they can be considered a basic element of more complex signals. We'll be discussing these more complicated signals later, but first let's look at sinusoids in some detail.

Frequency and period of sinusoids

There is, of course, not just one sinusoid but an infinite number of them. We need to specify what makes all these sinusoids different from one another. One crucial way in which they differ is in the time taken before they repeat. Here are two sinusoids which differ in this way:

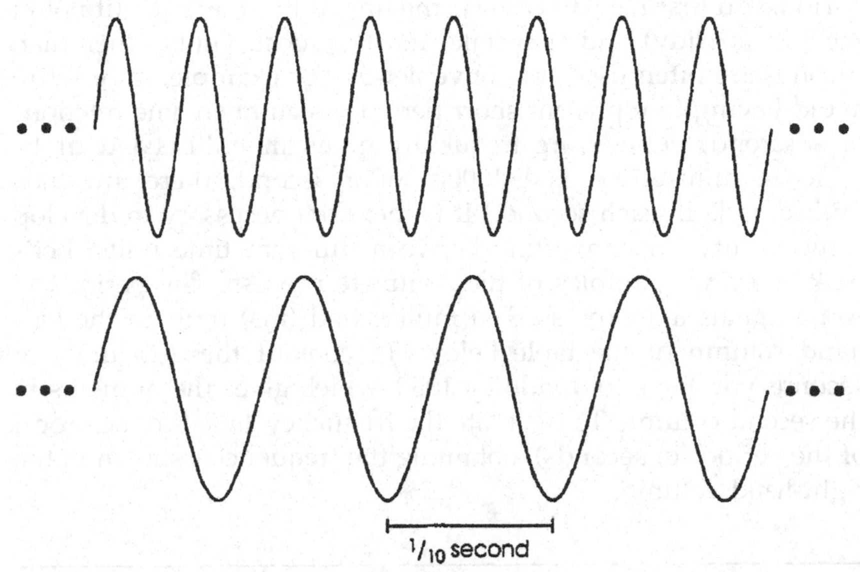

You can see that the lower sinusoid takes longer to repeat than the upper one. The shape that repeats is called a *cycle* and the time it takes to repeat is called the *period*. The period can be specified in any unit of time you wish, such as seconds or *milliseconds* (a thousandth of a second).

The same information can be specified by the frequency with which the cycle repeats, given by the number of cycles (or number of periods) that occur in a second. The unit of frequency is the *Hertz* (abbreviated to Hz), which is just another name for 'cycles per second'. Since the upper sinusoid in the example above has a shorter period than the lower, its frequency is greater.

It's easy to work out exactly the frequency of a signal from its period, or vice versa. For example, if, as in the bottom of the previous figure, a signal has a period of 1/10th of a second (translation: repeats every 1/10th of a second), obviously it has a frequency of 10 Hz (translation: repeats 10 times in a second). In general you can obtain the frequency of a periodic signal by taking the reciprocal of its period (that is, one over the period). So,

$$f = 1/p \qquad (1)$$

where f is frequency and p is period. Conversely, if you know the frequency of a signal, then its period (in seconds) is obtained by calculating one over its frequency:

$$p = 1/f \qquad (2)$$

You'll be encountering the frequency of signals expressed in Hz throughout this text. All that you've got to remember is that Hz refers to the number of times the signal repeats in one second. If you want to work out the frequency of a signal in Hz from its period, you first have to convert the time unit to seconds (if not in seconds already) and then take its reciprocal. Units other than seconds are often used for convenience. For example, in order to avoid having to represent short periods as cumbersome fractions of a second, periods are frequently given in millisecond units. (Since a millisecond is 1/1000th of a second, there are 1000 milliseconds in each second.) It is therefore necessary to develop some facility for converting between different time units. Let's look at a few examples of the complete process. The periods of some signals are expressed in millisecond (ms) units in the left-hand column of the table below. To convert these to units of seconds you have to divide by 1000, which gives the numbers in the second column. To calculate the frequency, take the reciprocal of the period (in seconds), obtaining the frequencies shown in the right-hand column.

Period (ms)	Period (s)	Reciprocal of period	Frequency (Hz)
1	0.001	1/0.001	1000
5	0.005	1/0.005	200
50	0.05	1/0.05	20
100	0.1	1/0.1	10
250	0.25	1/0.25	4
500	0.5	1/0.5	2

What if you want to do the reverse—you know a signal's frequency and you want to calculate its period? Well, all you do is reverse the process. For example, suppose a signal has a frequency of 160 Hz. What would its period be? First take its reciprocal, $1/160 = 0.00625$, which is its period in seconds. If you want to express this in milliseconds instead, then all you need do is multiply by a thousand ($0.00625 \times 1000 = 6.25$ ms).

So we can now describe one dimension along which periodic signals such as sinusoids vary—frequency. In the figure on the previous page, the period of the sinusoid at the top is half that of the sinusoid below. Or, to put it the other way round, the frequency of the top sinusoid is twice that of the one at the bottom.

If we put each of these signals through a loudspeaker, we would find that the upper signal sounds higher in pitch than the lower. The frequency of a sinusoid is the primary determinant of its pitch, with higher frequencies leading to higher pitches.

Constructing sinusoids

The fact that some basic shape is repeated over and over is characteristic of all periodic signals. The regularity arises in the case of auditory signals because the vibrating object causing the sound repeats its movement exactly. But the shape that repeats doesn't have to be sinusoidal. To show what makes a sinusoidal signal especially simple and worthy of such extended consideration, we'll show one way to construct one.

Consider a circle with a radius of one unit. Here's one with its horizontal diameter drawn in, and with a point placed on the circumference at the left-hand edge of the horizontal diameter (that is, at 'nine o'clock'):

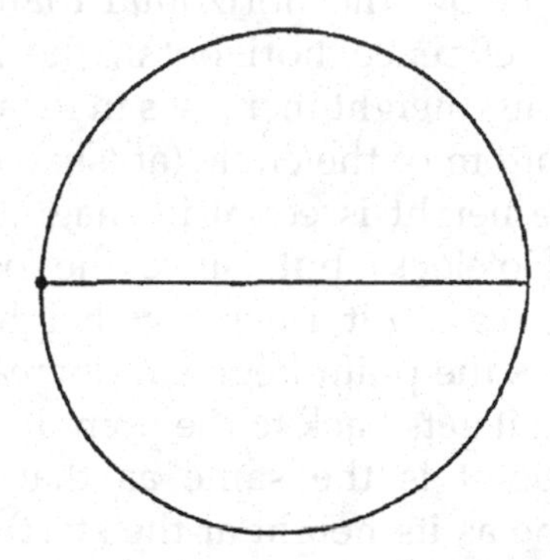

We'll now examine what happens to the height of this point above the horizontal diameter line as it moves at a constant speed around the circle. At the start of its travel (seen above) the point is on the horizontal diameter line and so its height is clearly zero.

Imagine that the point now begins to move in a clockwise direction. If it moves at constant speed, it will move the same distance round the circumference of the circle in equal units of time. Let's assume that the point takes 10 ms to go completely around the circle. It will then have passed 1/10th of the way round the circumference after 1 ms (shown as point B here):

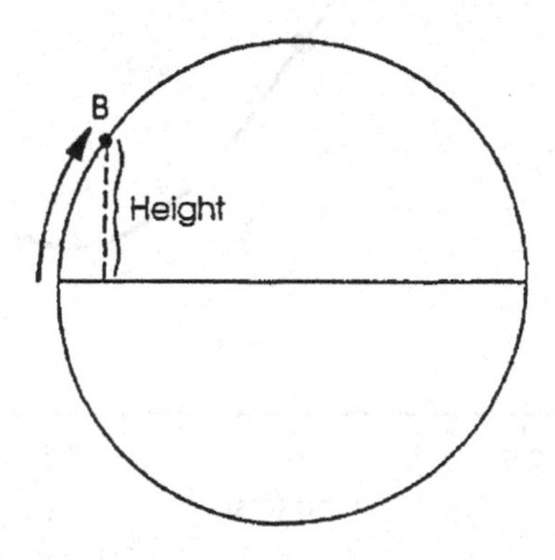

Now point B is above the horizontal diameter so its height is greater than zero. As the point travels further round the circle, its height continues to increase until it is right at the top of the circle, at 12 o'clock. Because this occurs when the point is a quarter of the way round the circle, the maximum height would be reached in our example after 2.5 ms. Here, the height of the point is the same as the radius (1, as we initially set the radius to this value).

From its position at 'high noon', as the point continues its travel, the height begins to decrease. When it's halfway round the circle (at 'three o'clock' or after 5 ms) it's at the same height as the horizontal diameter. So, at this point, the height is zero as it was at the start.

As the point rotates further around the circumference, it gets further and further below the horizontal diameter. Because the point is below the reference horizontal, the sign given to the height is negative. Thus, height increases in its 'negativeness' until the point is at the bottom of the circle (at 'six o'clock', 7.5 ms after it started). Here, the height is equal in magnitude to the radius (1, as it was at 12 o'clock) but, since the point is below the horizontal diameter, we call it a negative height (−1).

On further rotation, the point begins to increase its height from this value of −1 until it gets back to the point it started from. Now, the height of the point is the same as that of the horizontal diameter (0, the same as its height at the start).

If we plot these heights as a function of the time since the point started moving, a sinusoidally varying function arises. To help you relate the height derived from the movement of the point around the circle, insets show its position at selected times:

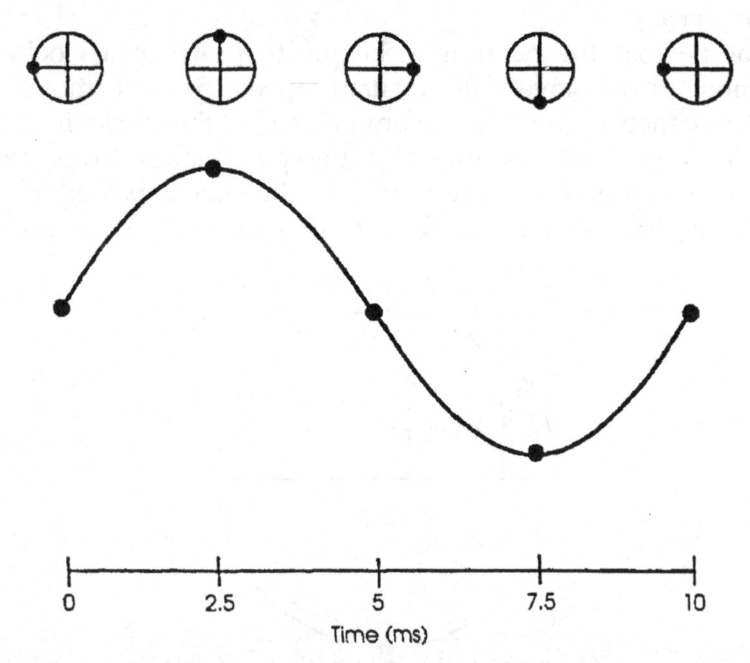

This exercise demonstrates that, as a point moves around a circle at constant speed, the height of the point measured relative to the horizontal diameter describes a sinusoidal function. That is, the smooth movement within each period of a sinusoid may be considered to derive from the constant distance that the point moves around the circle in each unit of time.

If the displacements of a point travelling at an irregular speed were plotted (in other words, if the point moves faster at some places than at others), the smooth up-and-down movement which is characteristic of a sinusoid would *not* occur. The only type of movement around a circle that produces a sinusoid is a point that moves at constant speed. For this reason, you will often see sinusoids referred to as arising from *uniform circular motion*. Not only does the point on the circumference of the circle move at constant (uniform) speed, but because of this, the point moves through the same angle with respect to the centre of the circle in each unit of time. As there are 360° in a circle, if you moved a point completely around a circle so that it ended up where it started, the point would have travelled through 360°. Thus, in our example where it takes 10 ms to move around the circle, the angular speed would be 36° per millisecond (that is, 360/10). The point would move 36° in the first millisecond going from point A to point B and 36° in the second (going from point B to point C) and so on:

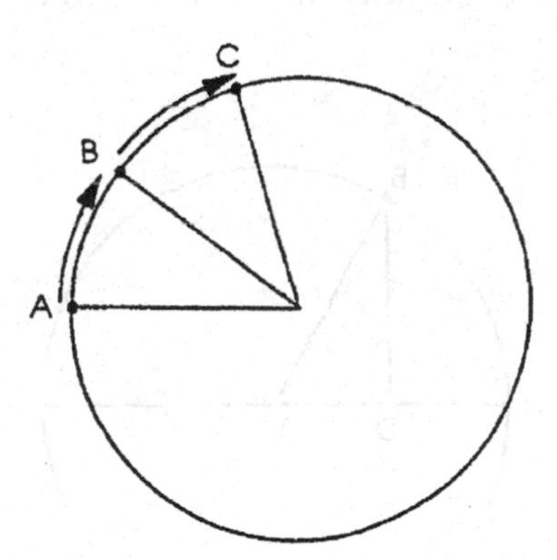

Though the rotations can be specified as so many degrees per millisecond, or degrees per second, it is more convenient to specify angular rotations in *radians* per second. Radians are another way to measure angles which relate more directly to the distance that the point travels on the circumference of the circle. Recall that the circumference of a circle has a length given by twice its radius times π ($2\pi r$). The radian measure indicates the number of radii that would be needed to go round the circumference once. There are $2\pi r$ ($\pi \approx 3.142$) radians in 360° (put another way, 2π radii are needed to wrap round the circumference). It might also prove useful to know that there are about 0.017 radians to a degree, since:

$$2\pi/360 \approx 0.017 \quad (3)$$

and about 57° in a radian, since:

$$360/2\pi \approx 57 \quad (4)$$

Using trigonometry to determine the shape of a sinusoid

So far, we know how to specify the time it takes a sinusoid to repeat—its period. If we knew the period of a sinusoid, we would also know that the height of the sinusoid measured at points in time exactly one period apart would all be the same. This results from the definition of period (the time for a complete rotation round the circle) since a point that had travelled back to where it started would have the same height as it did initially. However, as we've seen above, the height varies within each period. How can we conveniently work out the height at any particular moment within a period? One way is by using *sine* functions. We can show why this is so with some simple trigonometry using, principally, the definition of the sine of an angle. Let's go back to our circle of radius one again. We've labelled the centre of the circle as point C. The moving point is initially at A and after some time it has moved to B:

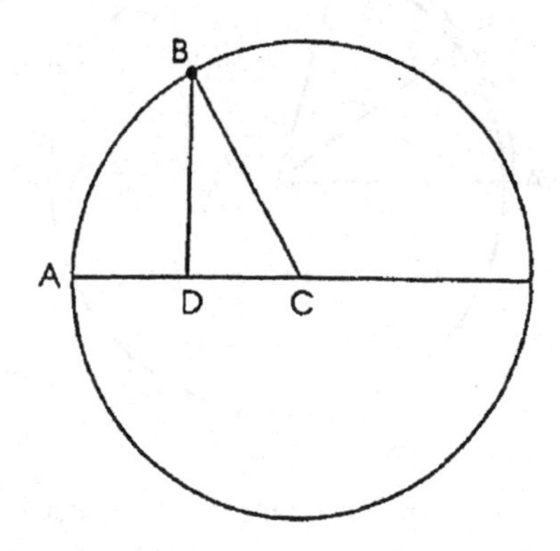

We want the height of the point above the horizontal diameter. If we drop a perpendicular from B to the line AC (let's call the point at which this line intersects the horizontal diameter D, as shown in the figure), what is the height BD? First note that BDC is a right angle triangle, since the angle described by those three points (which we will abbreviate as $\angle$BDC) is a right angle (90°). Now, the sine of an angle in a right angle triangle is given by the length of the line opposite it over the length of the hypotenuse. In our case:

$$\sin \angle DCB = BD/BC \tag{5}$$

If we multiply both sides of the equation by BC, then:

$$BD \text{ (what we want)} = BC \times \sin \angle DCB \tag{6}$$

BC is the radius of the circle and so doesn't change. If the circle has a radius of 1 unit as assumed here, we can forget about this (as anything multiplied by 1 is simply that number again). Therefore, the height of the sinusoid at any point in time is equal to the sine of the angle that the point has rotated through on the circle.

So, in order to find the height of the sinusoid at any point in time, all we need to determine is the angle the point has passed through and then take its sine. This is most conveniently done by using radians. We know that the moving point passes through 2π radians as it travels completely round a circle. If we consider sinusoids of different frequencies, then the sinusoid that has the higher frequency arises from a point that moves round the circle faster, or, to put it another way, travels through a bigger angle in any unit of time.

Thus the number of radians passed through in a second is given by $2 \times \pi \times f$ (where f is the frequency of the sinusoid). So, a sinusoid of frequency 1 Hz ($f = 1$) travels through $2 \times \pi \times 1$, or 6.284, radians in a second. A sinusoid of frequency 2 Hz ($f = 2$) travels through $2 \times \pi \times 2$, or approximately 12.6, radians in a second.

If we want to work out the number of radians passed through in some time less than a second, then we just multiply by whatever fraction of a second we're interested in. You can see this is so intuitively. If you consider a sinusoid of frequency 1 Hz which, as we've just said, passes through 6.284 radians in a second, how many radians would it pass through in half a second? Obviously 3.142. How did you get this number? Well, you just multiplied the number of radians passed through in a second ($2 \times \pi \times f$) by the fraction of a second you were interested in, here a half. This holds generally. So the angle passed through any time after the point has started travelling is given by $2 \times \pi \times f \times t$ radians (where t is the time interval). One other thing to note is that when the point has passed right round the circle (that is, when it has passed through 6.284 radians), measurements can start at zero radians again. This is simply another reflection of the fact that sinusoids are periodic.

Now that we have the angular speed, we can obtain the height of the sinusoid at any time by taking the sine of the angle that the point has rotated through in radians (using the fact that the height is given by the sine of the angle the point makes with its start). Thus, we can work out the value of the sinusoid by calculating $\sin (2 \times \pi \times f \times t)$. With the values that we set (0 at the origin, increasing to $+1$ at the top and decreasing to -1 at the bottom), the curve that is generated as the point moves around the circle is a *sine* function. You can see this if you consult tables of sinusoids; they begin at zero (where the angle is zero) and increase to $+1$ (where the angle is 90° or 1.571 radians).

Phase

Frequency is only one of the characteristics which specify a particular sinusoid. There are other ways they can vary as well. Consider these two sinusoids:

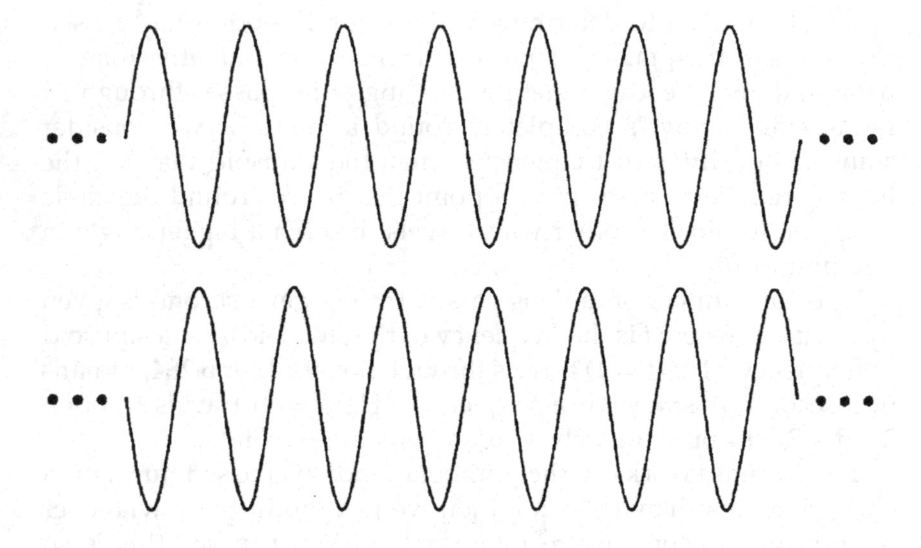

Both have the same frequency and were generated by uniform circular motion on the same circle. (You can convince yourself of this by measuring the periods with a ruler and comparing them.) Even though the two waveforms are sinusoids of the same frequency, they aren't identical. The difference—known as a difference in *phase*—arises from the fact that the waveforms start at different points in their cycles. The top one starts at zero and increases while the bottom one starts at zero and decreases. If the waveform at the bottom was shifted to the left a bit (a shift along the time axis here), it would look exactly like the waveform at the top.

You might get a clearer idea about what is meant by phase if we consider uniform circular movement of a point. If, as here, we're comparing two sinusoids, then we already know that the speed of movement around the circle is constant. If, as here again, the two waveforms have the same frequency, the speed at which the points rotate must be the same. However, the two points that are travelling around the circle need not start at the same place on the circle's circumference. In fact, sinusoids of the same frequency that differ in phase come from points that travel round the circle at the same rate but start at different places. So, in the example above, the points moving round the circle that produce the waveforms start opposite each other:

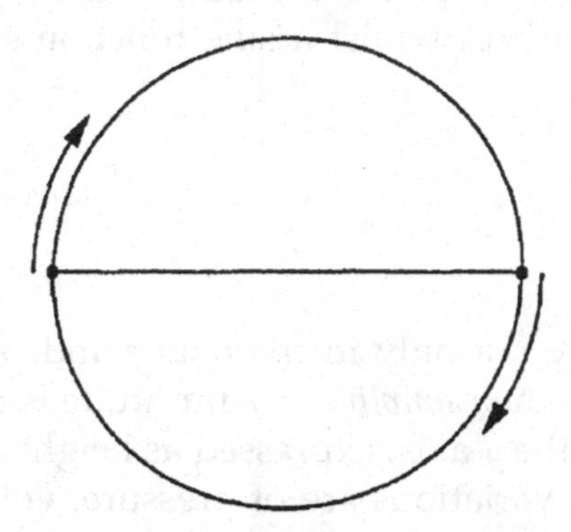

You should be able to see that, as the points move clockwise around the circle at the same rate, they will produce the two waveforms we started with. Another way to obtain the same result is to begin with the two points at the same place, but delay the start of one relative to the other. (In this case, the delayed point would not start to move until the first point was opposite it.) Two sinusoids of the same frequency but different phase can, then, be considered as time-shifted versions of each other (in our example, one-half of a cycle apart). This is another way of saying that two such sinusoids would look exactly the same if one was shifted along in time a bit (that is, horizontally).

The quantity that we use to specify differences in phase is the number of degrees (or number of radians) the two moving points are apart on the circle. We've already noted that there are 360° (or $2\pi = 6.284$ radians) in one complete rotation around a circle. If our two waveforms were generated by uniform circular motion of two points one-half of a cycle apart, the phase difference between the two waveforms would be $360°/2 = 180°$ (or π radians). Or, if the starting points on the circle were one-quarter of a circumference apart, the waveforms would differ by $360°/4 = 90°$ ($\pi/2$ radians). Two waveforms can vary in phase between 0° (that is, no difference in phase) and up to (but not including) 360°. Sinusoids of the same phase are said to be *in phase.* When two waveforms differ in phase by 360°, they will be one complete cycle apart and hence in phase again.

We can include phase in our general expression for calculating the value a sinusoid has at any point in time. If the phase is measured in radians, then the value at time t is given by $\sin(2 \times \pi \times f \times t + \varphi)$ where φ is the phase angle between the origin of the sine function and where the point starts (expressed in radians). To use this expression, you evaluate $2 \times \pi \times f \times t$ (as described earlier) but add in φ radians before you calculate the sine value. Thus, if φ is zero, the values produce the original sinusoid. A point that was advanced by $\pi/2$ radians (90°) would start with a value at its peak and begin decreasing (known as a *cosine* function). A cosine function starts at the top of its cycle and then moves downward, whilst a sine function starts at zero and initially increases.

Amplitude

Sinusoids can vary not only in frequency and phase, but also in how big they are—their *amplitude*. Amplitude is the magnitude of the excursions on the y-axis, expressed as height on our figures. In our examples, the variations are of pressure, voltage or displacement. Look at these two sinusoids:

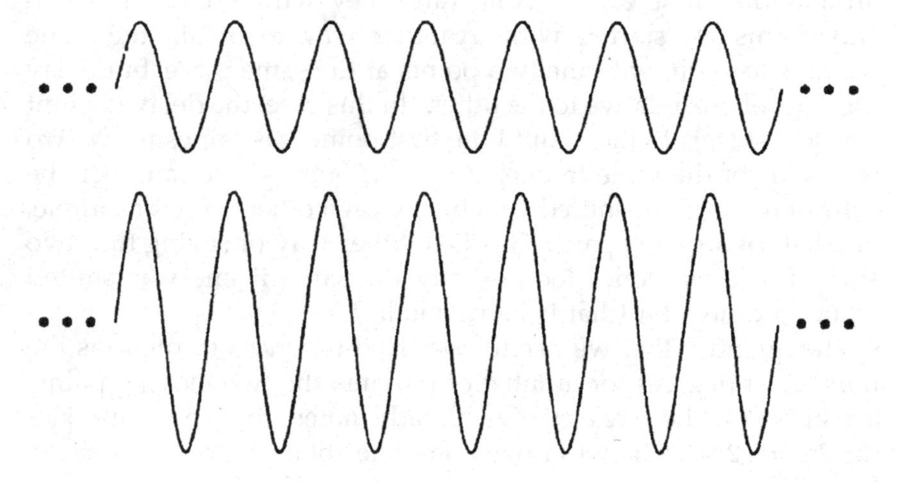

You can see that, although both have the same frequency and phase, they look different. The difference is in their amplitude. The sinusoid at the bottom is of greater amplitude than the sinusoid at the top. Differences in the amplitude of auditory signals typically arise from differences in the magnitude of displacement of the vibrating object. The amplitude of a sinusoid produced by a tuning fork will depend upon the force with which its prongs are 'twanged'. The time for the prongs of the tuning fork to move back and forth stays the same no matter how hard the prongs are 'twanged', so the frequency of the sounds is the same. A hard 'twang' will lead to bigger differences between maximum and minimum pressure than a soft 'twang', so the sound produced by the hard 'twang' has larger amplitude than that caused by the soft 'twang'. Signals that differ only in amplitude differ subjectively in their loudness.

Again, this factor can be included in our general expression for a sinusoid. Sinusoids with different amplitudes but the same frequency and phase rise and fall in unison but the extent of the excursions differs. Using A to indicate amplitude, our general expression for a sinusoid now becomes $A \sin(2 \times \pi \times f \times t + \varphi)$. If $A = 1$, we have the situation described previously. Any alteration to amplitude (alteration to A) is equivalent to considering our point moving round a circle with a radius different to 1. When A is 1, a curve is created which varies between $+1$ and -1. A value of A greater than 1 describes a curve with higher peaks and deeper troughs, whilst a signal with A less than 1 has lower peaks and shallower troughs.

The three factors that we've considered (frequency, phase and amplitude) are the only three attributes that sinusoids can vary on. Thus, any sinusoid can be uniquely specified by its frequency, amplitude and phase. Frequency refers to how fast the sinusoid moves up and down, amplitude to how big the up-and-down movements are, and phase to when the sinusoid starts.

Examples of other periodic signals

Of course, there are many, many periodic signals other than sinusoids, but only a few general cases need concern us now. Here is a periodic waveform that rises and falls linearly, but that falls more rapidly than it rises:

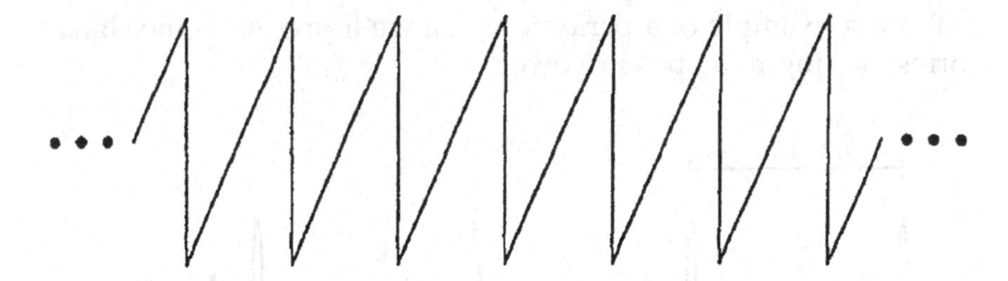

This is called a *sawtooth* waveform, since it is similar in shape to the teeth of a saw. You can see that even though the waveform is not sinusoidal, it repeats regularly. We can therefore define its period and, hence, its frequency in the same way as we did for sinusoids. Its amplitude, too, is often specified as with sinusoids. (We'll have more to say about the amplitude of complex periodic waveforms below.)

The sawtooth waveform is important in the study of speech production, as the vocal cords change the airflow from the lungs into a signal that looks similar to a sawtooth. The airflow increases as the cords open and decreases as the cords close. The reason for the sharp decrease in airflow in this case is that the vocal cords snap together faster than they open (see Chapter 13).

Here are some other examples of periodic waveforms:

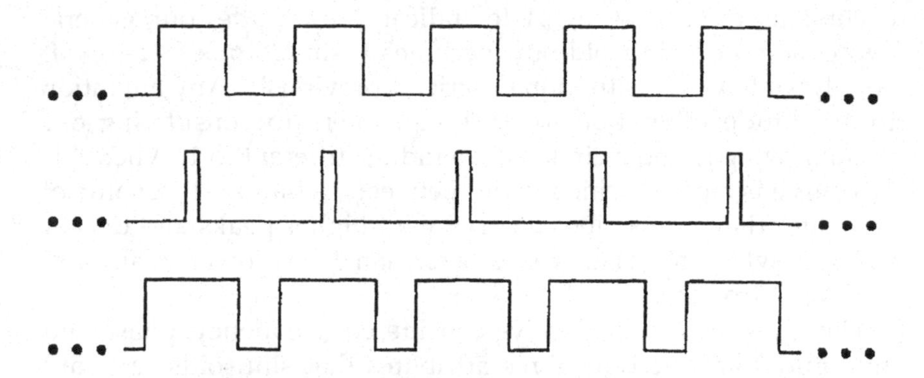

The waveform at the top is square in shape, and so it is known as a *square* wave. The other two waves are similar in certain respects to the square wave: they rise abruptly to their maximum, stay at this point for some time, fall abruptly to their minimum, and stay there for a while before the cycle repeats. Such waveforms are often called rectangular waves, or *pulse trains*. The signals differ in how long they stay at the top and bottom of their travel, their *duty cycle* (the proportion of the period in which they are at their maximum). Thus, a square wave would have a duty cycle of 0.5, as it is at its maximum for half of each period. Note that the period between repetitions (and therefore the frequency of all three waveforms) is the same.

The last example of a periodic signal we'll present is the signal corresponding to a spoken vowel:

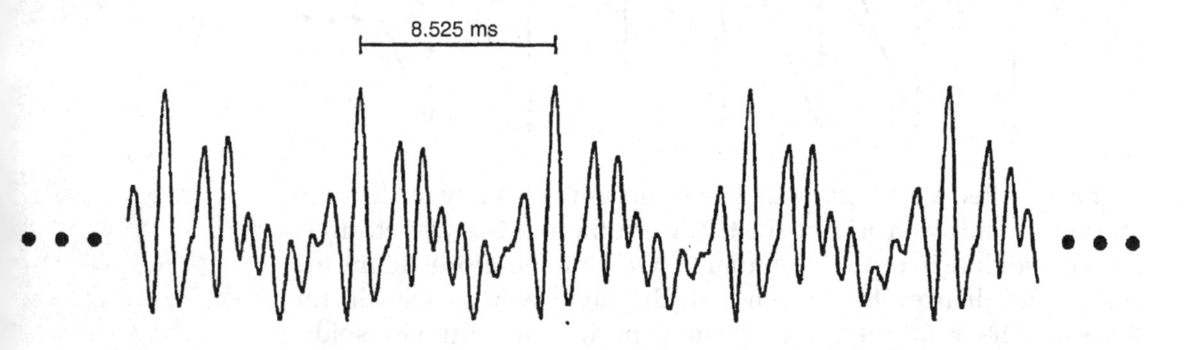

This is a trace derived from a real recording of the vowel 'ah'. You can see that it looks rather more complicated than the examples we have considered up until now, as it crosses the y-axis more than once within each period. However, it is still periodic in that it repeats regularly, at least over the time that we've measured it. In fact, as in all human affairs, this ideal never occurs. There will never be a strict periodicity, only a more or less good approximation to the ideal. We often refer to such signals as *quasi-periodic*.

One final (and related point) to emphasize is that truly periodic signals should last an infinite amount of time. You can appreciate this if you consider a signal that doesn't go on forever; the signal would have to be switched on and off, so it would only repeat exactly between these points in time. Usually, though, providing the period of the periodic signal is short relative to its total duration, this is sufficient for practical purposes for it to be treated as if it were truly periodic.

Aperiodic signals

Aperiodic signals, as you might expect, are not periodic. Two important types of such signals can be distinguished—random and transient.

Random signals

Here is one example of a random signal:

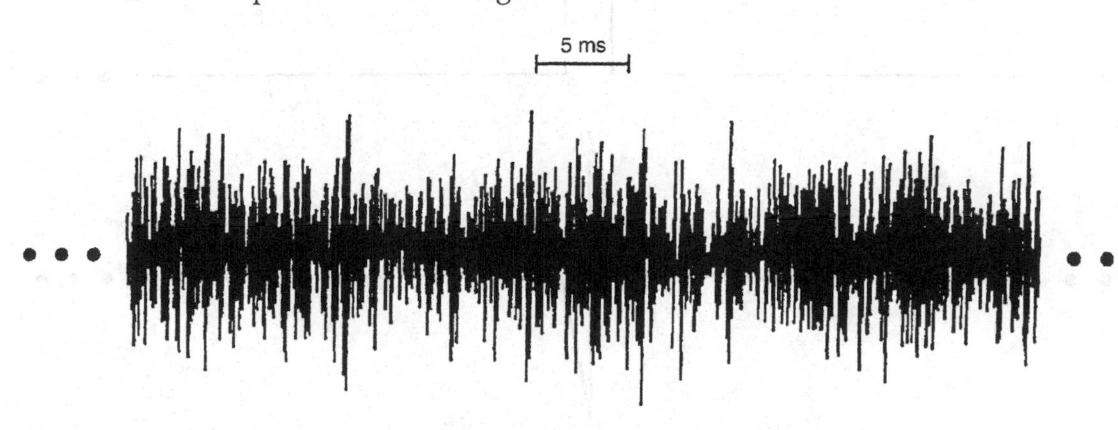

If you played this signal through a loudspeaker, it would sound 'noisy', like air escaping from a tyre. Signals like this are unpredictable, varying in a random manner over time.

There are different types of 'noise', usually distinguished by the mix of frequencies they contain. Noises which consist of a random combination of all frequencies at equal amplitude are called '*white*' by analogy with 'white' light (which can be considered as a mixture of all visible wavelengths). 'Pink' noise has more low-frequency energy, by analogy with reddish light, which is in the low-frequency end of the visible spectrum. Noise signals are often used in psychoacoustic experiments and, more importantly, occur naturally. Here, for example, is part of a 'sh' sound (as in the word 'ship') which, as you can see, varies randomly:

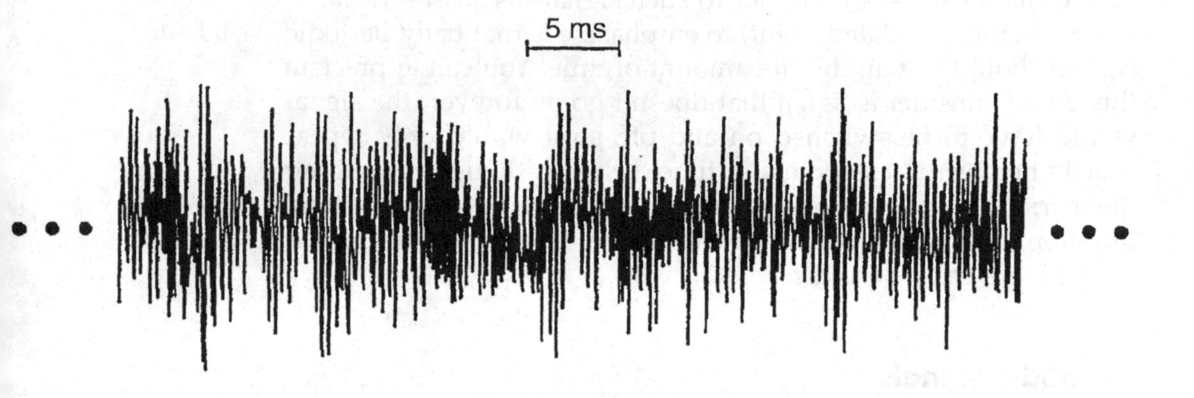

Transient signals

Here are examples of another sort of aperiodic signal:

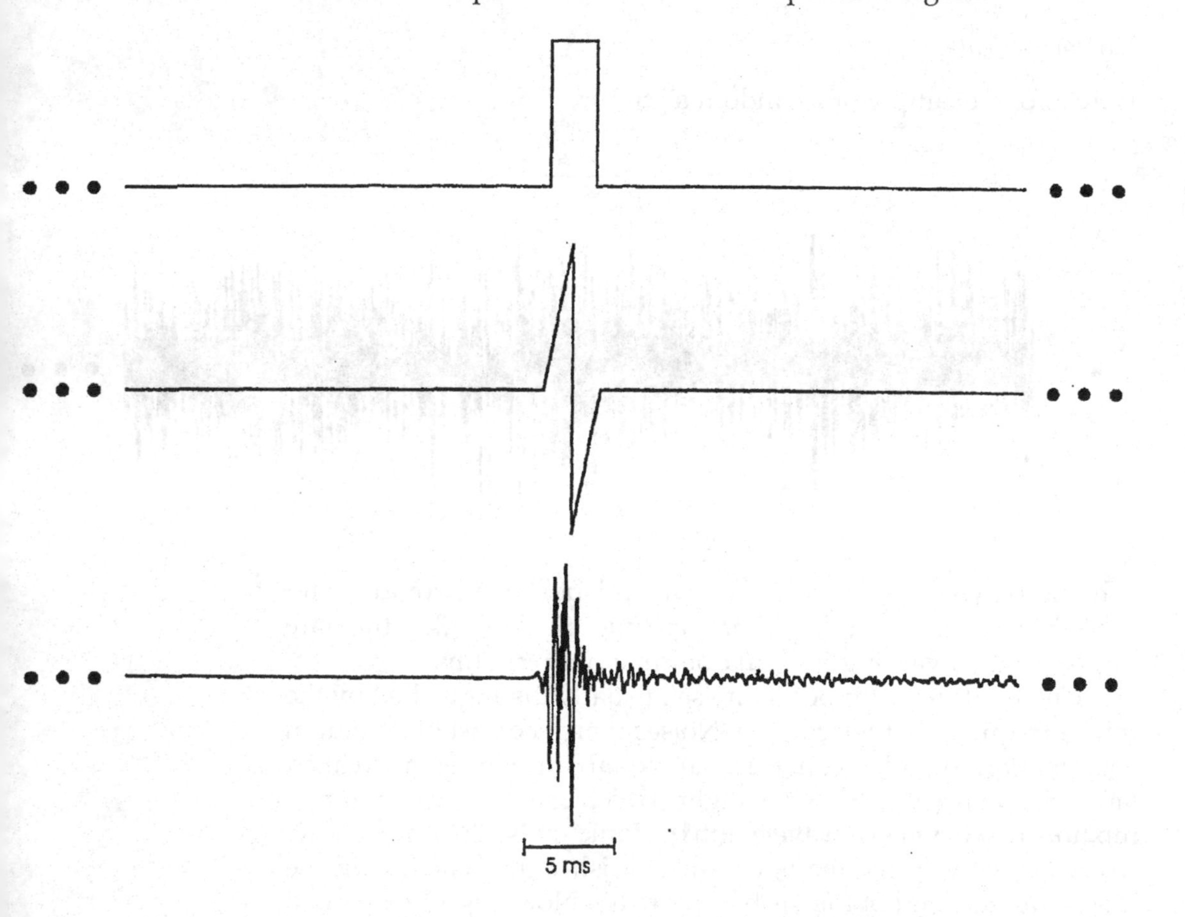

Each of these lasts for a short time only, and hence they are known as *transients* (since they don't hang around!) The two on the top are computer synthesized, and consist of single cycles of two of the periodic waveforms we described earlier (pulse train and sawtooth). The one on the bottom is a recording of a single handclap made in a small room.

Like random signals, transient signals are encountered in speech as well as in the laboratory. If you say the word 'tap', a little 'pop' or explosion, associated with the 't', occurs at the beginning of the word. This is sometimes called a *burst*, lasting only a brief period of time and not repeating. Bursts too are transient signals, as you can see from these three acoustic pressure waveforms (recorded with a microphone) of the bursts from the plosives p, t and k:

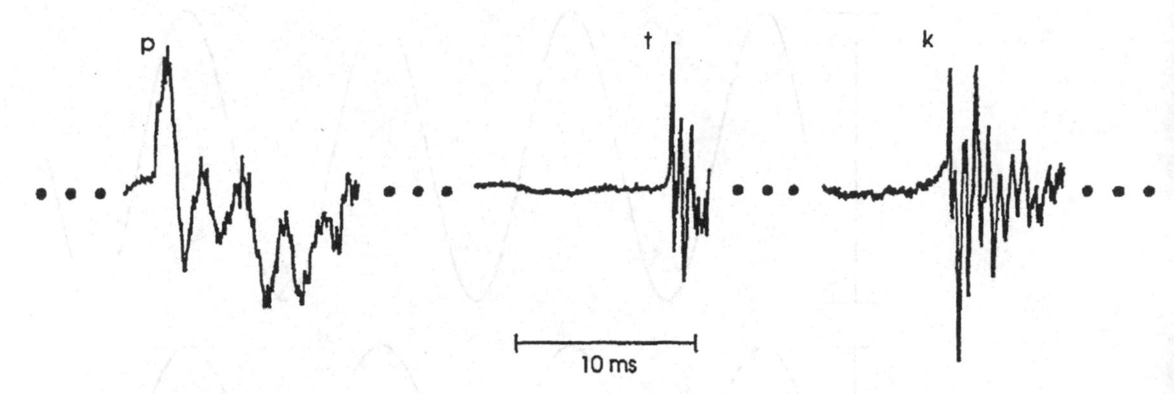

A complex speech signal can often be divided into sections of transient, random and periodic (or quasi-periodic) signals. Here, for example, is the first part of the word 'choose':

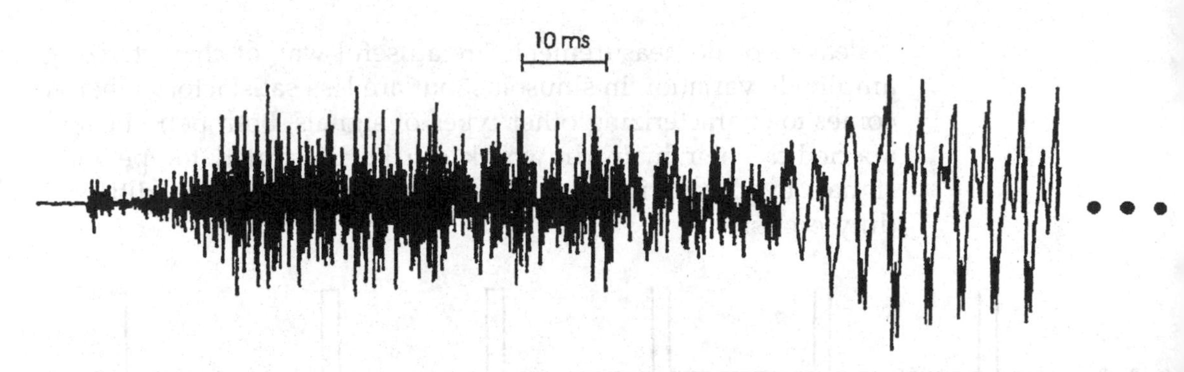

At the start of the word, you can see a small transient burst, followed by a signal which varies randomly for the remainder of the 'ch' sound (in fact, in a similar way as for a 'sh', shown above). This then gives way (at the vowel) to a quasi-periodic waveform.

Peak-to-peak and root-mean-square amplitude measurements

Before leaving our discussion of signals, we need to say a bit more about the general problem of defining appropriate amplitude measures, and, in particular, describe two useful ways in which amplitude can be measured.

One way that we could quantify the difference in amplitude between two signals is by measuring the distance between the peaks and troughs, known as the *peak-to-peak* value. Here are two sinusoids with their peak-to-peak values indicated by the vertical lines on the left. The sinusoid at the top has a peak-to-peak amplitude twice that of the sinusoid at the bottom:

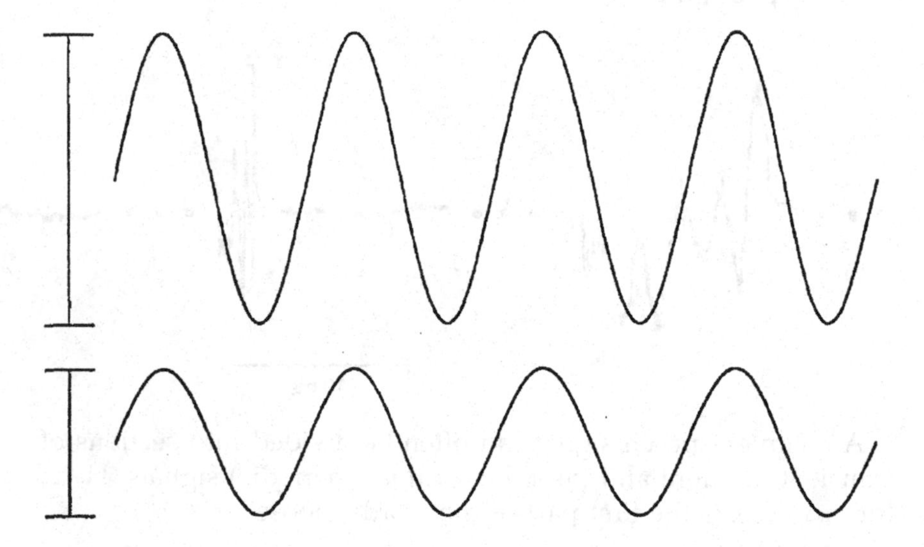

Peak-to-peak measurements are a useful way of characterizing amplitude variation in sinusoids, but are less satisfactory when it comes to characterizing other types of signals, both periodic and aperiodic. Their main drawback can be illustrated for periodic signals by comparing two electrical pulse trains with different duty cycles:

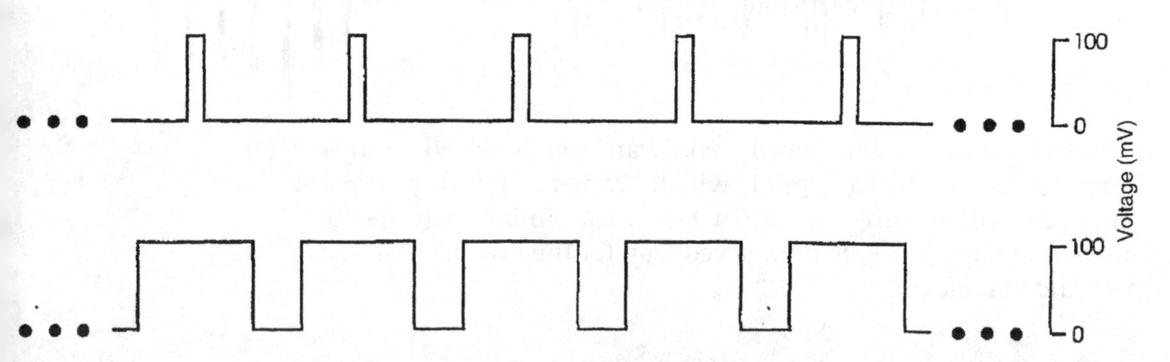

Although these two signals have the same peak-to-peak value, they differ in an important way. One of them (the lower) is at its maximum positive voltage for longer than the other. Or, put another way, the upper signal is at zero volts for a longer part of each cycle. Therefore, the two signals have different amounts of energy, as energy only exists when a signal has a non-zero voltage. Because two signals with very different energy content can have the same peak-to-peak amplitude (as this example shows), it is necessary to have other ways of characterizing amplitude—ways which characterize variation in energy content because of variation in the shape of a cycle.

We'll illustrate the workings of one such method with a sinusoidal pressure variation, as we'd expect to find from 'twanging' our tuning fork:

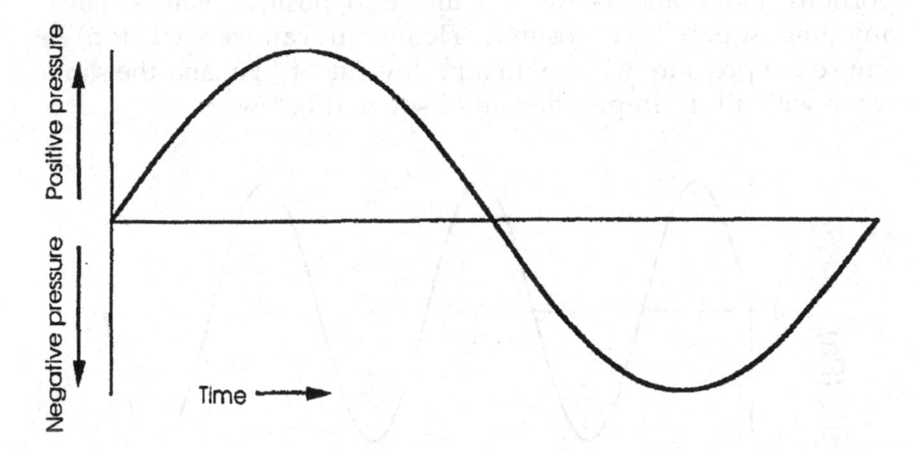

As we noted in Chapter 2, it is not the absolute atmospheric pressure that is of importance in measuring sound, only the relatively rapid fluctuations around this ambient pressure (represented in the figure above as a horizontal line). The bits of the waveform above this line have a positive pressure relative to that of the atmosphere and the bits below are at a pressure below atmospheric pressure. If we increased the amplitude of the sinusoid, we would increase the extent to which the curve goes above and below the line.

Let's consider the bits that are above atmospheric pressure for a moment. At all points along this first portion of the sinusoid, the height of the curve above atmospheric pressure characterizes the amplitude of the signal. Signals with higher amplitude would deviate more from the line than signals with lower amplitude. Thus, if we summed up our pressure measurements in the section where they are positive, we would have a way of characterizing the amplitude of this part of the signal. This would be the same as summing up the area between the positive pressures and atmospheric pressure. By doing this, we would end up with a value that indicates that the signal has a certain amount of positive pressure relative to atmospheric. The same could be done with the section of the signal below the line (but, here, bigger negative values would be associated with signals of higher amplitude). If we added together the deviations of the portions above and below the line, we would find that they summed to zero for a sinusoid (which is to say that the portion of the curve below the line has the same shape as that above the line, but that they are opposite in sign). This *cannot* be a good amplitude measure, as we would end up saying that all sinusoids, no matter what the extent of their excursions, have an amplitude of zero.

How do we take account of positive and negative pressure at the same time? It turns out to be useful to *square* the pressures measured relative to atmospheric pressure. This converts the portions above and below the line into positive values, since anything squared is positive. Here you can see (at top) a sinusoidal pressure wave with a peak value of 1 Pa, and the same wave with all its amplitude values squared below.

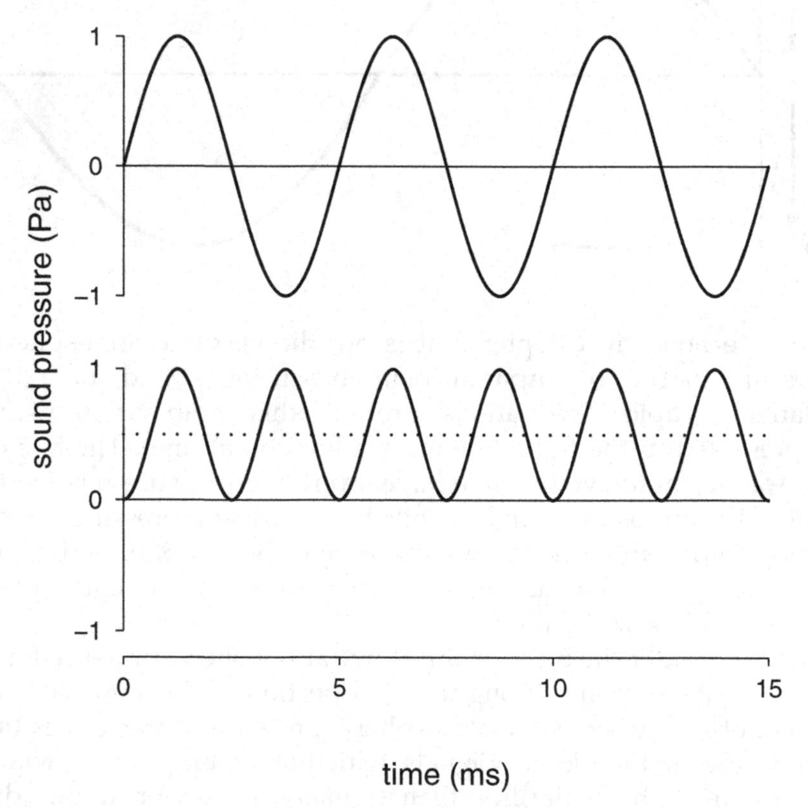

It's interesting to note that a sinusoid squared has the same shape as a sinusoid of twice the frequency, as can be gathered from the trigonometric identity $\sin^2(x) = 1/2 - \cos(2x)/2$. The crucial factor is that the original x which determines the frequency of the sine wave has become $2x$ in the new cosine term.

The total area under this squared curve is then added up, and the average value across one cycle calculated (by dividing by the period). This value (0.5 Pa here because the original peak value of the sinusoid was 1 Pa) is drawn in as a dotted line along with the squared sinusoid. Finally, the square root of this value is taken to compensate for the initial squaring, giving a value of 0.707 Pa—this quantity is then known as the *root-mean-square* (RMS) amplitude.

To summarize how an RMS amplitude measure is obtained: the momentary pressures about atmospheric pressure are squared (thus eliminating negative numbers), averaged and the square root taken. RMS amplitude measurements are useful quantities because they relate directly to the power in the signal. In fact, the RMS level of a signal is the value that a constant wave (in other words, a waveform consisting of a single horizontal line) would need to be in order to give the same power as the original signal. (You should be able to convince yourselves that a square wave with the same peak-to-peak amplitude and periodicity as a sinusoid, also going between equal magnitude positive and negative values, would have a larger RMS amplitude.) The RMS and peak-to-peak amplitudes are related in an exact way for sinusoids—the RMS amplitude measurement is 0.707 of the peak value. Though this relationship only holds for sinusoids, other relationships hold for other specific types of periodic waveforms.

Note that real RMS amplitude measurement devices take an average over several periods. If the averaging time is long relative to the period, it doesn't matter whether the averaging is done over a whole number of cycles, so the real measurement device doesn't have to determine when a period starts and finishes.

Measuring amplitude—the relationship between amplitude and intensity

The RMS amplitude is the most important way of specifying the amplitude of a signal because it presupposes the use of a related quantity—*intensity*. For a sound wave, intensity in a free or completely diffuse field is proportional to amplitude squared. Thus, if you have a pressure measurement (either at one point in time or averaged as in an RMS amplitude value), you can convert it into an intensity measure by squaring (giving an instantaneous or averaged intensity measurement, respectively) and multiplying by an appropriate constant. Or, if you want to convert an intensity measure into an amplitude measurement, divide by the appropriate constant and take the square root. The reason we need to be able to perform these conversions is that the scale used in acoustic work is really an intensity scale. But, as we've already seen, most acoustic instruments measure pressure. As you'll see in more detail below, for convenience we'll make pressure measurements and assume that this simple conversion between pressure and intensity always holds, even in situations where it does not.

Defining scales

Once we have a measure of intensity, there is another complication. The scale that is normally used is more complex than those you are more familiar with, for example length. Every time the length of an object increases by some agreed-upon physical quantity, the length increases by an extra unit of that measure. Thus, an increment of 1 inch is the same size whether the extra inch is added to a line of 5 inches or 5000 inches. So, to measure length, the number of 1-inch units of length just have to be counted up. However, the units on intensity scales normally used in acoustics are not physically equal like this, so we don't count up the number of physically equal units. To explain exactly what we do, we need to consider what is meant by a scale of measurement in some detail.

All scales of measurement need a starting point and the size of the unit of measurement to be defined. Let us consider temperature scales to illustrate how these apply. You are probably all familiar with the centigrade and fahrenheit scales. The starting point on the centigrade scale is defined as the melting point of pure water, and this is given the arbitrary value of 0°C. The unit of measurement is given by defining the boiling point of pure water as 100°C and splitting the scale between 0°C and 100°C into 100 equal units. So, for example, the mercury in a thermometer expands by the same amount in going from 0°C to 1°C as it does in going from 50°C to 51°C, in a similar way to the units of length discussed above. Note that both of the predetermined values on the centigrade scale are arbitrary and chosen for convenience. This arbitrariness can be appreciated by noting that the melting and boiling points of water are different on the fahrenheit scale (32°F and 212°F). Similarly, the division of the scale is arbitrary too—a degree on the fahrenheit scale is smaller than a degree on the centigrade scale (there are more one-degree steps going from the melting point to the boiling point of water on the fahrenheit than on the centigrade scale). Although both types of temperature scales partition the steps between two defined points into equal-sized units, we'll see in a moment that there is nothing obligatory about such an equal-increment partition.

To summarize, temperature scales illustrate the two important aspects of scales of measurement—the need for a reference point and a method of dividing up the scale. After we've defined our intensity scale, we'll illustrate how these ideas apply to it.

Definition of the intensity scale—dB values

To define an intensity scale, then, the things we need are a reference point and a method of determining the size of the steps. The intensity scale used in acoustic work is called a *bel* scale, after Alexander Graham Bell, the inventor of the telephone and a teacher of the deaf. The bel scale is defined as the logarithm to the base 10 of the ratio of two acoustic intensities:

$$\text{bels} = \log(I/I_{\text{ref}}) \qquad (7)$$

(Note that whenever we use 'log' on its own, base 10 will be understood. For ease of reference, the crucial properties of logs are summarized in the Appendix to this chapter.)

I is the intensity of the sound that has been measured and I_{ref} is a reference intensity (which we'll have more to say about later). The bel unit is a bit too big for convenience, and so we normally deal in decibels (1/10th of a bel, abbreviated dB—note the lower case 'd' and upper case 'B'):

$$\text{decibels} = 10 \log(I/I_{\text{ref}}) \qquad (8)$$

As we've mentioned before, it is much more convenient and practical to work with pressures than with intensities. Therefore, we want to define a scale that works directly with pressures. We will, as is usual, assume that intensity is proportional to pressure squared in order to change the scale from an intensity-based one to a pressure-based one. Strictly speaking, this relationship is only true for free-field or diffuse-field sounds (and not, for example, for speech in an ordinary room). However, we will use the scale to express pressure values without necessarily implying anything about what the intensity would be.

It is thus possible to incorporate the conversion of pressure to intensity in our scaling by assuming intensity to be proportional (symbolized by the factor k) to the square of pressure:

$$I = kp^2 \qquad (9)$$

and:

$$I_{\text{ref}} = kp_{\text{ref}}^2 \qquad (10)$$

As you will soon see, the constant of proportionality k will cancel out in the ratio of I to I_{ref}. In more detailed formulae:

$$\text{decibels} = 10 \log(I/I_{\text{ref}}) \qquad (11)$$

First substituting equations (9) and (10) into equation (11) we obtain:

$$\text{dB} = 10 \log(kp^2/kp_{\text{ref}}^2) \qquad (12)$$

and then

$$dB = 10 \log(p^2/p_{ref}^2) \quad (13)$$

by cancelling the factors of k on top and on bottom.

Since p and p_{ref} are both squared, the square can be taken outside of the bracket:

$$dB = 10 \log (p/p_{ref})^2 \quad (14)$$

Finally, using the fact that the log of something raised to the power of 2 is equal to twice the log (a particular instance of law III of logarithms presented in the Appendix), we obtain the standard expression:

$$dB = 2 \times 10 \log(p/p_{ref}) = 20 \log(p/p_{ref}) \quad (15)$$

dB are also often used in expressing voltages, the electrical quantity analogous to acoustic pressure (seen in Chapter 2). In that case, we use:

$$dB = 20 \log(V/V_{ref}) \quad (16)$$

where V_{ref} is the reference voltage.

Scale reference points

One essential feature of dB scales is that they require a reference value. To make matters more concrete, for the moment we'll restrict consideration to the measurement of pressure only. We do not specify any single value of p_{ref} for a good reason—in acoustic work many different references are used. We'll say more about the most commonly used one shortly. First, let's discuss what a variable reference point does, as this quantity makes the scale look very different from those we discussed earlier when we considered the measurement of temperature.

Consider only the quantity inside the brackets in the definition of the dB scale given by equation (15)—p/p_{ref}. Let's work out some values, assuming, first of all, that p_{ref} is equal to 2 Pa. When p is equal to p_{ref} (that is, when p is also 2 Pa in this case), the value of p/p_{ref} is 1. If we had used a reference of 4 Pa, the ratio would have been 1 when p was 4 Pa as well. Thus, by choosing different values of p_{ref}, we can shift the pressure that is given the value of '1' up or down. This is just another way of setting our starting point, similar to what we saw with temperature. The difference between temperature and dB scales is that we are able to vary the reference on dB scales more or less willy nilly.

But why do we need to do this? We managed perfectly well by defining a single reference on the centigrade scale (which also had the advantage that we didn't need to do all these calculations!)

One reason is that it is often convenient to be able to use different reference pressures depending on the task involved. For example, when testing the ability of hearing-impaired listeners to discriminate between two sounds, it may be better to present them at some constant level relative to the threshold of the individual. Then, even if the listeners have different hearing losses, there should be more comparability in what they hear than if sounds are presented to everyone at the same absolute level. Measurements based on references using individual thresholds like these are known as *sensation levels*.

In other situations (say, in quantifying the sensitivity of a microphone, or the maximum output level of a loudspeaker), a standard reference level is desirable. The most important pressure reference used in working with sound is 20 micropascals (20 μPa—a micropascal is 10^{-6} Pa, or one-millionth of a pascal). When this reference is used, the levels are known as *sound pressure levels* (abbreviated to SPL).

A crucial point about reference levels is that they *must* be specified, otherwise the measurements are meaningless. This is true of temperature, too. If you said that the weather forecaster predicted a temperature of, say, 10° for the weekend, no one would know whether to wear heavy gloves or not unless they knew the temperature scale used (and, therefore, the reference). Usually, though it's pretty clear which temperature scale is meant, as only two are used regularly, and we always know, at least roughly speaking, what temperatures to expect for a given time of the year. (Readers in England may doubt this, and quote in evidence the saying 'If you don't like the weather, wait a minute'.)

Reference levels for measurements on dB scales are indicated in two ways. Firstly, there is a standard set of abbreviations which imply the use of a particular reference. For example, if a sound is at 39 dB SPL, the use of SPL indicates that the reference 20 μPa has been used. Secondly, for nonstandard reference values (and all voltage measurements), the reference used is given explicitly, preceded by the abbreviation *re* (for 'referred to'). Thus, 39 dB re 20 μPa is exactly the same as 39 dB SPL.

Taking the logarithm of the ratio

If we simply employed the quantity p/p_{ref}, the scale of pressure would be a linear scale, as with temperature. That is, some increase in p would always produce the same increment in the quantity p/p_{ref}. For example, if p_{ref} is 3 Pa and, initially, p is 3 Pa too, the ratio is 1. If we add 3 to the numerator, the ratio is now 6/3 or 2, an increase of 1. Add another 3 Pa to the numerator and we go to 9/3 or 3. Every time you add 3 Pa to the numerator, the ratio increases by 1. Thus, every time we add a constant amount, the measure increases by a constant absolute amount, just as with temperature. Every time the mercury in a thermometer increases in length by some specified amount, the temperature is said to have increased by one degree. In both cases, the same physical increment causes the same increase on the scale.

Now we'll show how the use of logarithms makes a dB scale so different from the temperature scale. You should all be familiar with taking logs (if you need to revise them, see the Appendix to this chapter). However, you may not be aware of what the process of taking logs actually does. We'll try to explain this by starting with some examples. The log of 10 is 1, the log of 100 is 2 and the log of 1000 is 3. Going from log(10) to log(100) results in an increase of 1 (=2−1). Going from log(100) to log(1000) also gives an increment of 1 (=3−2). The same would happen in going from 1000 to 10,000, and so on. But the number we logged isn't increasing by the same amount. An increase of 1 log step going from 10 to 100 Pa (assuming we're still measuring pressures) only extends 90 Pa, while an increase of 1 log step going from 100 to 1000 covers 900 Pa. This contrasts with temperature scales. At the risk of boring you silly, let's just repeat that a specified increase in the length of mercury in a thermometer equals one degree, whatever point the mercury is at in the thermometer (and, therefore, whatever the temperature is). This is not so on a log scale—an increment of one log unit covers a smaller range at the bottom end of the scale (for example, going from 10 to 100) than at the top (going from 1000 to 10,000). Note that an increase of a log unit always produces a 10-fold increase in the numerator (100 is 10 times 10; 1000 is 10 times 100 etc.). This is the critical feature of log scales—*equal ratios are transformed into equal distances.* On linear scales, equal absolute increments take up equal distances.

Now let's consider what this achieves. First, the range of values is compressed, since a log increment covers a wide range at the top end of the scale. Think about the log steps as distances in miles. The first log step takes up 90 miles, the next 900 miles, the next 9000 miles—you don't have to go very many log steps before you're dealing with enormous distances. A log scale deals with this range of numbers conveniently.

Second, it turns out that measurements in log units correlate more closely with our subjective perceptual experiences. For example, we're more sensitive to a given increase in pressure (heard as a change in loudness) at low levels than at high levels. To a first approximation, the minimal detectable change in pressure is a constant *proportion* of the overall level (a rule known as Weber's law). It's clear that our measurement scale ought to correspond with our sensitivity. A log scale does exactly this—a constant *proportional* change becomes a constant *absolute* change in dB.

A related feature of logarithmic units is that they determine a scale that varies in its accuracy in something like the way that our sensitivity to pressure differences does. There is no sense in being able to measure pressure changes accurately that we are not sensitive to, while being unable to resolve differences that are clearly discernible. A micrometer is of little use in measuring the distance between London and Scunthorpe or London and Brooklyn, as is a ruler in measuring distances in subatomic physics. Since a log scale is more finely graded at low levels than at higher levels, it provides units that are matched to perceptual sensitivity at *both* ends of the scale.

These arguments should prevent you getting the idea that log scales are 'unnatural' and linear scales 'natural', just because the latter are more familiar. Breaking up the temperature scale between 0 and 100 into 100 equal-sized steps is just as arbitrary as it would have been if we had broken it up into 100 logarithmically spaced units.

Further features of dB scales

There are some important properties of dB scales that have only been implicit in our discussion.

Firstly, 0 dB does not mean no sound at all. Since $\log 1 = 0$, 0 dB SPL means the sound has the same pressure as the reference (20 μPa for dB SPL).

Secondly, since log 0 does not exist (or is equal to minus infinity), it is not possible to express a level of zero (or no sound) in dB.

Thirdly, a negative value of dB does not mean negative sound! Since $20 \log(p/p_{ref})$ is less than 0 when $p<p_{ref}$ (that is, when $p/p_{ref}<1$), a negative dB value simply means that the specified level is smaller than the reference.

Fourthly, dB values don't add and subtract in the normal way. Two sinusoids at 1 kHz each at a level of 94 dB SPL, when added together in phase don't give 188 dB SPL! It's necessary to convert back to pressure values to do the addition and then recalculate the value in dBs. So, to find what pressure corresponds to 94 dB SPL + 94 dB SPL, first put 94 dB into the formula used to define the dB scale with respect to pressure measurements ($20 \log(p/p_{ref})$ with p_{ref} equal to 20 μPa):

$$94 \text{ dB SPL} = 20 \log(p/20\ \mu\text{Pa}) \qquad (17)$$

Next, divide both sides by 20:

$$94/20 = \log(p/20\ \mu\text{Pa}) \qquad (18)$$

Now we need to get rid of the log on the right-hand side (RHS)—in other words, undo the work of the log. This is done by raising 10 to the powers indicated by both sides of the equation:

$$10^{94/20} = 10^{\log(p/20\ \mu\text{Pa})} \tag{19}$$

leaving us with:

$$10^{94/20} = p/20\ \mu\text{Pa} \tag{20}$$

as a log to base 10 is the power 10 has to be raised to. A few presses on a calculator shows that:

$$10^{94/20} = 50118.7 \tag{21}$$

so:

$$50118.7 = p/20\ \mu\text{Pa} \tag{22}$$

Cross-multiplying by 20 μPa gives:

$$20\ \mu\text{Pa} \times 50118.7 = p \tag{23}$$

which is about 1,000,000 μPa or 1 Pa.

We can now add 1 Pa + 1 Pa = 2 Pa and convert back to dB SPL to find the intensity of a signal formed by the addition of two signals of 94 dB SPL:

$$20\ \log(2\ \text{Pa}/20\ \mu\text{Pa}) = 100\ \text{dB SPL} \tag{24}$$

not 188 dB SPL. This example illustrates the conversion of dBs back to pressure and vice versa, and that a doubling of amplitude is equivalent to an increase of 6 dB. The general equation for converting into pressure (or voltages) from dB is:

$$p = p_{\text{ref}} 10^{\text{dB}/20}$$

Finally, the laws of logarithms can be used to make quick 'off-the-cuff' dB calculations. Because $\log(xy) = \log(x) + \log(y)$, if we can express the initial ratio to be converted into dB with some simple multiplicative factors, all we need to do is add together the dB values for each factor.

For example, how many dB above 1 V is 4 V, a ratio of 4/1 (= 4)? Note first that 4 is 2 × 2, and, since each factor of 2 gives an increase of about 6 dB, the overall increase is 6 + 6 = 12 dB re 1 V. Here are some convenient ratios to remember for rough calculations:

Ratio	2	3	5	10
dB (approx.)	6	10	14	20

So, a ratio of 100 (= 10 × 10) = 40 dB, because each factor of 10 adds 20 dB. A ratio of 1000 = 60 dB, because 1000 is 100 multiplied by another factor of 10 which adds another 20 dB, and so on.

Exercises

1. Fill in the missing entries in the following table, which shows the relationships of the period of a sinusoid (in two different units) to its frequency (you should not need a calculator):

Period (ms)	Period (s)	Frequency (Hz)
1	0.001	1000
?	0.05	?
10	?	?
?	?	50
?	1	?
2	?	?
?	?	200
0.5	?	?
?	0.1	?
?	0.02	?

2. Fill in the missing entries in the following table (you should not need a calculator):

Radians	Degrees
π	180
4π	?
?	360
$\pi/2$	?
?	540
$\pi/4$	?

3. On page 15, you were shown two sinusoids. The upper sinusoid had a frequency twice that of the lower. Given that the period of the lower sinusoid is 1/10 s, what is the frequency of the upper sinusoid?

4. The waveform of the vowel shown on page 26 (lower) had a period of 8.525 ms. First give a rough estimate of its frequency (greater or less than 100 Hz, or 200 Hz?), then calculate it exactly. Find out what musical note this corresponds to. Is such a frequency most typical of a man, woman or child speaker?

5. What is the RMS amplitude of this square wave:

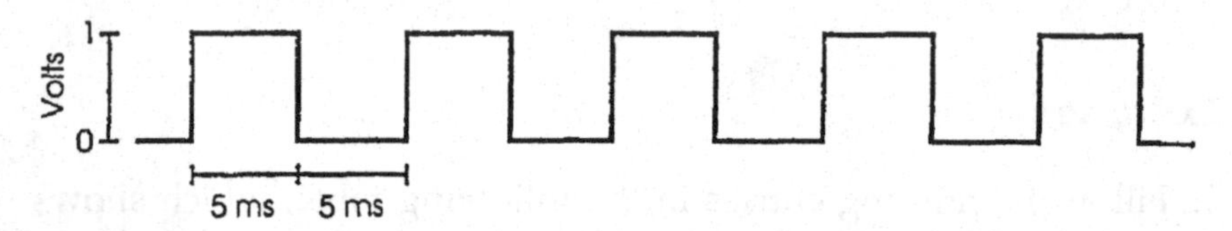

What would be the peak-to-peak amplitude of a sinusoid with the same RMS amplitude?

6. Here are two pulse trains with the same periodicity as the previous square wave (10 ms). Calculate their duty cycles and RMS amplitudes:

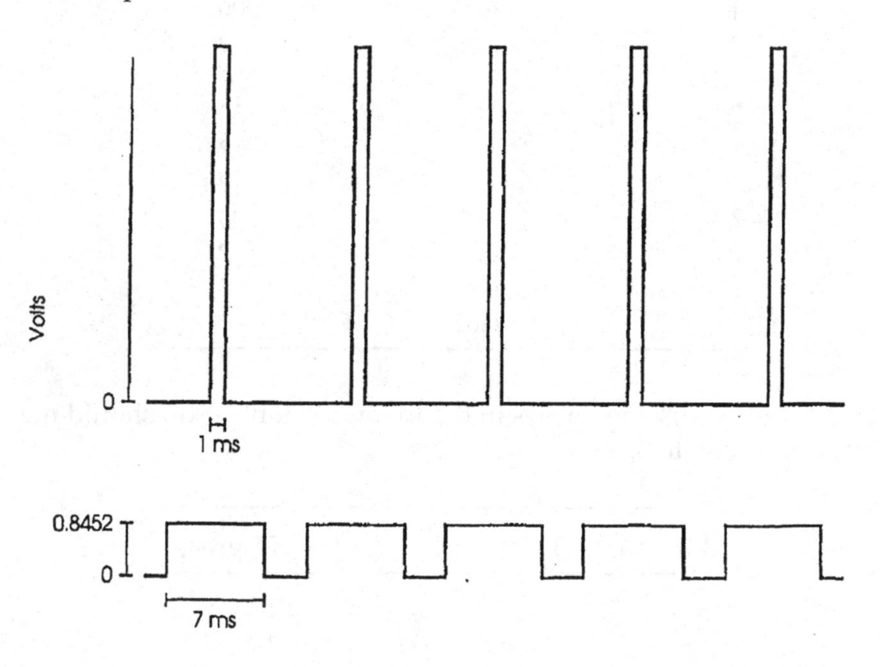

7. Determine the general relationship between the RMS and peak-to-peak values of a square wave (50% duty cycle) which goes from 0 V to some arbitrary voltage, *V*.

8. Determine the general relationship between the RMS and peak-to-peak amplitudes of a pulse train with arbitrary duty cycle ($d\%$) which goes from 0 V to some arbitrary voltage, *V*. Check the results of your formula with your answers to exercises 5, 6 and 7. (Hint: Do not forget that d must be divided by 100 if it is to represent a proportion and not a percentage.)

9. (For readers with a basic knowledge of calculus.) Show that the RMS amplitude of sinusoid is 0.707 of its peak value. (Hint: Start by squaring a sine wave and then taking its integral over one period.)

10. Fill in the missing values in the following table (a calculator should not be needed):

Volts	dB re 1 V
2	6
1	?
0.5	?
?	20
?	−12
0.2	?
?	60
8	?

11. Suppose that two sinusoids with the same frequency are added together in phase, one of 94 dB SPL, and one of 74 dB SPL. What is the resulting SPL? Explain this (perhaps surprising) finding.

12. (More difficult.) What is the minimum dB difference that must exist between two sinusoids such that adding them together in phase does not make the level of the sum more than 2 dB greater than the larger of the two signals? (Hint: Let l be the linearly expressed amplitude of the lower amplitude signal, and g be the linearly expressed amplitude of the greater amplitude signal. The condition stated above is satisfied by solving:

$$20\ \log\left\{\frac{g+l}{\text{ref}}\right\} - 20\ \log\left\{\frac{g}{\text{ref}}\right\} = 2$$

Make use of the laws of logarithms to simplify this equation, and solve for l in terms of g. Finally, calculate the dB value of l relative to the level of g. If you find this too difficult, get the answer anyway by trying some different differences in dB and estimating the value empirically.) Find the same condition for a maximum change of 1 dB in the sum.

Appendix: Exponents and logarithms (to the base 10)

The log to the base 10 of a number is the number of times 10 has to be multiplied by itself to produce the original number. So, the log of 100 to the base 10 is 2, because $10 \times 10 = 100$. 100 can also be represented as 10^2. '2' is known as the *exponent* of 10, and is the same value as the log. Though we have given a whole-number log to keep things easy, exponents aren't always whole numbers. A log of 2.5 means that 2.5 is the exponent. So the number that was logged is $10^{2.5}$, which is 316.23 (as you would expect, somewhere between $10^2 = 100$ and $10^3 = 1000$). Logs need not always be positive, either. A negative log (or exponent) simply means that the original number is less than 1.

A summary of important aspects of exponents and logs follow. For ease of explanation, only whole numbers will be used here, but the results obtained are true for fractional numbers as well.

Exponents

A. Assume that m is a positive integer (1, 2, 3, 4, 5, ...).

10^m is defined as:

$$\underbrace{10 \times 10 \times \cdots \times 10}_{m \text{ times}}$$

B. This leads to three laws of exponents.

I. $10^m \times 10^n = 10^{m+n}$

Proof:

$$10^m \times 10^n = \underbrace{10 \times 10 \times \cdots \times 10}_{m \text{ times}} \times \underbrace{10 \times 10 \times \cdots \times 10}_{n \text{ times}}$$

Thus, on the right we have:

$$\underbrace{10 \times 10 \times \cdots \times 10}_{m+n \text{ times}}$$

Which, by definition, is 10^{m+n}

II. $10^m/10^n =$ (i) 1 if $m = n$

(ii) 10^{m-n} if $m > n$

(iii) $1/10^{n-m}$ if $m < n$

Proof:

(i) if $m = n$, $10^m/10^n = 10^m/10^m = 1$

$$\text{if } m > n,\ 10^m/10^n = \frac{\overbrace{10 \times 10 \times \ldots \times 10}^{m \text{ times}}}{\underbrace{10 \times 10 \times \ldots \times 10}_{n \text{ times}}}$$

$$= \frac{\overbrace{10 \times 10 \times \ldots \times 10}^{n \text{ times}} \times \overbrace{10 \times 10 \times \ldots \times 10}^{m-n \text{ times}}}{\underbrace{10 \times 10 \times \ldots \times 10}_{n \text{ times}}}$$

Cancelling out n terms from the top and bottom, we have:

$$= \underbrace{10 \times 10 \times \cdots \times 10}_{m-n \text{ times}} = 10^{m-n}$$

(ii) if $m < n$,

$$10^m/10^n = \frac{\overbrace{10 \times 10 \times \cdots \times 10}^{m \text{ times}}}{\underbrace{10 \times 10 \times \cdots \times 10}_{m \text{ times}} \times \underbrace{10 \times 10 \times \cdots \times 10}_{n-m \text{ times}}}$$

Cancelling out terms:

$$= \frac{1}{\underbrace{10 \times 10 \times \cdots \times 10}_{n-m \text{ times}}}$$

$$= \frac{1}{10^{n-m}}$$

III. $(10^m)^n = 10^{mn}$

Proof:

$$(10^m)^n = \underbrace{(10^m) \times (10^m) \times \cdots \times (10^m)}_{n \text{ times}}$$

$$= \overbrace{\underbrace{(10 \times 10 \times \cdots \times 10)}_{m \text{ times}} \times \underbrace{(10 \times 10 \times \cdots \times 10)}_{m \text{ times}} \times \cdots \times \underbrace{(10 \times 10 \times \cdots \times 10)}_{m \text{ times}}}^{n \text{ times}}$$

$$= 10^{\overbrace{m+m+m+\cdots+m}^{n \text{ times}}} = 10^{mn}$$

C. To extend these results to any values of n and m, we need to make the following definitions:

$$10^{-n} = 1/10^n$$

$$10^0 = 1$$

This means law II can be expressed simply as:

$$10^m/10^n = 10^{m-n}$$

which will now hold true for all values of m and n.

Logarithms to the base 10

By definition, if $x = 10^y$, then $y = \log(x)$. The three laws of exponents can then be translated into three laws of logarithms:

I. $\log(xy) = \log(x) + \log(y)$

Let $x = 10^m$, $y = 10^n$.

$$\begin{aligned}\log(xy) &= \log(10^m \times 10^n) = \log(10^{m+n}) \text{ by law I of exponents} \\ &= m + n = \log(10^m) + \log(10^n) = \log(x) + \log(y)\end{aligned}$$

II. $\log(x/y) = \log(x) - \log(y)$

Again, let $x = 10^m$, $y = 10^n$.

$$\begin{aligned}\log(xy) &= \log(10^m/10^n) = \log(10^{m-n}) \text{ by law II of exponents} \\ &= m - n = \log(10^m) - \log(10^n) = \log(x) - \log(y)\end{aligned}$$

III. $\log(x^n) = n \log(x)$

Let $x = 10^m$.

$$\begin{aligned}\log(x^n) &= \log[(10^m)^n] = \log(10^{m\,n}) \text{ by law III of exponents} \\ &= mn = nm = n \log(10^m) = n \log(x)\end{aligned}$$

CHAPTER 4

Introduction to Systems

Signals are, of course, only half of our story. It is now time to consider *systems*. Simply put, a system is something which performs some operation on, or transformation of, an *input* signal to produce an *output* signal. Let's start by looking at a system that is easy to describe in a few words, one that is used to increase the amplitude of a signal. Such devices are commonly known as *amplifiers*.

The particular amplifier we will discuss takes the value at its input, multiplies it by two and presents it at the output. We can draw a diagram of this system transforming a particular input signal to get a better feel about what's going on:

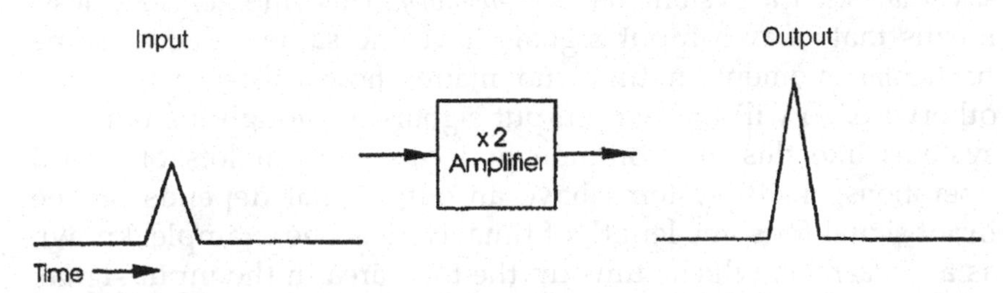

On the left is the input signal. Because we are showing the waveform directly, the x-axis represents time. The specific meaning of the y-axis depends upon the type of input signal. If the input is an electrical signal, the height of the curve will typically indicate the voltage at any particular time; if a sound, the height will indicate sound pressure. In all cases, however, the y-axis indicates some measure of amplitude. Here we have chosen a triangular waveform as the input. The arrow on the left indicates that the input signal is fed in to the system, represented by the rectangular box. The arrow at the right of the box represents the passage of the signal out of the system. At the far right is seen the output signal. As expected from our earlier description of this amplifier, the output signal looks just like the input, except that it is twice as big.

We can also represent this system by a formula:

$$\text{output} = 2 \times \text{input} \tag{1}$$

Remember that 'output' and 'input' represent waveforms over some stretch of time, not just single numbers. To make this explicit, we will write both signals as functions of time, represented by t:

$$\text{output}(t) = 2 \times \text{input}(t) \tag{2}$$

or slightly abbreviated as:

$$\text{outp}(t) = 2 \times \text{inp}(t) \tag{3}$$

We can translate this formula back into words as: 'The value of the output at time t is given by the value of the input at time t multiplied by 2'.

Don't get the impression that all systems can be described in such a simple way. This system is rather special in that its output at any particular moment depends on the value of its input at a single moment in time. So, for example, in order to know the *output* of the system at some time, we only need to know its *input* at that same time. Put another way, the output of the system doesn't depend in any way on what has happened to it in the past. Because it only needs to look at one moment of the input signal to determine its output, it doesn't need to store any information: we say that such a system has no *memory.* This 'memoryless-ness' means that, if two input signals have the same value at some particular moment in time, no matter how different they are otherwise, so will the two output signals. Although memoryless systems like this are common and can perform lots of useful operations, many systems have an output that depends on the input signal for some length of time. Here is one example, known as an *integrator,* which sums up the total area in the input signal:

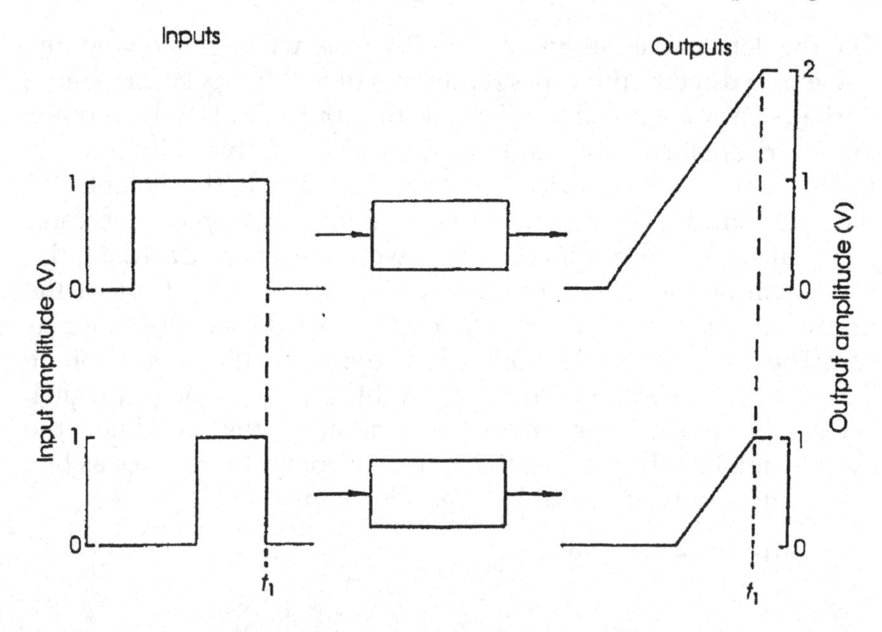

Two input waveforms are shown, both transient rectangular pulses but of different durations. Although they have the same value at time t_1 having come back down to zero, the outputs at that time are *not* the same. This is unlike the amplifier, where, if two inputs were the same at some moment, the outputs would also be the same. In fact, most of the systems we will discuss throughout this book will be ones *with* memory.

In the real-life examples given in Chapter 1, it should be easy for you to recognize what the systems are, since you already know the signals. For the discussion here, we will use a dated piece of equipment (a 'Walkman'), which is convenient because it plays music from a physical storage device (magnetic recording tape contained in a cassette with two spools) quite separate from the rest of the systems involved. In playing a cassette tape, the first system is the tape-head and associated circuitry which converts the magnetic information recorded on the tape into an electrical signal. This signal then enters an amplifier, which boosts the power of the signal. Finally, this boosted signal goes to yet another pair of systems, the headphones, which convert the voltage variations of the signal into the air pressure variations we hear as sound. We could also consider all these systems connected together as one big system, with the magnetic information on the tape as the input, and the sound as the output.

Note that our analysis of systems puts no constraints on the form the input or output signals take. A system can have electrical signals as input and output (as in an amplifier) or an electrical signal as input and sound as output (as in a headphone) or vice versa (as in a microphone) or any other combination. Similarly for signals—they can be electrical, mechanical or acoustic. The methods of analysis we will develop describe signals and systems in ways that make it possible to use techniques which do not depend on the particular physical form that the signals or systems take.

However, it would be too much to expect to be able to analyze *every* system with the same techniques. It turns out that there is a large and important subset of systems that *can* be approached with one set of ideas. These are known as *linear time-invariant* (LTI) systems, and they will be the focus of our discussions. For the rest of this chapter, we'll detail just what is meant when we say that a system is LTI—beginning with the notion of *linearity.*

What, then, makes a system 'linear'? A system is said to be linear if it fulfills two requirements: that it is both *homogeneous* and *additive*. We'll deal with homogeneity first, as it is a property that is frequently investigated in real-life systems.

Homogeneity

Homogeneity means that if an input signal 'inp(t)' applied to some system gives an output 'outp(t)' then the input signal multiplied by some arbitrary constant value will give an output of the same constant times 'outp(t)'. In a simple formula, if:

$$\text{inp}(t) \rightarrow \text{outp}(t) \tag{4}$$

(where we translate '→' as meaning 'is transformed by a particular system into'), then:

$$k \times \text{inp}(t) \rightarrow k \times \text{outp}(t) \tag{5}$$

where k is an arbitrary number.

So, for example, if we know the output of a homogeneous system to a sinusoid of amplitude 1 V at 300 Hz and wanted to know the output to a 300-Hz sinusoid of amplitude 2 V, all we would do is multiply the amplitude of the previous output signal by a factor of 2. Homogeneity, then, implies a *proportionality* between the level of the input and the level of the output. As in the example just given, if the input level is doubled, so is the output level. If the input level is halved, so is the output level. Any expansion or reduction in the level of the input will lead to an equivalent change in the level of the output.

How does one go about testing a system for homogeneity? Although homogeneity applies to any signal, a sinusoid of a particular frequency is almost always used in practical tests. One then investigates how the amplitude of the output changes with changes in the amplitude of the input. The easiest way to do this is to plot the amplitude of the output versus the amplitude of the input for a number of levels of the input signal. A system that is homogeneous will give a result like this:

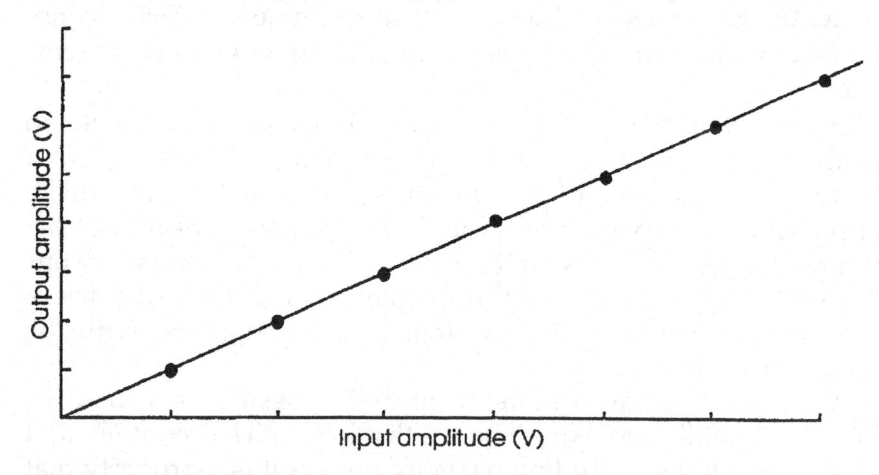

As stated above, homogeneity implies that the output amplitude must be proportional to the input amplitude, reflected here as a linear relationship between the two variables. The slope of the line can vary but it must pass through the origin, as can be seen by setting the arbitrary constant k equal to zero in equation (5). Put into words, a homogeneous system will have zero output for zero input. This last property, although true for every homogeneous system, is not a very good way of differentiating homogeneous from non-homogeneous systems. Many non-homogeneous systems show zero output for zero input.

Note, finally, that the shape of the output signal, whatever it happens to be, cannot change with changes in the level of the input. This can be inferred from the original definition of homogeneity, since the output signal can only change by a multiplicative factor.

To summarize, a system can be tested for homogeneity by putting in a sinusoid at a particular frequency and then plotting the amplitude of the output versus the amplitude of the input for a number of different input levels. If a straight line passing through the origin results, while the output wave shape remains the same, the system is homogeneous.

Testing homogeneity in the peripheral auditory system

We'll now look at some experiments that sought to test the homogeneity of parts of the peripheral auditory system, crucial to the understanding of hearing, since all airborne sounds pass through it:

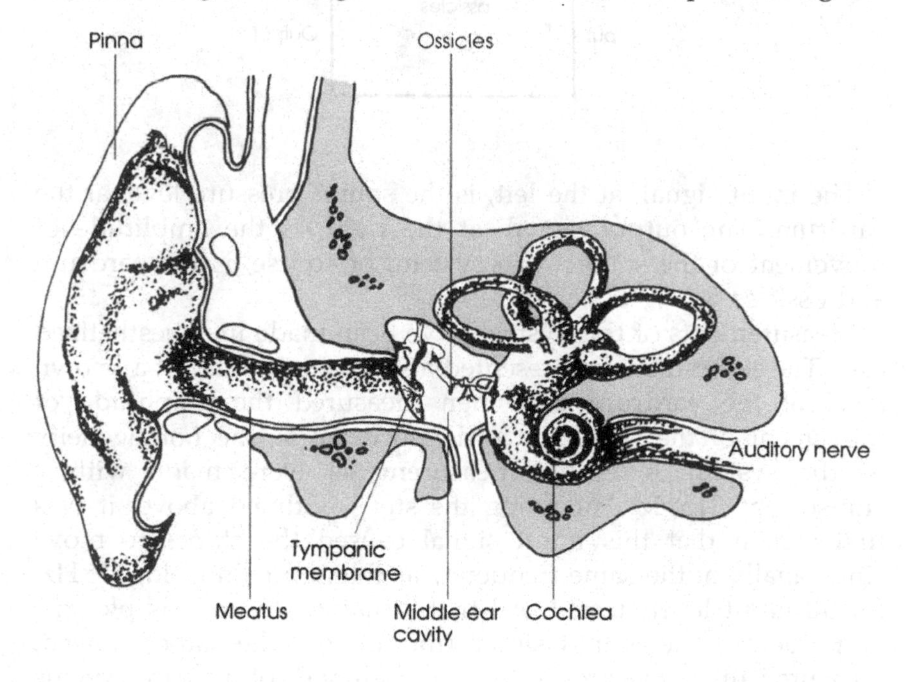

You probably already have a pretty good idea of the general functioning of this system, but a quick review won't hurt. Sound travels through the air, is collected by the *pinna* (in common language, the ear) and then passes into the ear canal or *meatus*. These two structures, the only ones readily visible, make up the *outer ear*. Stretched across the bottom of the meatus is the eardrum or *tympanic membrane*. Sound pressure variations in the meatus cause this membrane to vibrate. Behind the tympanic membrane is a set of three small bones, the *ossicles*. They are known as the *malleus* (hammer), the *incus* (anvil) and the *stapes* (stirrups). The tympanic membrane and the ossicles together are known as the *middle ear*. The malleus is attached to the tympanic membrane, while the stapes is inserted into a hole in the inner ear or *cochlea*, known as the *oval window*. As the tympanic membrane moves, so does the malleus, incus and stapes in turn, resulting in a movement of the stapes in the oval window. This is only the first stage in the process that leads to a perception of sound, but we'll stop here for a moment to investigate more thoroughly the conversion of sound at the eardrum into a movement of the stapes. In a 'box-and-arrows' representation, here is our system:

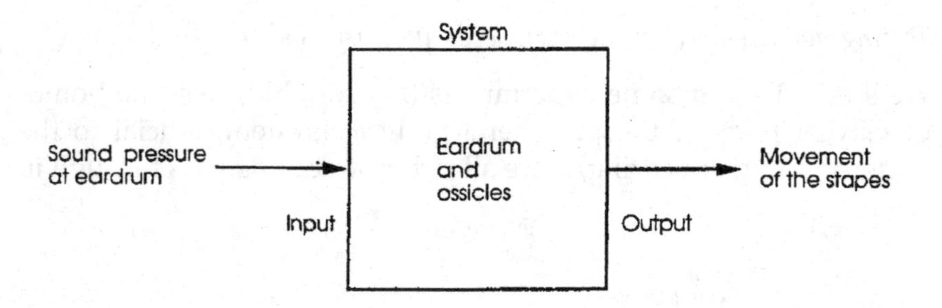

The input signal, at the left, is the sound pressure level at the eardrum. The output signal, at the right, is the amplitude of movement of the stapes. The system, of course, is the eardrum and ossicles.

Measurements of this system have been made in anaesthetized cats. The experimenters presented a sinusoidal sound of a known level at the eardrum and then measured the amplitude of movement of the stapes. In order to examine the homogeneity of the system, extensive measurements were made with a sinusoid at 315 Hz. Following the steps outlined above, it was first shown that this input signal caused the stapes to move sinusoidally at the same frequency as the input sinusoid, 315 Hz, for all input levels tested. All that remained to do was plot the amplitude of the output signal (the amount the stapes moved, measured in *micrometres* (μm)—one millionth of a metre) versus the amplitude of the input signal (the sound pressure level expressed in pascals):

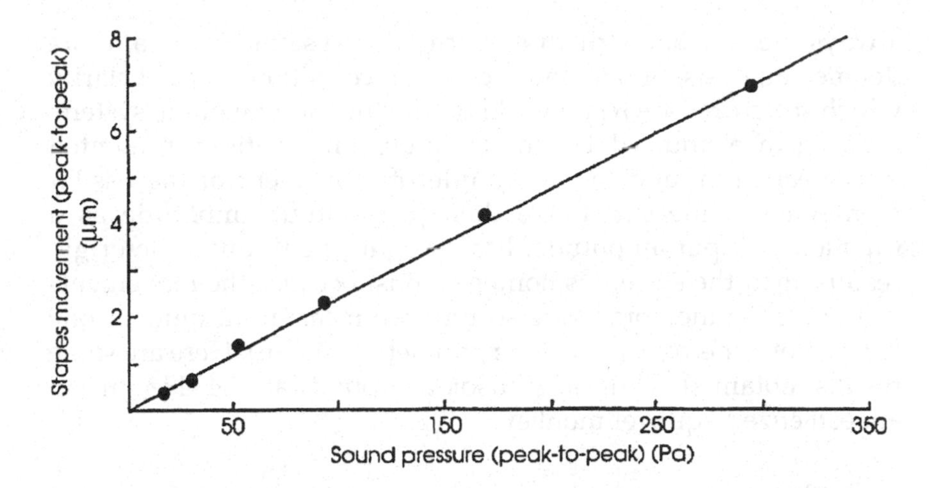

The circles mark the actual measurements made. As is clearly seen, the points describe a straight line that intersects with the origin. Therefore, the system is homogeneous. You may notice that a linear scale is used for both the input (sound pressure in pascals) and output (stapes displacement in μm), as tests of homogeneity are easier to appreciate in this form than when dB scales are used.

So, we now know that the system comprising the eardrum and ossicles is homogeneous. Things don't always work out this way, though, as consideration of another part of the peripheral auditory system demonstrates. We'll continue our discussion where we left off before, at the movement of the stapes in and out of the oval window of the cochlea.

The cochlea itself is a fluid-filled tube divided down almost its entire length by the *basilar membrane.* The movement of the stapes causes waves in the fluids of the cochlea which in turn cause the basilar membrane to vibrate. The movement of the basilar membrane is *transduced* (changed) into firing patterns on the auditory nerve by the *hair cells.* The nerve spikes are then transmitted to the brain, where they are interpreted as sound.

Because the basilar membrane plays such an essential role in the chain of events that eventually leads to a sound being heard, it is important to know how it vibrates in response to sound. Therefore, the output signal we want to study is the amplitude of vibration of a point on the basilar membrane. We'll choose the sound pressure level at the eardrum as the input signal, as we did before for the middle ear. The system, then, can be roughly labelled as eardrum-plus-ossicles-plus-cochlear-fluids-plus basilar membrane. We say roughly, because there are other structures present (e.g., Reissner's membrane and the hair cells) which may affect the vibration of the basilar membrane. Even though the input and output signals are unambiguously defined, it's not always clear what structures form the system. Only a more detailed analysis, or one under different conditions, would clarify which structures are responsible for the functioning of the system.

So, again, a sinusoidal sound at different intensities is presented at the eardrum, and the amplitude of movement of the basilar membrane is measured. We then plot output amplitude as a function of input amplitude. If a straight line through the origin results, then the system is homogeneous. Because the movements of the basilar membrane are so tiny, we measure in units of one *billionth* of a metre, known as a nanometre (10^{-9} m). Here are some results obtained with a sinusoidal sound at 7.4 kHz in an anaesthetized squirrel monkey:

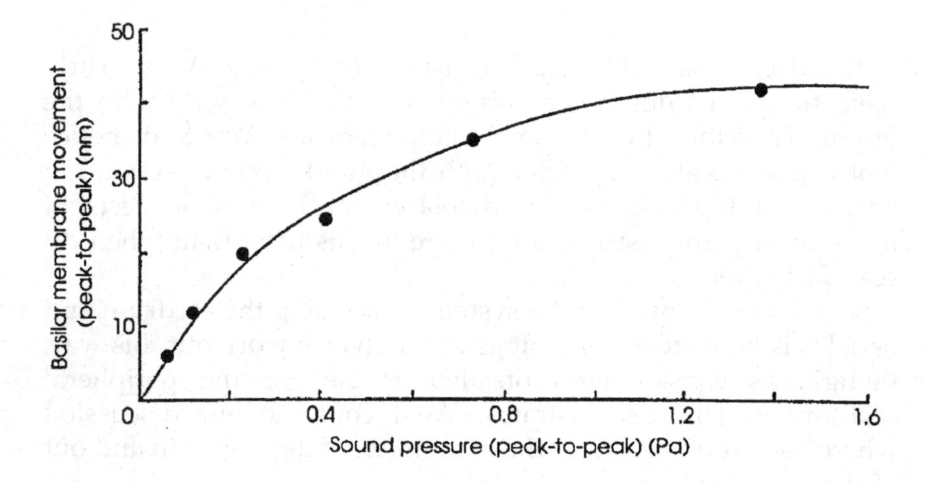

Although the curve does seem to pass through the origin, it does not describe a straight line. The amplitude of movement of the basilar membrane, once it reaches a particular amplitude, increases at a slower rate than the increase in the sound pressure level. In other words, the relationship between output and input amplitude is *not* proportional. Therefore, the system is not homogeneous and is hence non-linear. A system which exhibits this property is often said to show a *saturating non-linearity,* because the amplitude of the output levels off, or saturates, with increases in the input. This non-linearity in basilar membrane movement has wide-ranging implications for the understanding of human auditory perception.

To what part of the system do we ascribe the non-linearity? This experiment on its own only shows that there is a non-linearity somewhere between the eardrum and the basilar membrane, but doesn't allow us to specify exactly where. However, the previous results showed the 'eardrum-plus-ossicles' system to be linear. Therefore, the non-linearity must be 'between' the stapes and the basilar membrane, somewhere in the inner ear.

Testing homogeneity 'back in the day': a cassette recorder

Saturating non-linearities must have their uses, or they presumably wouldn't be found in natural systems. It should be clear, however, that they would be highly undesirable in systems that were concerned with the *fidelity* of reproduction, for example, in any kind of recording system. We would therefore expect systems for reproduction to be strictly homogeneous, at least for a reasonable range of input levels.

Let's perform such a test on a particular example of a recording system, technically far easier than measuring stapes and basilar membrane movement! Before the advent of modern digital equipment, it was quite common to record music on cassette tapes, so they could be played on a 'Walkman' of the type described earlier. In order to test homogeneity on such a cassette recorder, all we need do is pick a sinusoid of some frequency, record it at various levels and then play back the tape, examining and measuring the output signal. We'll use an input electrical sinusoid of 1 kHz (as this is a frequency typically used for testing audio equipment), starting with an input level of 1 V peak-to-peak (as this is typical of the size of signals one would want to record). Before we can make the recording, though, we need to set the *record level* knob. This adjusts the sensitivity of the recorder in order to permit its optimal operation over a wide variety of situations. Some tape-recorders do this automatically, adjusting themselves continuously to the level of the input signal. This has the advantage of being easy to use, but will seriously distort many aspects of the signal. Therefore, although automatic recording level control is adequate for recordings of lectures and 'low-fi' music, tape recorders for serious purposes should always be equipped with a *manual* control and VU ('view') meter which allows the control to be set appropriately. Even up-to-date high quality audio recorders will have a manual volume control, although the meter is typically in the form of a set of LEDs.

We've now got the 1-V, 1-kHz sinusoid being fed into our 'very serious' (if seriously outdated) tape recorder. We first adjust the level of the recording control so that the meter reads 0 dB VU (i.e., 0 dB on the 'view' meter). We then record short stretches of the 1-kHz tone at this and a variety of other levels, both above and below 1 V. We can then play them back, and inspect the output.

Let's look first at our standard input signal of 1 V, and the output it leads to:

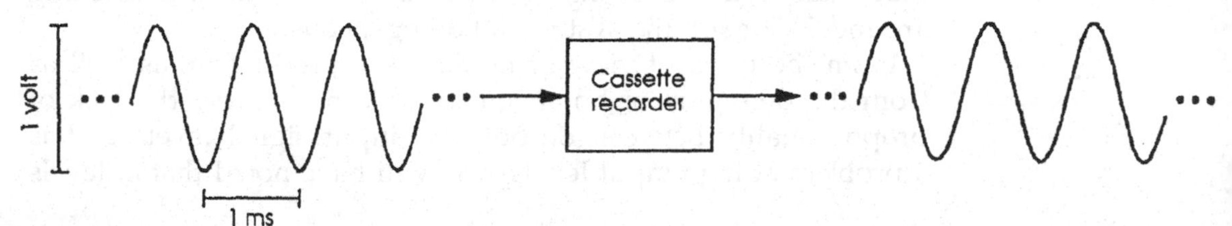

You can see that the output looks pretty much just like the input sinusoid, and has the same level. If we now look at an input a bit lower in level, at 0.6 V, you can see a similar outcome:

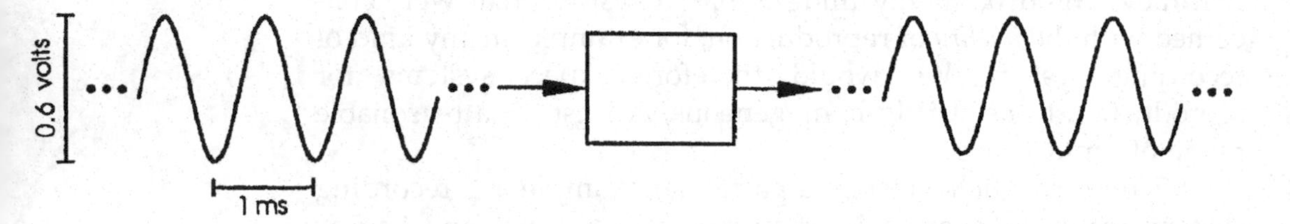

From the point of view of homogeneity, things are looking pretty good so far for this cassette recorder. This impression is strengthened if we now plot output level versus input level for input voltages at 1 V and below:

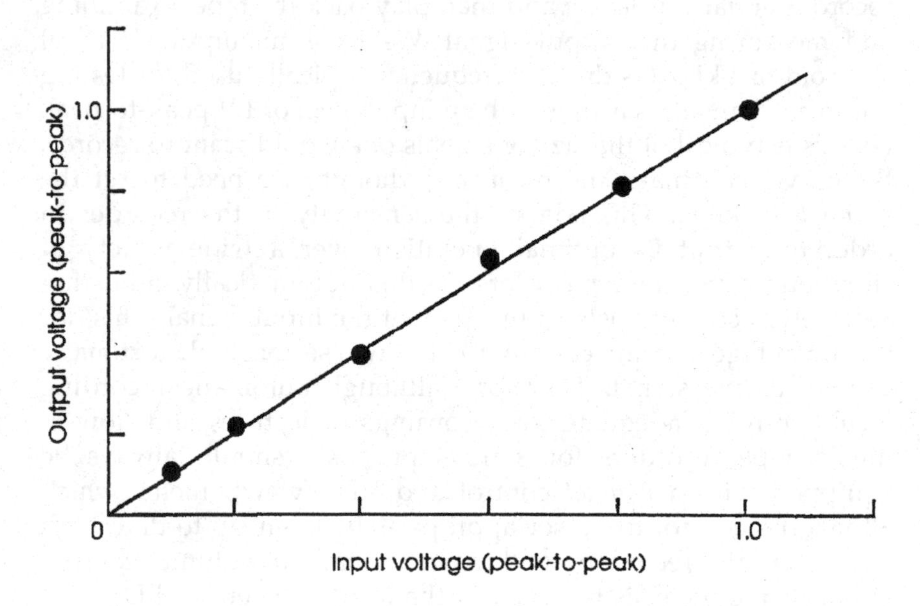

As you can see, the measured points (indicated by the filled circles) lie quite close to a straight line passing right through the origin. The system looks to be homogeneous.

However, if we extend our measurements *upwards* in amplitude from 1 V, we get the result seen on top of the next page—a saturating non-linearity.

Therefore, it appears that as long as our input signal doesn't go much above the level that gives 0 dB VU (the limit here being around 2 V or so), the system is homogeneous.

From the point of view of fidelity of reproduction, as well as from the definition of homogeneity, it is not simply the lack of proportionality between output and input signal levels that is a problem at high input levels. You will have noted that at levels

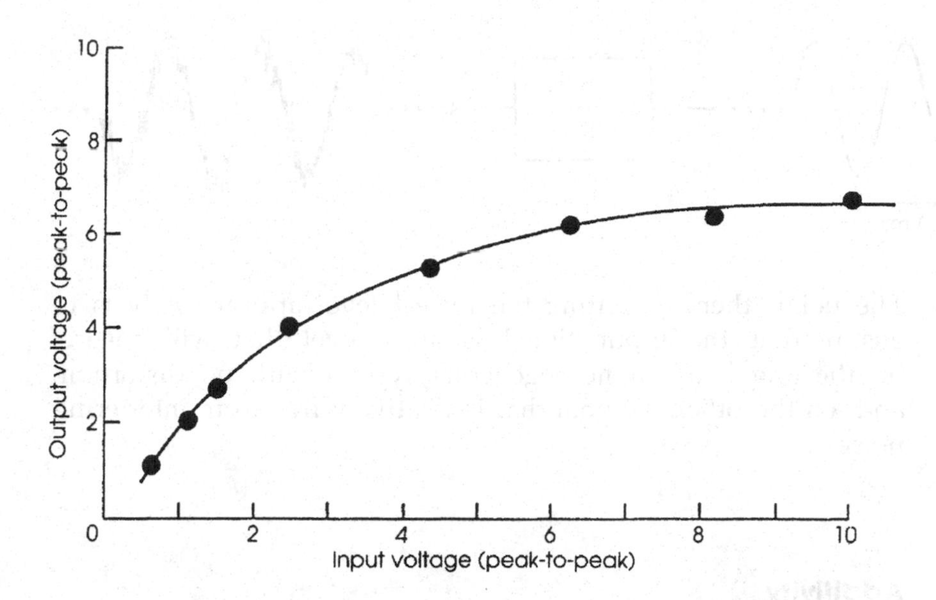

near 1 V the shape of the output signal mirrored that of the input signal quite closely—this is what we mean by 'good reproduction'. At high signal levels, this is not so, as can be seen for the output signal resulting from a 10 V input:

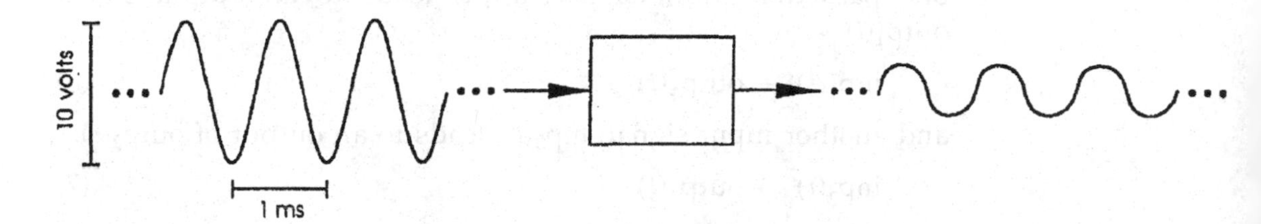

Here a sinusoidal input does *not* lead to a sinusoidal output—the output is *distorted*. Clearly, this is not what we want of a high-fidelity system. Incidentally, current-day digital recorders distort waveforms that are too high even more severely than this, 'clipping' off the tops and bottoms of the waves. Older systems are often said to fail more 'gracefully'.

Don't get the idea that only signal levels that are too high are a problem. Although our tape-recorder is probably homogeneous right down to levels near zero, a different problem arises when the input gets too small. All systems have a certain amount of 'noise' in them (here mainly from the cassette tape itself) that tends to stay relatively fixed in amplitude. Therefore, as the input level gets smaller and smaller (and so does the output level from the tape), the level of the noise starts to approach that of the signal. This can make the output signal look very messy, as can be seen from what comes out of our tape recorder for an input sinusoid of 20 mV (0.02 V):

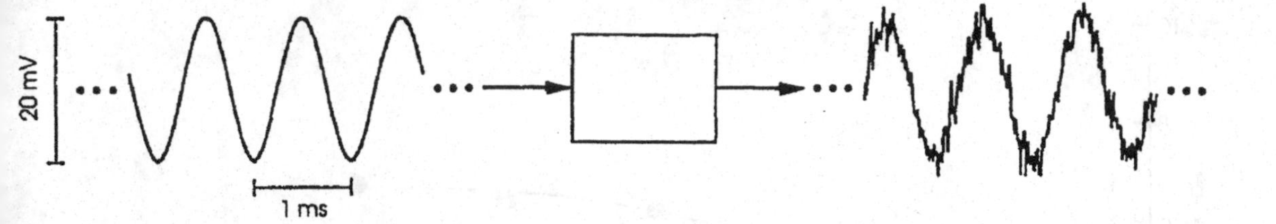

The point, then, of setting the record level appropriately is to ensure that the input signal is at a level that will ensure, on the one hand, a homogeneous system with no distortion, and, on the other, a signal that is relatively free from interfering noise.

Additivity

So much for homogeneity. Additivity, the other half of linearity, means that the output of a system to the sum of two input signals can be predicted from the sum of the outputs to each of the two inputs on their own. Put in a more convenient form, if one particular input signal $\mathrm{inp}_1(t)$ leads to an output signal $\mathrm{outp}_1(t)$:

$$\mathrm{inp}_1(t) \rightarrow \mathrm{outp}_1(t) \tag{6}$$

and another input signal, $\mathrm{inp}_2(t)$, leads to an output of $\mathrm{outp}_2(t)$:

$$\mathrm{inp}_2(t) \rightarrow \mathrm{outp}_2(t) \tag{7}$$

then:

$$\mathrm{inp}_1(t) + \mathrm{inp}_2(t) \rightarrow \mathrm{outp}_1(t) + \mathrm{outp}_2(t) \tag{8}$$

To summarize, if a system is additive, and if the output to each of two inputs is known, then the output to the sum of the two inputs is given simply by the sum of the two separate outputs.

The next figure illustrates how this property works for the amplifier discussed at the beginning of this chapter. Shown at the top left are two separate input signals, and their corresponding outputs (at right) after being multiplied by a factor of 2. If we know that a system is linear (as we will later show our amplifier to be), then in order to find its output to a sum of the two input signals, all we need do is sum the two output signals. The crucial point here is that we get the same answer whether or not we add the two input signals together, and then put them through the system, or first put the two inputs through the system separately, and *then* add the results:

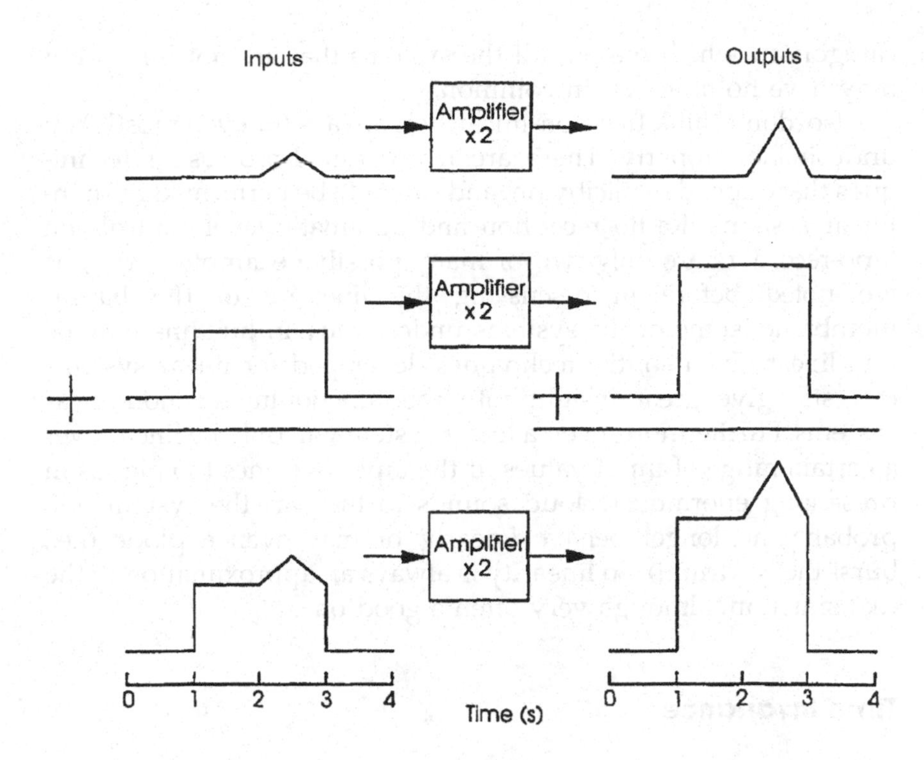

Linearity = homogeneity + additivity

In order to show that a system is linear, then, we need to show that it is both homogeneous and additive. Although this can be done separately, it is often more convenient to use a condition that includes *both* criteria in one simple formula.

Suppose we have two arbitrary input signals, $\mathrm{inp}_1(t)$ and $\mathrm{inp}_2(t)$, that lead to the outputs $\mathrm{outp}_1(t)$ and $\mathrm{outp}_2(t)$, respectively:

$$\mathrm{inp}_1(t) \rightarrow \mathrm{outp}_1(t) \tag{9}$$

$$\mathrm{inp}_2(t) \rightarrow \mathrm{outp}_2(t) \tag{10}$$

By a straightforward combination of the properties of additivity and homogeneity, we can say that a system is linear if and only if, for arbitrary constants a and b:

$$(a \times \mathrm{inp}_1(t) + (b \times \mathrm{inp}_2(t)) \rightarrow (a \times \mathrm{outp}_1(t) + (b \times \mathrm{outp}_2(t)) \tag{11}$$

At this point, let us just point out that linear systems are only a small subset of the possible systems, one could imagine. To say a system is non-linear is really to say very little about it. It is something like, when describing an animal, calling it a non-mammal. In this sense, 'non-linear systems' is a waste-basket category in which we put all the systems that are not linear, but may have nothing else in common.

Also, don't think that non-linearity is always (or even mostly) an undesirable property. There are many signal-processing techniques that depend explicitly on, and can only be performed by, non-linear systems. Radio reception and automatic level controls on tape-recorders are only two of many possible examples. Also, as we noted before in discussing the linearity of the basilar membrane, some of the systems under study in humans may be non-linear. Even so, the techniques developed for linear systems can still give great insight into the functioning of non-linear systems. Furthermore, even a linear system will only be linear over a certain range of input values. If the input becomes too big (as in presenting enormously loud sounds to the ear) the system will probably no longer behave linearly, or may even explode (i.e., burst the eardrum!). So linearity is always an approximation to the real situation, although very often a good one.

Time invariance

One further property, apart from linearity, will characterize many of the systems we will deal with: *time invariance.* Intuitively speaking, a system that is time invariant is one that does not change over time. It doesn't matter *when* a particular signal is input to a time-invariant system; the output signal will always be the same. The only thing that will change is the absolute time at which the output signal appears, since the time between the application of the input and the appearance of the output will remain the same. This property can also be written down using the notation we have developed previously. Given

$$\text{inp}(t) \rightarrow \text{outp}(t) \tag{12}$$

then

$$\text{inp}(t) \text{ delayed by } d \text{ seconds} \rightarrow \text{outp}(t) \text{ delayed by } d \text{ seconds} \tag{13}$$

There is a less wordy way to write this down. Consider the input signals inp(t) and inp($t-3$). These are clearly the same signal, only shifted in time relative to one another. At time 3 s, say, inp($t-3$) has the same value of inp(t) at time $t = 0$. At time 6 s, inp($t-3$) has the same value as inp(t) at time $t = 3$. In other words, inp($t-3$) is 3 s behind inp(t). Similarly, inp($t-d$) is simply the signal inp(t) delayed by d s. You have already encountered this idea in a slightly different form in Chapter 3, where it was pointed out that a sine wave and a cosine wave differed only in phase, which in that case was equivalent to a shift in time.

We can therefore write:

$$\sin(2\pi ft) = \cos(2\pi ft - \pi/2) \tag{14}$$

This is just another way of saying that a sine wave of frequency f Hz has exactly the same shape as a cosine wave of f Hz, except that the sine wave is 1/4 of a cycle behind the cosine wave (or equivalently, that the cosine is 1/4 of a cycle ahead of the sine).

This type of notation can be used to simplify the definition for time invariance. A system is time invariant if, given:

$$\text{inp}(t) \rightarrow \text{outp}(t) \tag{15}$$

then

$$\text{inp}(t - d) \rightarrow \text{outp}(t - d) \tag{16}$$

for any value of d.

Our 'two-times amplifier' is one example of a time-invariant system, as also can be seen from the following example. If a triangle is input to the system at time $t = 3$ s, then the amplified triangle also appears at 3 s. If, however, the triangle is delayed for 3 s, not being input to the system until $t = 6$ s, the output, too, is delayed by 3 s, not beginning its appearance until $t = 6$ s:

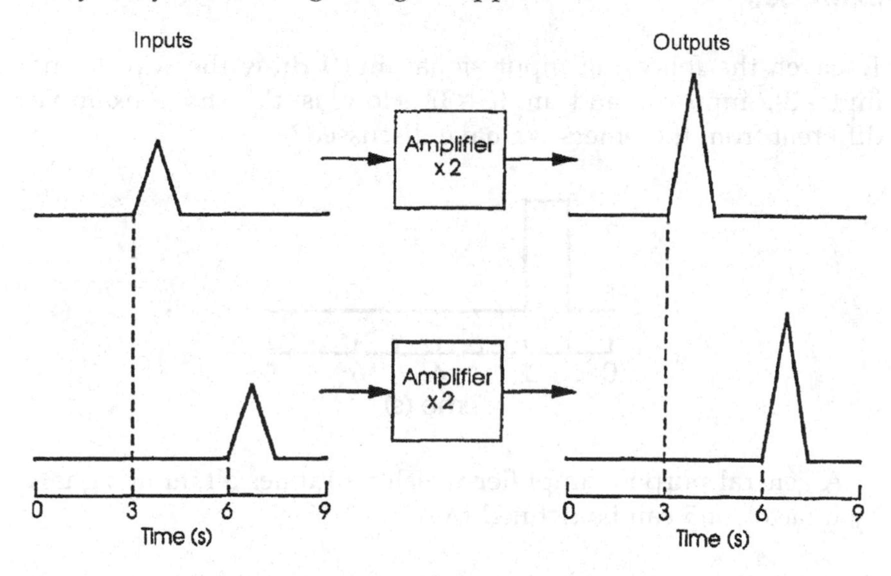

Although we will normally assume that the systems we discuss are time invariant, some of them will definitely not be. Often it is enough for time invariance to hold only over some short length of time. Consider, for instance, the continuous production of a series of different vowels, like 'ee-ah-oo'. The input signal is the sound coming from the vibration of the vocal folds and the output is the sound radiating from the lips. The vocal tract, then, forms the system, which, by changing shape, causes an alteration of the sounds coming from the larynx which we hear as changes in vowel quality (this will be explained much more fully in Chapters 6 and 7). The larynx activity can remain the same during these vowel changes so it is clearly changes in the system (the vocal tract) that are causing changes in the output. Since these changes are occurring in time, the system is said to be *time-varying*. We can usually treat such systems, though, as time invariant as long as we look at their output over sufficiently short periods of time. Some systems, like hi-fi stereo amplifiers, can be considered time invariant over the course of years (unless they break down) while the vocal tract can, for some sounds like plosive consonants, only be considered time invariant for some milliseconds, since it is changing shape so fast.

In summary, we will primarily be considering the behaviour of LTI systems. Although these form only a small subset of the systems imaginable, the applicable techniques can describe a large number of useful systems. Even systems with some non-LTI aspects can be usefully analyzed over part of their range of operation.

Exercises

1. Given the following input signal inp(t) draw the waveforms inp(t−3), inp(t−6) and inp($t+3$). How is this final example different from the others we have discussed?

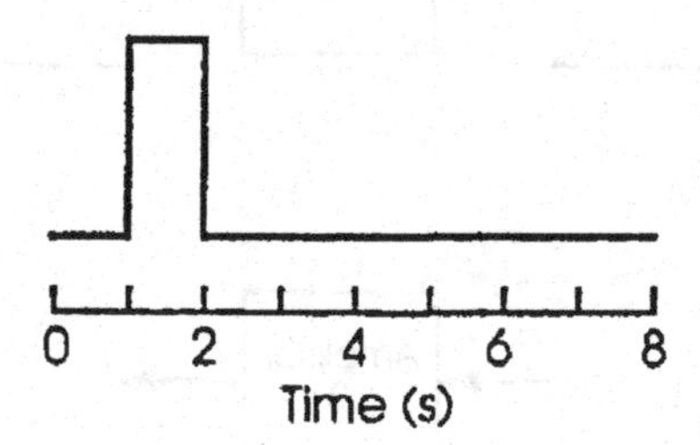

2. A general-purpose amplifier which multiplies all input signals by a factor of 5 can be defined by:

$$\text{outp}(t) = 5 \times \text{inp}(t) \tag{17}$$

Make some arguments that this system is linear. Is it time invariant? What about an amplifier that multiplies all input signals by a factor of m:

$$\text{outp}(t) = m \times \text{inp}(t) \tag{18}$$

Is it LTI? What conditions on m would make it time invariant? (Hint: Consider what would happen if m changed with time.)

3. Describe the way you would test whether or not a telephone system is homogeneous.

4. Show that a system which squares the value of the input is not LTI.

5. Describe the changes caused by each of the following systems to their input signals. Which show input/output pairs that are consistent with a system that is homogeneous? Why?

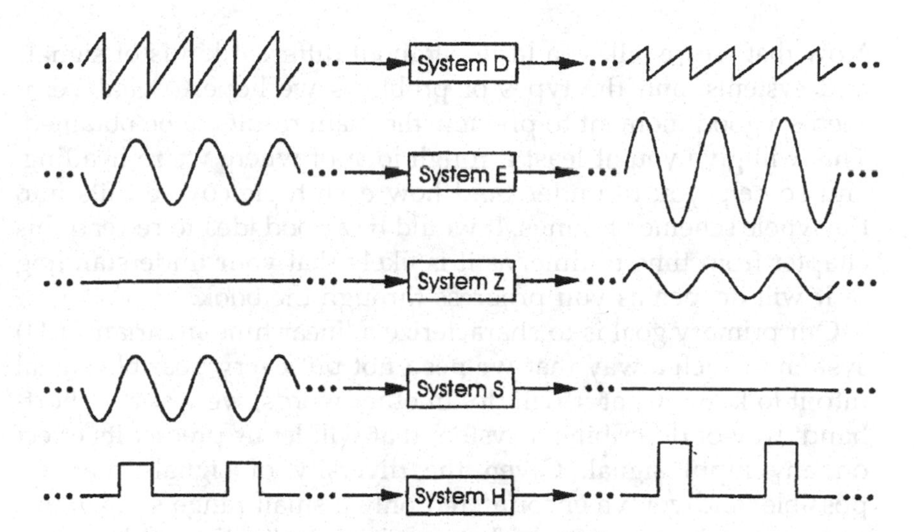

6. Describe the changes caused by each of the following systems to their common sinusoidal input. Write a formula describing the changes for those systems that appear to be LTI.

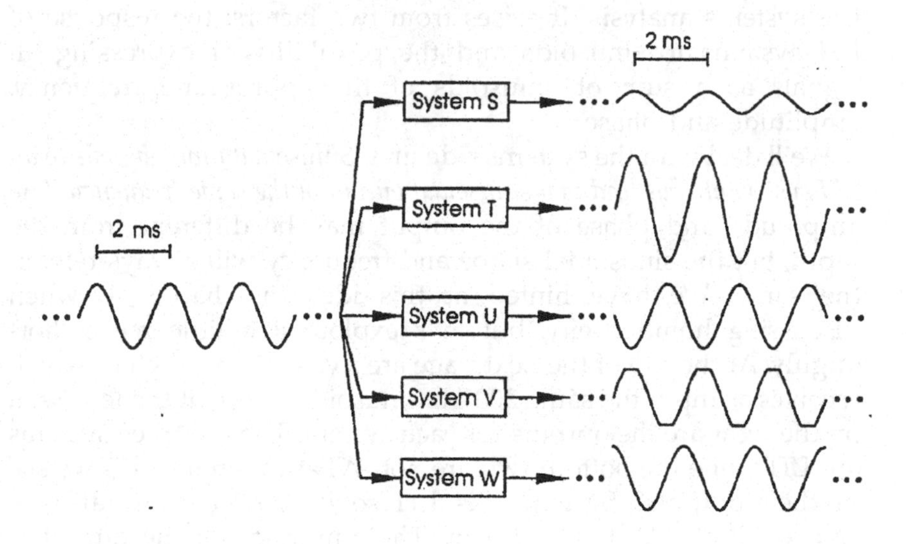

CHAPTER 5

A Preview

Now that we've talked a little bit about different kinds of signals and systems, and the types of problems we hope to analyze, it seems a good moment to preview the main results to be obtained. This will give you at least a rough idea of where we're heading, and so help you to understand how each topic covered fits into the whole scheme of things. It would be a good idea to re-read this chapter from time to time, as it is likely that your understanding of it will deepen as you progress through the book.

Our primary goal is to characterize a linear time-invariant (LTI) system in such a way that we need not put every possible signal into it to know what it will do. In other words, we want a 'short-hand' way of describing a system that will let us predict its effect on any input signal. Given the diversity of signals that are possible (and you've encountered only a small range so far), this may seem a hopeless task. The surprising fact is that, as long as a system is LTI, we only need to know its response to *sinusoidal* inputs in order to predict its response to *any* input.

This property holds only for LTI systems and is the very heart of LTI systems analysis. It arises from two factors: the response of LTI systems to sinusoids and the possibility of expressing all signals as a sum of sinusoids of the appropriate frequency, amplitude and phase.

We'll deal with the systems side first. *Sinusoidal input signals to an LTI system always lead to a sinusoidal output of the same frequency.* The amplitude and phase of the output may be different from the input, but the sinusoidal shape and frequency will always remain the same. We have hinted at this idea in Chapter 4, when discussing homogeneity, but let's explore it a little more thoroughly. At the top of the next page are five systems, each of which receives as input the same 200-Hz sinusoid, shown at the left. Seen on the right are the outputs for each system. The top three systems are LTI while the bottom two are not. What differences do we see in their outputs? System 1 is LTI so its output must also be sinusoidal at 200 Hz, as shown. The amplitude of the output is somewhat reduced from that of the input, but LTI systems can alter the amplitudes of sinusoids. Of course, the amplitude of the output could be zero, so that some sinusoidal inputs give no output, but

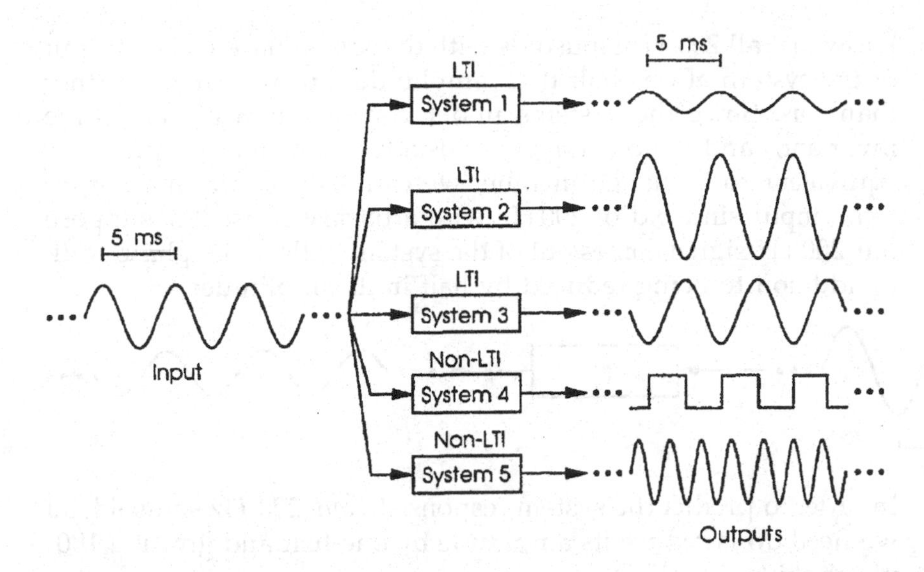

no other sinusoids will be created. LTI System 2 also changes the level of the input signal, here amplifying it. Again, since the system is LTI, the output is still sinusoidal at 200 Hz. LTI System 3 must, by definition, also have a 200-Hz sinusoidal output, but here the signal is altered in phase, and not in amplitude.

Now we come to the non-LTI systems. The output of System 4 is clearly not sinusoidal, so it cannot be LTI even though the output is at the same frequency as the input. System 5 has a sinusoidal output, but at a different frequency than the input, so it is not LTI either.

To look at this another way, if we feed sinusoidal input signals to LTI systems, we already know quite a lot about the outputs that will arise. They will be sinusoidal and at the same frequency as the input. To specify the exact output of the system to a particular sinusoid, we still must determine the output's amplitude and phase, usually by making an empirical measurement. Once we've done this, however, we need not do it again for sinusoids of other amplitudes and phases. As long as we stick to a sinusoid of one frequency, we need only make one measurement, and use this information, along with the properties of linearity and time invariance, to predict the exact outputs in other cases.

The detailed reasoning goes like this. Suppose we know the response of an LTI system to a 200-Hz sinusoid of peak-to-peak amplitude 1 V and phase angle of 180°. The homogeneity aspect of linearity means we can predict the system response to a sinusoid of the same phase and frequency but arbitrary amplitude. This follows from the proportionality between input and output amplitudes in homogeneous systems. For example, if our 200-Hz, 1-V input sinusoid comes out of a system at 0.5 V, we know that all 200-Hz sinusoids with the same phase will come out of the system at one-half their amplitude. But we can go further than this. Since the systems under consideration are also time invariant, and since, for a sinusoid, a change in phase is equivalent to a change in time, we can also predict the output to an input sinusoid of 200 Hz with arbitrary phase. So, suppose our 200-Hz signal comes out of the system with a 180° phase shift, in addition to being reduced by half in its amplitude:

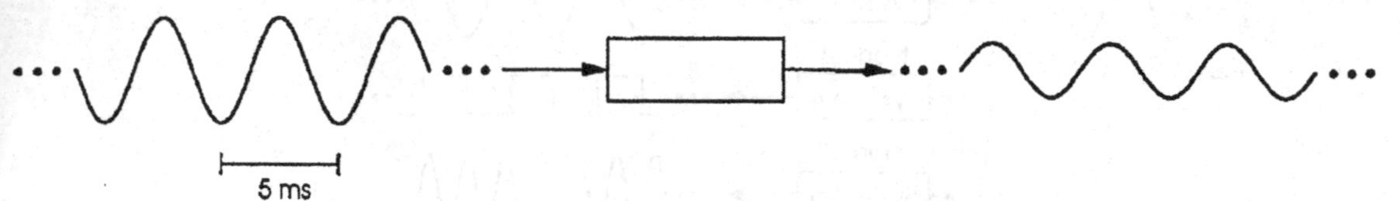

In order to predict the system response to *any* 200-Hz sinusoid, all we need do is reduce its amplitude by one-half and give it a 180° phase shift.

In short, given the response of an LTI system to a sinusoid of a given frequency, we can predict the response to any sinusoid of the *same frequency* (that is, of an arbitrary amplitude and phase).

Of course, the system response to this one sinusoid tells us nothing of the response of the system to sinusoidal inputs of *other* frequencies. These must be specified separately.

This may not seem to have got us very far in our goal of efficient characterization of a system. Sinusoids are only a very tiny fraction of all the signals we may want to pass through a system. We have yet, however, to make use of the additivity property of LTI systems. By our definition of additivity, if we know the response of the system to two signals then we can predict the system response to the sum of the two signals. So, if we know the response of a system to, say, sinusoids of 200 and 400 Hz, separately, we know the response to the two added together.

Let's determine the output of the system that we've just discussed to such an input signal. We already know that, at 200 Hz, sinusoids are reduced in amplitude by one-half and phase-shifted by 180°. The response to a 400-Hz sinusoid must be determined independently. Suppose a 1-V 400-Hz sinusoid comes out of the system with an amplitude of 2 V and with no phase change:

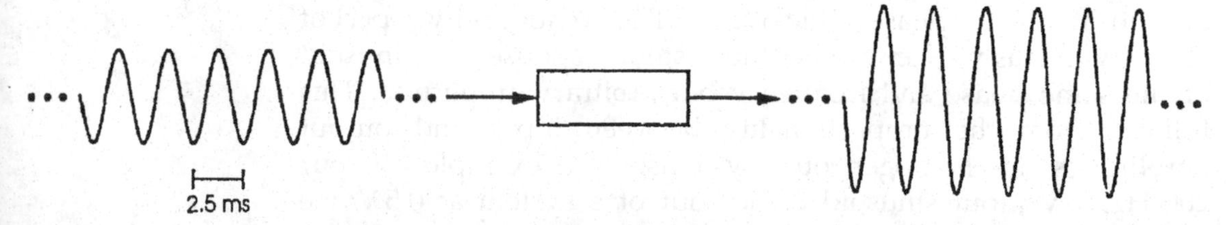

We can now predict the system response to *any* 400-Hz sinusoid—just double the amplitude of the input. In order to obtain the system output to the sum of these two inputs, we simply sum the respective outputs:

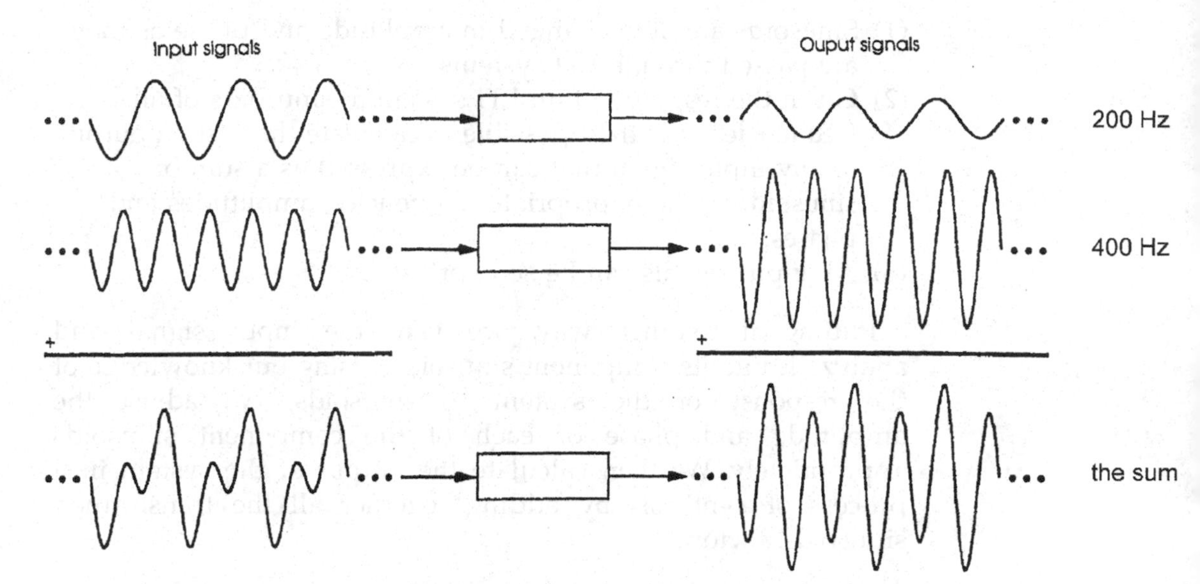

The 200 Hz and 400 Hz input signals at the left have to be added together moment by moment in time. Starting at the far left, where both the sinusoids have zero amplitude, $0+0=0$, which is the value shown in that position in the summed wave. Conducting this procedure over the entire waveform (explained much more thoroughly in Chapter 7), gives the desired input signal wave shown at the bottom left of the figure. The addition of the 200 Hz and 400 Hz output signals (after transformation by the system) is shown at the right.

Although this output is for a particular pair of sinusoids at 200 and 400 Hz, we need not be restricted to just these two. Using homogeneity and time invariance in the way discussed above, we can predict the output to 200- or 400-Hz sinusoids with any amplitude and phase. By then invoking additivity, we can predict the result for a sum of 200- and 400-Hz sinusoids of arbitrary amplitudes and phases. And there is no reason to stop at the sum of two sinusoids. We could add up three, four, ten or even an infinite number of them.

In short, given the response of an LTI system to sinusoidal stimuli, we can predict the system output to any input signal that can be expressed as a sum of sinusoids of the appropriate frequencies, amplitudes and phases. This result would be of little import were it not for a remarkable proof known as Fourier's theorem, which shows it is possible to express *all* signals as a sum of sinusoids of the appropriate frequencies, amplitudes and phases. This is why it is only necessary to know the response of an LTI system to sinusoids.

Because these concepts are so important, let's summarize the argument:

(1) Sinusoids are only changed in amplitude and phase as they are passed through LTI systems.
(2) Given the response of an LTI system to sinusoids of all frequencies, it is then possible to calculate the system output to any input signal that can be expressed as a sum of sinusoids of the appropriate frequencies, amplitudes and phases.
(3) All input signals can be so expressed.

Putting it another way, we take the input signal and analyze it into its component sinusoids. Using our knowledge of the response of the system to sinusoids, we adjust the amplitude and phase of each of the component sinusoids appropriately. We then calculate the output of the system in a process of synthesis by adding together all the transformed sinusoids. Pictorially:

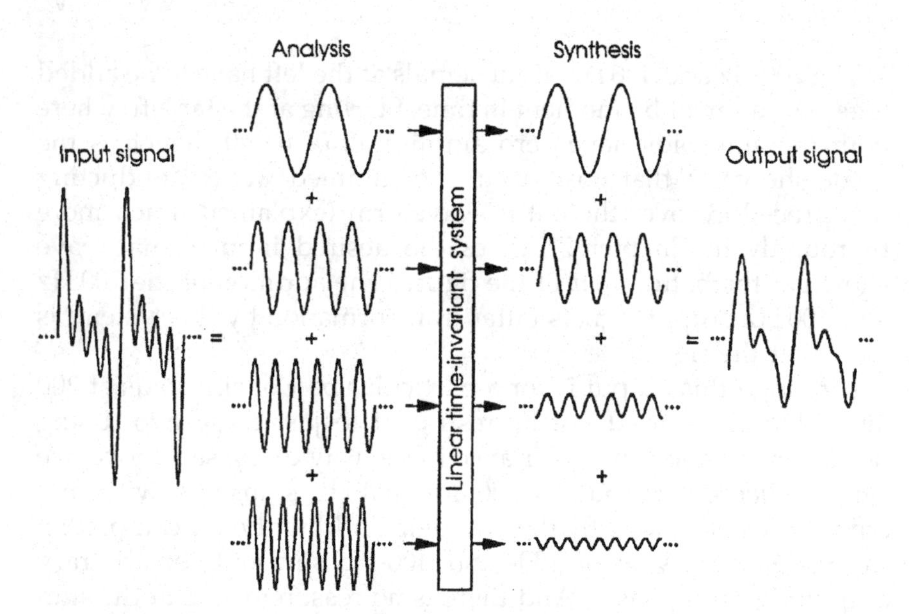

Let's go through this picture a little more slowly. The input signal, at far left, is analyzed into its sinusoidal components. The column headed 'Analysis' shows the constituent sinusoids of the original input wave. These are then put through the LTI system. This system has different effects for each of the four different frequency sinusoids. Generally speaking, the higher in frequency the sinusoid, the more this system reduces its level, as you can see from each of the output waves in the column headed 'Synthesis'. Finally, the output signal at the far right column is obtained when these four sinusoids are summed at corresponding points in time.

In this way, we need not put every signal through a system to see what output we would get. All we need to know is how the system treats sinusoids. This makes life a lot simpler than it would otherwise be. Say, for instance, we were interested in the operation of a hearing aid, and what it did to speech sounds. It wouldn't be necessary to put through it every different kind of speech sound to know what it would do. All that would be necessary, given that the aid could be treated as LTI, is a measurement of its response to sinusoids, a much more manageable task. This information could then be used to predict the output of the aid to *any* sound, speech or otherwise.

Thus, the response of an LTI system to sinusoids is the 'shorthand' way we need to characterize systems efficiently. In Chapter 6, we explore the nature of this characterization in some detail.

Exercises

1. For a sinusoidal input signal of 450 Hz and 0° phase, explain what outputs are possible for an LTI system.

2. Consider a system that alternates between letting an input signal through unchanged, and letting nothing through at 3-ms intervals. For example, there could be a simple switch between input and output that was turned on and off repeatedly. Sketch the input and output signal of this system for a 1-kHz sinusoid, and for a single pulse 1-ms wide. Is this system homogeneous? Additive? Linear? Time invariant? Explain the reasons for your answers, preferably with accompanying sketches.

3. Describe (in a fairly informal way) what an LTI system could do to an input signal made by adding together two sinusoids, one of 100 Hz and one of 200 Hz. What forms could the output signal take? (Hint: There are four main possibilities, given that each sinusoid can be passed through the system or not.)

4. Which of the following systems could be LTI? Explain your reasons. Of the LTI systems, what change(s) in the input signal has the system made?

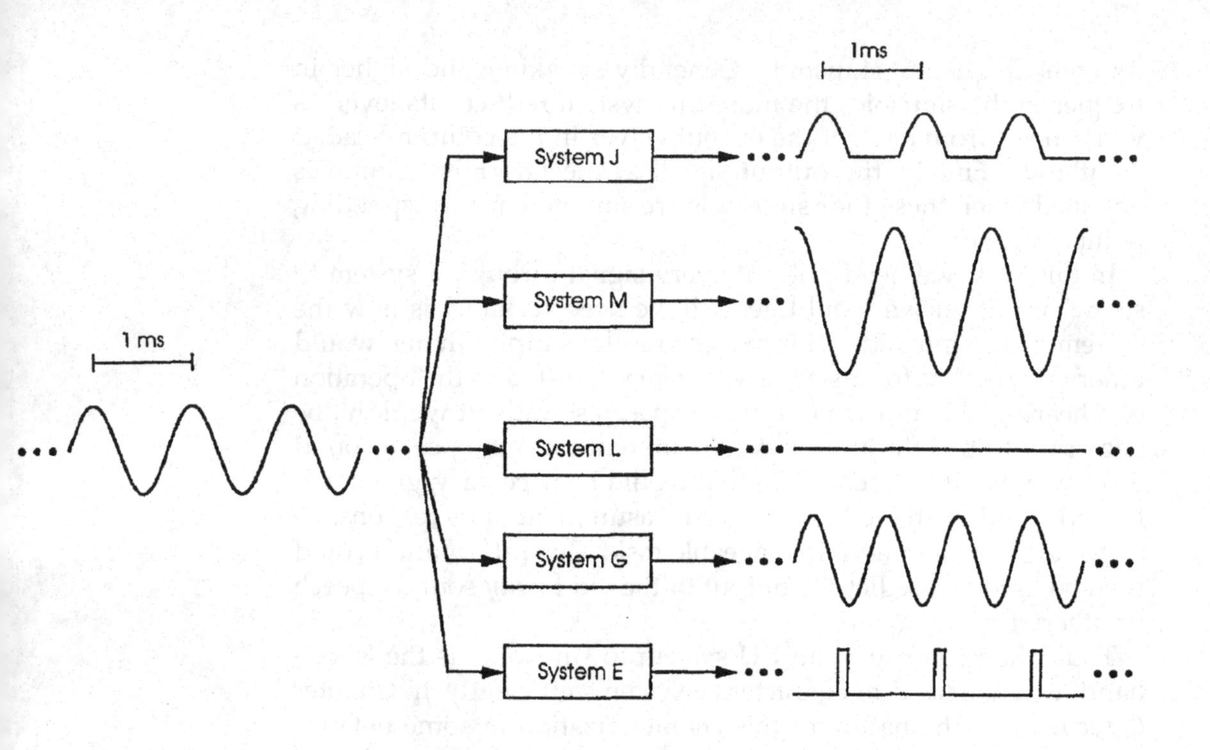

5. Explain how all three of the following systems could be LTI:

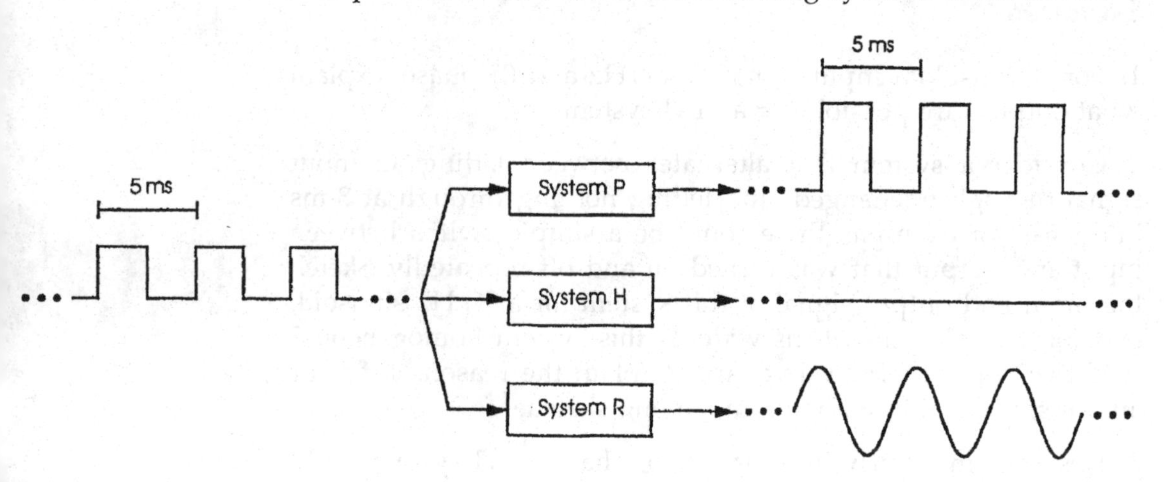

CHAPTER 6

The Frequency Response of Systems

We'll now spend a good deal of time describing the response of linear time-invariant (LTI) systems to sinusoids alone. At first sight, this might seem a waste of time, since it is only in rather unnatural experimental situations that one is faced with sinusoidal inputs. Just keep in mind the point made in Chapter 5—that LTI systems are completely characterized by their response to sinusoids.

This characterization is known as a *transfer* function, as it describes what happens to sinusoidal signals as they are *transferred* through the system. Since sinusoids can be changed in two ways by LTI systems (in their amplitude and phase), it is convenient to consider the transfer function as consisting of two parts—the *amplitude response* and the *phase response*. As the amplitude response of a system is usually considered more important (since the ear is relatively insensitive to phase changes in sounds), you will often see the term *frequency response* used to mean only the amplitude response. In fact, transfer function and frequency response mean exactly the same thing. The phase response of a system can have a big effect on what its output *looks* like (although not necessarily on what it *sounds* like), so a complete characterization of a system requires *both* amplitude and phase responses. Let's discuss amplitude responses first.

Amplitude responses—the basic concept

Up until now, we've only talked about the response of systems to a relatively small number of sinusoids. We could have entered these numbers into a table if we wanted to. For example, suppose we were studying a particular electrical device (call it System X) and were trying to determine its effect on the amplitude of sinusoidal inputs (that is, its amplitude response). A measurement must be made for each frequency of interest, a simple task. Normally we choose a level for the input signal that will remain constant. Here we'll use 2 V. Then we pick a frequency, say 500 Hz, put this into the system, and measure the level of the output. Say we get nearly 2 V out (1.98 V to be exact). Then we go on to a sinusoid of a different frequency, say 1 kHz, send it through the system and measure the output, now 1.42 V.

Continuing this process for whatever frequencies we're interested in, we might end up with a table like this:

Frequency of input sinusoid (Hz)	Amplitude of output sinusoid (V)
125	2
250	2
500	1.98
1000	1.42
1500	0.56
2000	0.24
3000	0.08

So far so good. But what if we wanted to know what happened to sinusoids at 400 Hz, or 1733 Hz? In order to be able to predict the response to a sinusoid of *any* frequency, we would need a table with an entry for every possible frequency. This would mean an infinite number of entries, which would take a pretty long time to write down! Therefore, we will always give system responses as a *graph*, with frequency running along the x-axis and some measure related to output amplitude along the y-axis. Here's the amplitude response of System X in a graphical form:

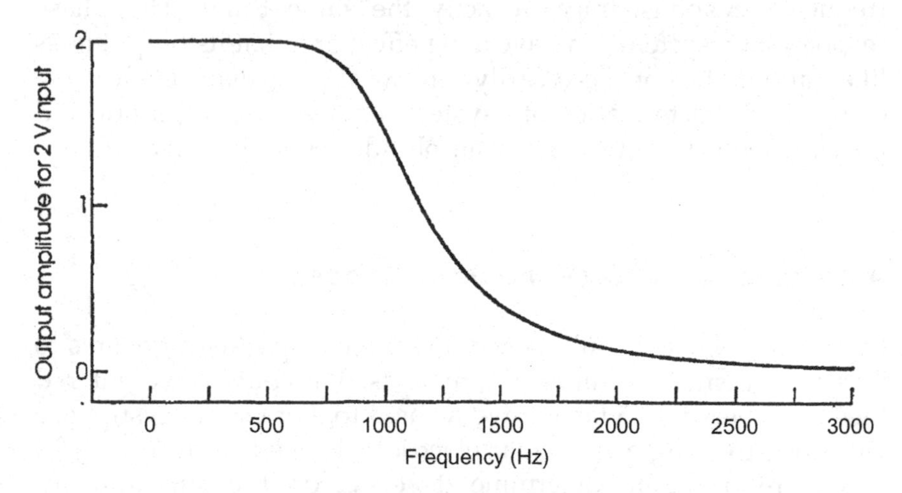

Of course we can't possibly measure the response at every frequency, but we can usually make enough measurements to get a smooth curve. You can see that the graph not only contains the information in the table, but also gives the output of the system for *every* frequency in the range pictured, due to interpolation between the measured points. Thus a 2-V sinusoidal input at 400 Hz leads to a 2-V output, and one at 1733 Hz gives about 0.4 V out.

Graphs have another important advantage over tables in displaying information of this sort. A graph gives a much better indication of the pattern of the amplitude response curve. System X above, for instance, shows a simple and commonly occurring pattern. For sinusoids below some frequency (here about 600 Hz), the output amplitude is the same as the input amplitude, at 2 V. Above this frequency, however, sinusoids are reduced in amplitude in their passage through the system. The degree of reduction, or *attenuation,* increases as the input sinusoid increases in frequency. A 2-V 2-kHz sinusoid appears at the output at an amplitude of only 0.24 V, while a 3-kHz input sinusoid of 2 V has an output amplitude of only about 0.08 V (80 mV). An amplitude response of this basic shape (essentially decreasing from low to high frequencies) is known as *low-pass,* since all frequencies *lower* than some given frequency are *passed* through the system equally well, while frequencies higher than the given frequency are attenuated relative to the lower ones.

Many natural systems exhibit a low-pass characteristic, including one we have discussed before: the middle-ear system consisting of eardrum plus ossicles. In order to construct its amplitude response, we present a sinusoidal sound of a constant amplitude (here, 120 dB SPL at the eardrum) but at a number of different frequencies, and measure the amplitude of the movement of the stapes for each frequency. Plotting each point at the appropriate place on the graph, we end up with this:

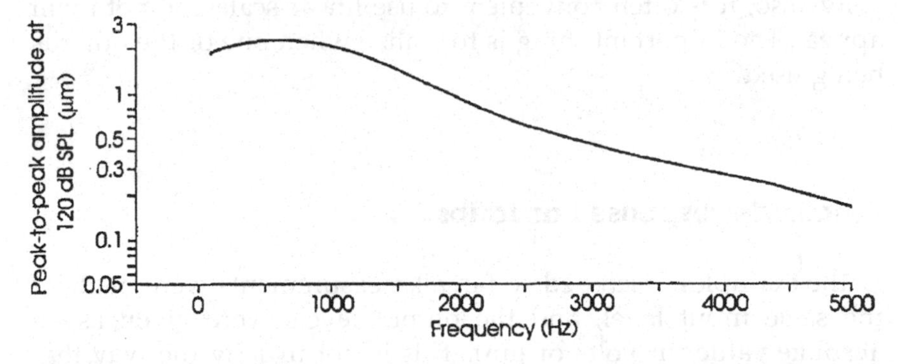

Note the similarity between this curve and the one for System X. Again there is a fairly flat line up to some frequency (here about 800 Hz), beyond which the amplitude of the output decreases, further reductions in amplitude occurring for further increases in frequency. Clearly, this is also a low-pass system. Note that the *y*-axis (giving the peak-to-peak amplitude of the stapes movement) is logarithmic. This is for essentially the same reasons (discussed in Chapter 3) that the sound pressure level expressed in dB (a logarithmic scale) is more convenient than one expressed in pascals (a linear scale).

The range of stapes movements we want to represent can be a bit too large on a linear scale to allow important differences in measurements at low amplitudes to be evident in any graph. For example, even a doubling in the amplitude of the stapes movement may only represent a very small absolute change. Since the middle-ear system, at least in these experiments, seems to be LTI, the 1,000,000:1 dynamic range in sound pressure level for normal hearing will translate into a 1,000,000:1 range for the amplitude of movement of the stapes. Such large ranges are more conveniently represented on logarithmic scales.

There is another related reason for using logarithmic scales. In practical experiments, it is almost always the case that the error in measurement is proportional to the absolute value of the measurement being made. In other words, the bigger the thing we're measuring, the bigger the absolute error. (Consider the size of errors that might be made in measuring the width of a bookshelf versus the distance between two cities.) In the context of measuring stapes movements, we expect the error measured at 1 kHz (at a peak-to-peak amplitude of about 2 μm) to be about 10 times larger than the error at 4.5 kHz (because the peak-to-peak amplitude here is about 10 times smaller at about 0.2 μm). A logarithmic scale makes these errors take up the same distance when plotted, since constant ratios are transformed into constant differences. If a linear scale was used, the higher values would seem much more scattered than the lower values, even though the error in measurement was proportionately equal.

Even so, it is often convenient to use linear scales, so both will appear. The important thing is to realize when one or the other is being used.

Amplitude responses as ratios

In the examples discussed so far, all measurements were made at the same input level, and the output levels were given as an absolute value (in volts or μm). This is not usually the way that amplitude response curves are reported, as this restricts their direct applicability to one particular measurement. It would be more convenient if amplitude responses could be expressed in a way that made them independent of the particular input and output amplitudes measured. This is done by defining the amplitude response as the ratio between the level of the output and the level of the input, both as a function of frequency. Put in symbols, let $A(f)$ be the value of the amplitude response, and

Input(*f*) and Output(*f*) indicate the amplitudes of the input and output sinusoids, all as a function of frequency, *f*. Then:

$$A(f) = \frac{\text{Output}(f)}{\text{Input}(f)} \tag{1}$$

Let's calculate and draw the amplitude response for System X. Starting with a sinusoidal frequency of 250 Hz, we simply divide the level of the output (2 V) by the level of the input (also 2 V) to get 1. At 1 kHz, we would be dividing 1.42 V (the output amplitude) by 2 V (the input amplitude) to obtain 0.71. The following table shows a more complete set of calculations:

Frequency of sinusoid (Hz)	Output amplitude (V)	Input amplitude (V)	Output/ Input
125	2	2	1
250	2	2	1
500	1.98	2	0.99
1000	1.42	2	0.71
1500	0.56	2	0.28
2000	0.24	2	0.12
3000	0.08	2	0.04

Again, it is usually more informative to graphically display such information:

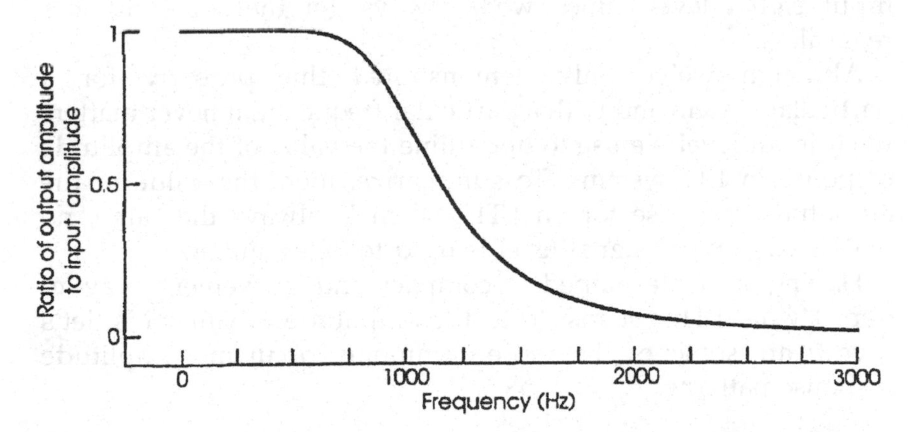

This looks just like the first curve we drew for System X when we plotted output amplitude for 2 V input as a function of frequency, instead of the ratio between output and input amplitudes. So it should. All we've done is to divide the *y*-values by 2, which is equivalent to a simple re-scaling of the *y*-axis, or change of units (for example, plotting voltage in terms of millivolts instead of volts). The advantage of using this ratio instead of absolute values is that the amplitude response is then defined on the basis of an input with an amplitude value of one unit (here, 1 V). This makes calculations of output amplitudes easier, since if:

$$A(f) = \frac{\text{Output}(f)}{\text{Input}(f)} \tag{2}$$

and we multiply this equation by Input(f) on both sides, then:

$$\text{Input}(f) \times A(f) = \text{Output}(f) \tag{3}$$

To put this in words, if we know the value of the amplitude response and the level of the input signal, we can predict the amplitude of the output by a straightforward multiplication of the values.

The other advantage of referring all measurements to the signal input level by taking ratios is that we need not make all measurements at the same input level. Although the absolute amplitude of the output will change with changes in input level, the value of the amplitude response (output over input) will not. This property follows directly from the homogeneity of LTI systems. For example, if we measure the output of System X to a 2-V 250-Hz input signal, we get 2 V out. If we measure the output to a 4-V signal at the same frequency, we'll get 4 V out. (Remember, homogeneity means that doubling the input amplitude results in a doubling of the output amplitude). If we report either one of these values, we also have to report the input level used to measure it. If we report the ratio between output and input, however (here $2/2 = 1$ and $4/4 = 1$) we need not specify input signal levels since we'll always get the same number regardless.

Although we've only demonstrated this property for a particular system and with a particular frequency, it never matters what input level we use to determine the value of the amplitude response in LTI systems. To summarize, then, the value of the amplitude response for an LTI system is always the same, no matter what input signal level is used to determine it.

Having now developed a compact and convenient way to depict what LTI systems do to the amplitude of sinusoids, let's investigate some of the more commonly occurring amplitude response patterns.

Filters

Both of the systems whose amplitude responses we have looked at so far (System X and the middle ear) pass low frequencies better than high frequencies. Systems which let some band of frequencies pass better than others are in general known as *filters*. There are two main simple kinds of filters, named according to the relationship between the range of frequencies over which sinusoids are transmitted relatively well and the range of frequencies over which they are attenuated.

Let's examine the case of *low-pass filters* first, two examples of which we've already met. Generally speaking, low-pass filters attenuate high-frequency sinusoids more than they do low-frequency sinusoids. An idealized low-pass filter would pass all sinusoids below some frequency (known as the *cutoff frequency*) equally well while completely stopping the passage of sinusoids above this frequency. What would the values for the amplitude response be for these two regions? Let's go back to our original definition:

$$A(f) = \frac{\text{Output}(f)}{\text{Input}(f)} \tag{4}$$

For frequencies above the cutoff frequency, we want the output amplitude to always be zero, no matter what the level of the input is. In other words, nothing will come out of the system. In this case, substituting $\text{Output}(f) = 0$ into equation (4), we note that $A(f) = 0$ no matter what $\text{Input}(f)$ is.

For frequencies below the cutoff, we want the sinusoids to come out at the same level as they went in, so that $\text{Output}(f) = \text{Input}(f)$. Substituting this into equation (4), we get:

$$A(f) = \frac{\text{Output}(f)}{\text{Input}(f)} = 1 \tag{5}$$

The amplitude response curve for an ideal low-pass filter is therefore pretty simple. It has a value of 1 for all frequencies below the cutoff frequency (abbreviated as f_c) and a value of 0 for all frequencies above:

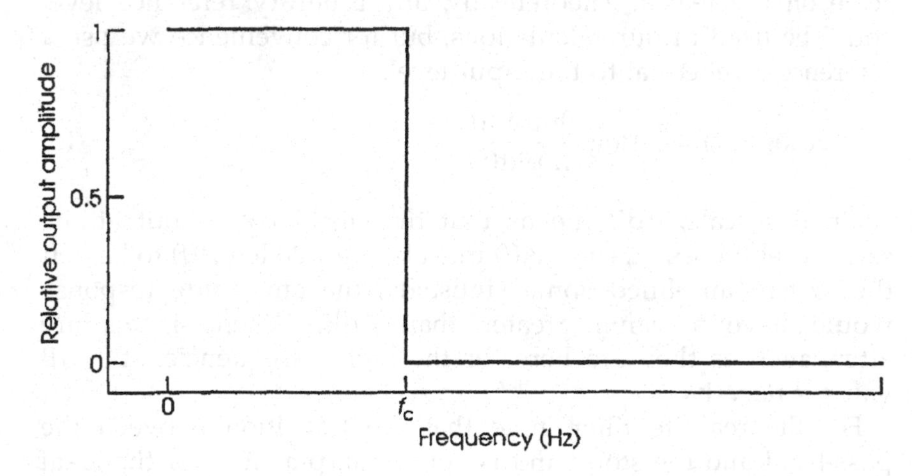

Note that we've labelled the y-axis as 'relative output amplitude' just to stress that the output amplitudes have been divided by the input amplitudes to obtain the given values of the amplitude response. You will also often see the word 'gain' used to refer to the same concept. In this sense, 'gain' simply means the factor by which the input signal is to be multiplied.

The output of an ideal low-pass filter for any sinusoidal input is easy to predict. An input sinusoid that has a frequency lower than f_c, the cutoff frequency, will come out of the system at the same level as it went in. For such signals, it is as if the system wasn't there. For sinusoids above the frequency f_c, there will be no output at all. We call the region over which the sinusoids go through the system well the *pass-band*, and the region where they are attenuated the *stop-band*.

As with most things you are taught, however, what the ideal is and what actually happens are two different matters! Real-life filters are never ideal. An amplitude response for a real-life filter might look more like this:

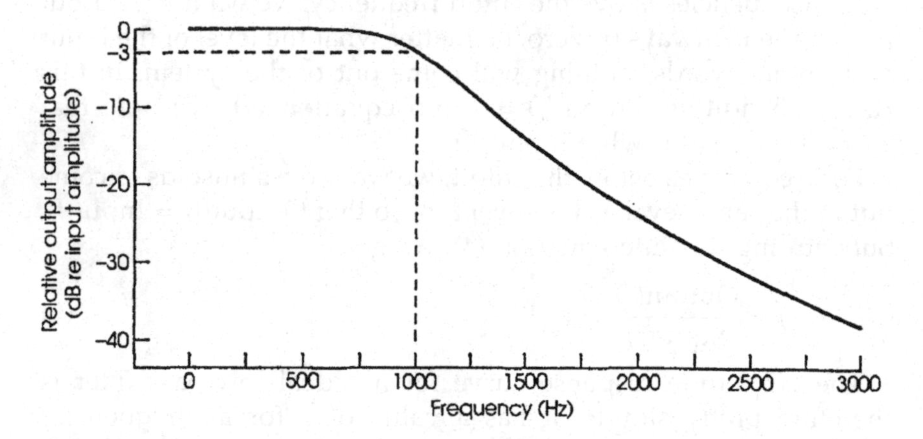

In fact, this is just the amplitude response for System X with a dB scale on the y-axis. Theoretically, any arbitrary reference level could be used in our calculations, but for convenience we use a reference level equal to the input level:

$$20 \log A(f) = 20 \log \frac{\text{Output}(f)}{\text{Input}(f)} \tag{6}$$

On this scale, 0 dB means that the signal comes out at the same level it goes in, since $A(f)$ must be 1 for 20 log $A(f)$ to be 0. If the system amplified some sinusoids, the amplitude response would have a value greater than 0 dB. When signals are attenuated, as they are here for the higher frequencies, the dB value is negative.

For the real-life filter, note that the transition between the pass-band and the stop-band is not as sharp as it is for the ideal filter. Therefore, there will be sinusoids not in the pass-band which will be attenuated relatively little as they pass through the system.

This makes the specification of a cutoff frequency a little more difficult than it is for an ideal filter, as there is no single frequency which uniquely gives the boundary between the pass-band and the stop-band. To get around this problem, we take as the cutoff frequency the point at which the amplitude response has decreased by 3 dB from its maximum. You can see this marked by the dashed lines on the previous figure. This 3-dB point is also known as the *half-power point* because a signal that is 3 dB less than some reference signal has one-half of the power (see the exercises). You will often see low-pass filters described as being, say 'low-pass at 1 kHz', as System X is. Yet this is not really enough information about what the filter does, as these two amplitude response curves demonstrate:

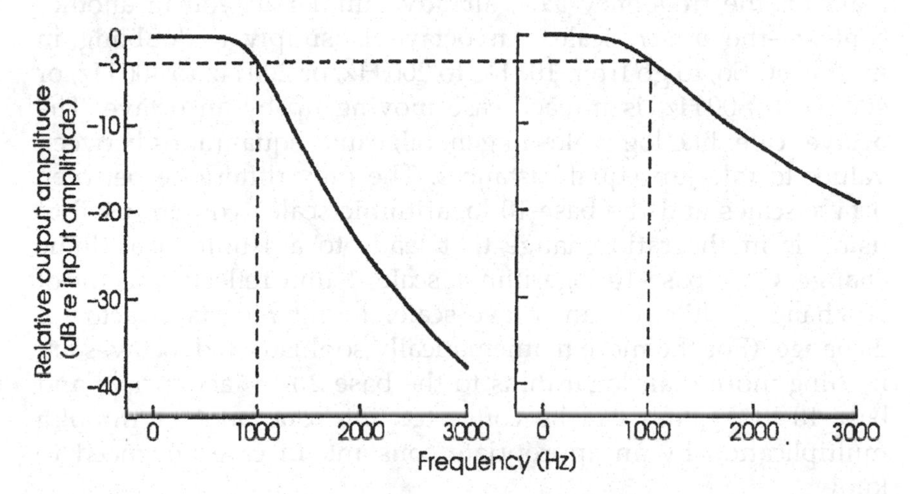

Even though these two filters have the same cutoff frequency, they differ in the slope at which the amplitude response decreases once the cutoff frequency is reached. As the *roll-off* becomes steeper and steeper, the filter's amplitude response more closely approximates that of an ideal filter. The less sharp the roll-off, the better that sinusoids with frequencies not in the pass-band of the filter will be represented in the output. So, for a better characterization of the functioning of a filter, we need to know the rate at which the amplitude response decreases. Many filters have an amplitude response that decreases roughly linearly when plotted on scales that not only have dB on the y-axis, but also the logarithm of frequency on the x-axis. At the top of the next page are the two amplitude responses shown above re-plotted in this way.

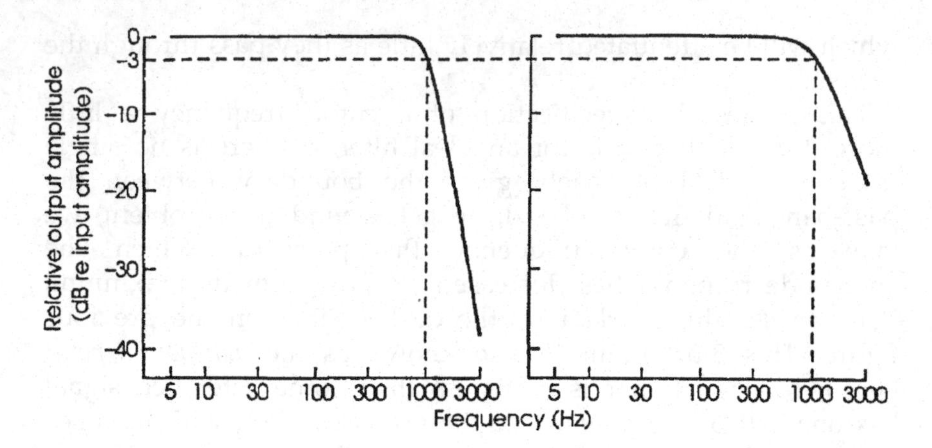

Because the filter shape, once it begins its roll-off, is roughly linear, it can be specified by its slope. As usual, we need a way to quantify this slope. To do so, we use a very special logarithmic scale for the frequency axis, already familiar to you in another context—the *octave* scale. An octave is simply a doubling in frequency. So, to go from 100 Hz to 200 Hz, or 200 Hz to 400 Hz, or 400 Hz to 800 Hz, is in each case moving up by an octave. The octave scale, like log scales in general, cause equal ratios between values to take up equal distances. The only difference between octave scales and the base-10 logarithmic scales you are used to using is in the ratio change that leads to a 1-unit logarithmic change. On a base-10 logarithmic scale, 1 unit reflects a factor of 10 change, while, on an octave scale, 1 unit reflects a factor of 2 change. (For the more mathematically sophisticated, octaves are nothing more than logarithms to the base 2.) In fact, octave and base-10 log values can be converted into one another through multiplication by an appropriate constant. In case you need to know:

$$\log_2[x] = \frac{\log_{10}[x]}{\log_{10}[2]} = \frac{\log_{10}[x]}{0.301} = 3.32 \log_{10}[x] \tag{7}$$

Therefore, the two scales are essentially the same thing, differing only by a multiplying constant. Therefore, since filter roll-offs are often linear on dB versus log frequency scales, they will also be linear on dB versus octave frequency scales. This is why a single value designating the slope is all that is usually needed. Roll-offs are usually expressed, then, as so many dB per octave. They are calculated by noting the difference in attenuation in dB at two frequencies, one octave apart, in the region where the amplitude response seems to be linear. This gives the roll-off directly in dB per octave. Take the example at the top of the following page, System X in yet another guise—on dB versus log frequency scales.

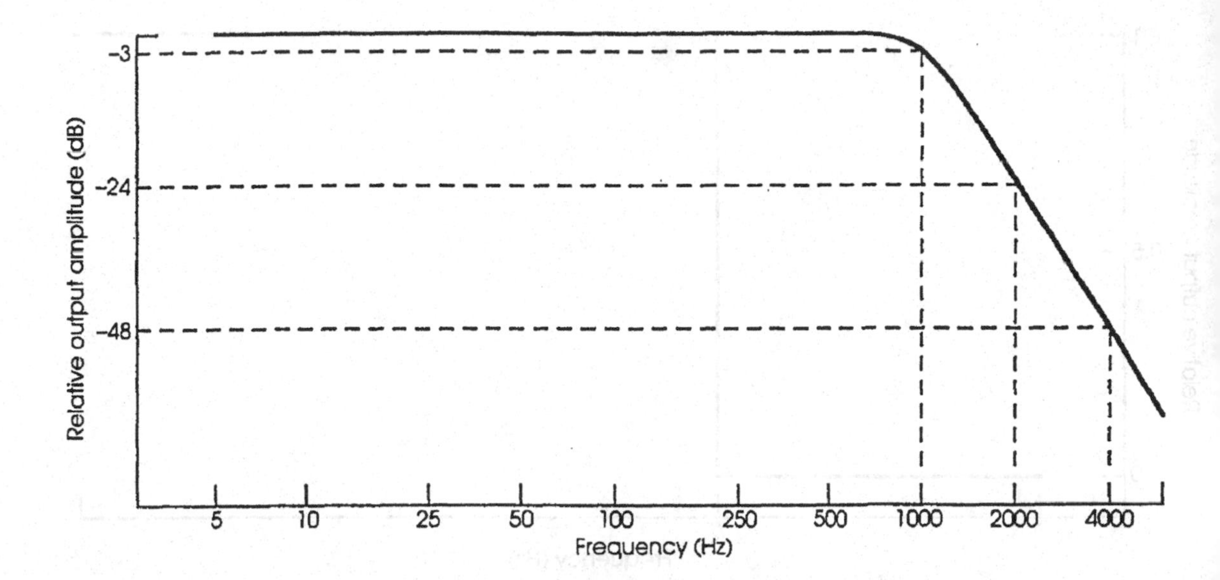

At 2 kHz, System X attenuates the signal by 24 dB. At one octave above this (4 kHz), the attenuation is 48 dB. Therefore, this filter has a roll-off of 48–24 = 24 dB/octave. Note that if we had calculated the roll-off from 1 to 2 kHz, before the amplitude response reached its maximal slope, we would have obtained a figure of 21 dB/octave. (How is this figure arrived at?)

Filters come in a wide variety of possible slopes. In order to give you some idea of the range possible, slopes of less than 18 dB/octave are considered fairly shallow, and those between 18 and 48 dB/octave moderately steep, while filters with cutoffs greater than 90 dB/octave are considered very steep indeed. Different applications require different specifications, and just because a filter is steeper doesn't necessarily mean that it's better suited to the task in hand.

You will notice that although we use linear amplitude scales for drawing the amplitude response of an ideal filter, we tend to use dB scales for real filters. As we've already discussed, dB scales are more convenient for representing the vast range of measurements that is necessary. But they are less convenient for ideal filters because we cannot properly draw on dB scales an amplitude response that is zero at some frequencies, since the logarithm of zero is not a finite number (see the Appendix of Chapter 3 for details). As ideal filters cannot exist in the real world, this restriction on representing values of zero is not much of a constraint.

By analogy with low-pass filters, there are filters that only let sinusoids at or above a certain frequency through. These are known as *high-pass* filters, and ideally have an amplitude response like this:

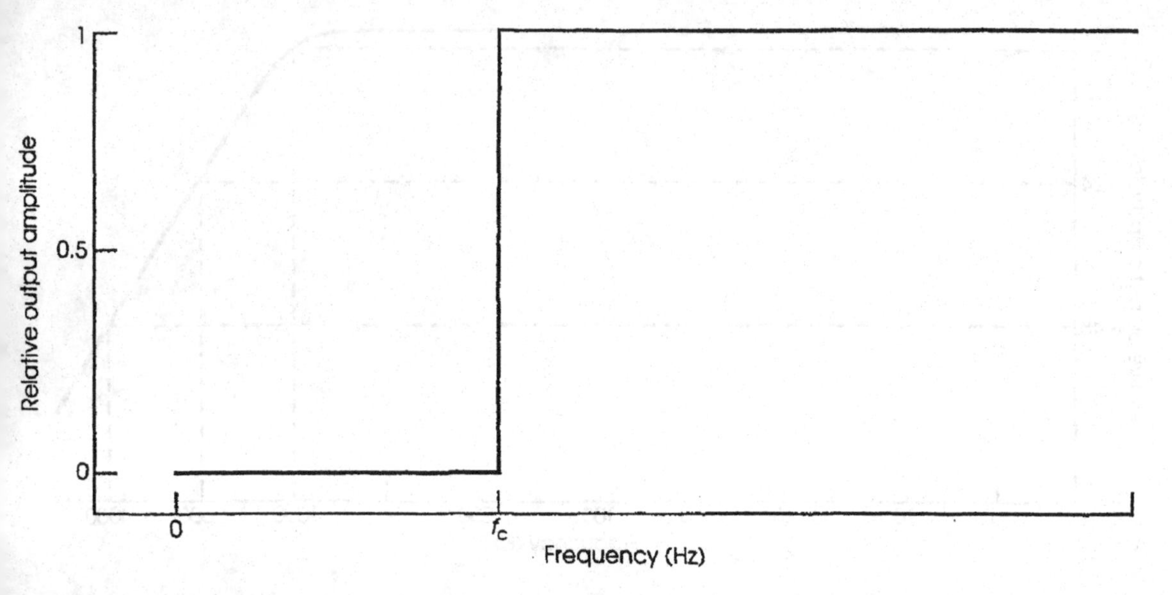

but in real life more like this:

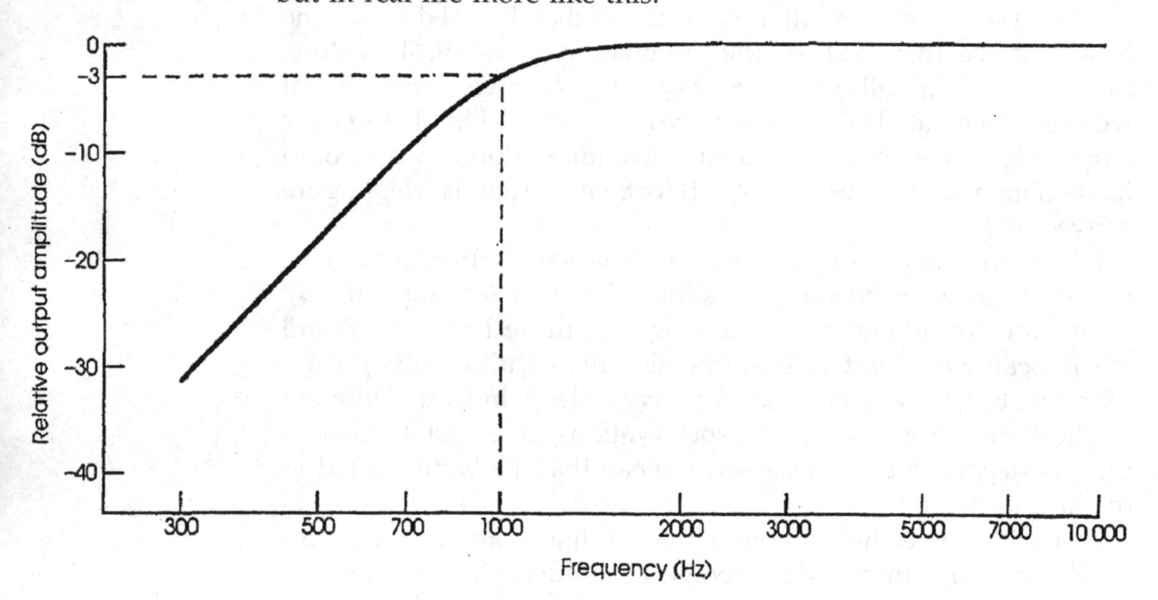

Again, as you can see here, we specify the characteristics of such a filter by its cutoff frequency and roll-off in exactly the same way as for a low-pass filter.

Systems in parallel

Low-pass and high-pass filters exhibit the two simplest response shapes, and it is convenient to distinguish two further filter types that can be considered as a combination of one high-pass and one low-pass filter. Let's imagine that we have the cutoff frequency of a high-pass filter ($f_{c\text{-}hi}$) set at a higher frequency than that of a low-pass filter ($f_{c\text{-}lo}$), as in these two ideal filters:

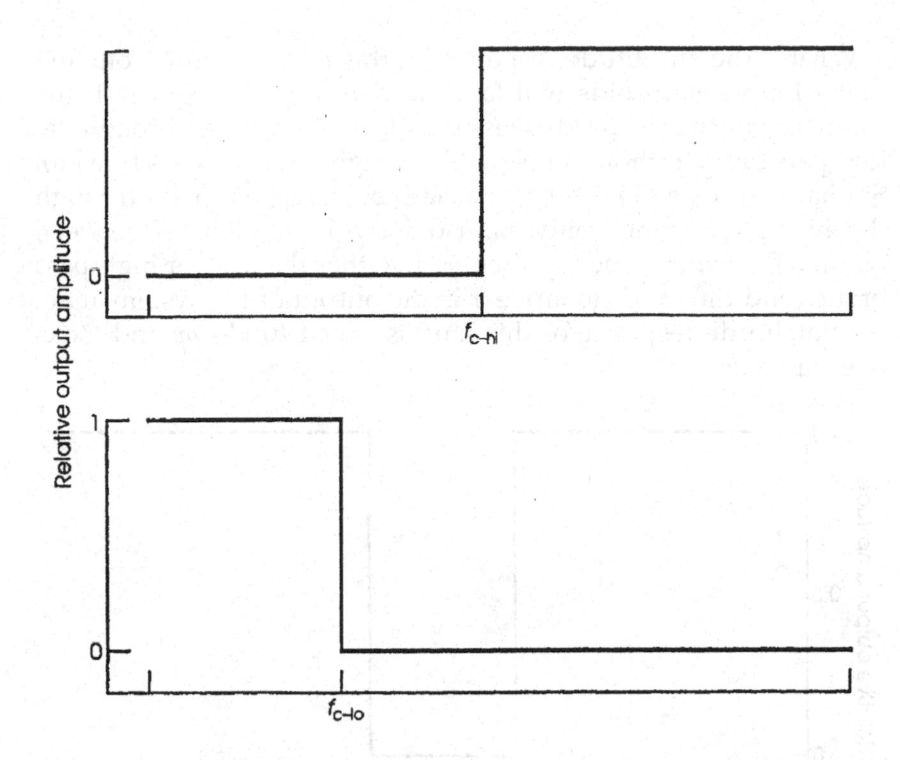

(Notice that subscripts 'c-lo' and 'c-hi' refer to the type of filter, low- or high-pass.)

We can construct a new type of filter from these two filters by connecting them in parallel, thus feeding the two filters independently with the same input signal and then adding the two outputs together:

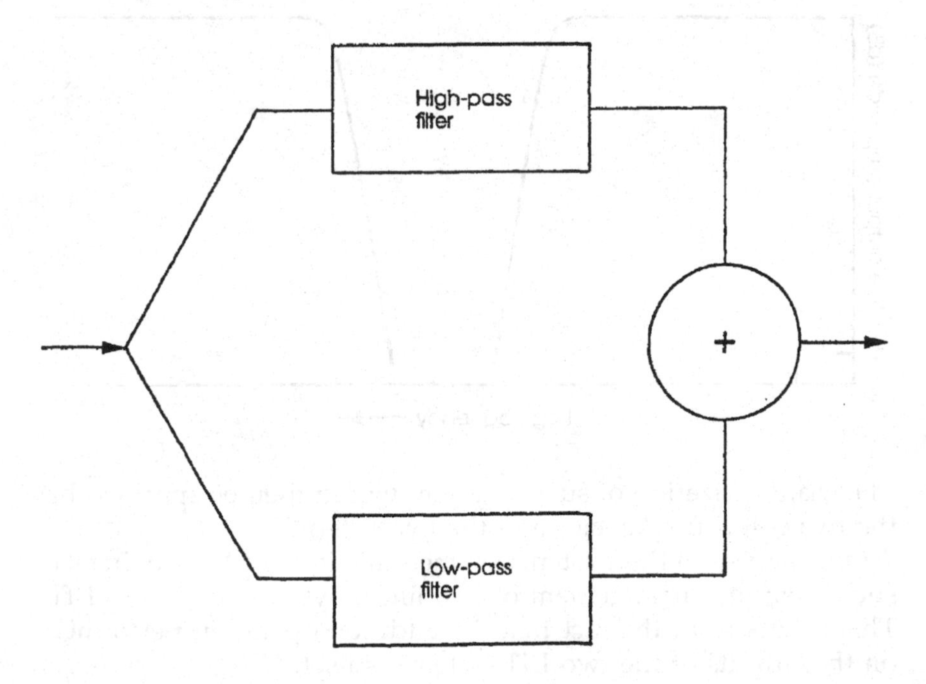

What is the amplitude response for this new system? Note first that all input sinusoids will fall into one of three regions. If the frequency of the sinusoid is less than $f_{c\text{-}lo}$, it will pass through the low-pass branch (though not at all through the high-pass branch). Similarly, sinusoids with frequencies above $f_{c\text{-}hi}$ will pass through the high-pass branch only. Sinusoids with frequencies between $f_{c\text{-}lo}$ and $f_{c\text{-}hi}$ will not be represented in either the low- or high-pass branch and thus will not appear in the output of the system at all. An amplitude response of this kind is called *band-stop* and looks like this:

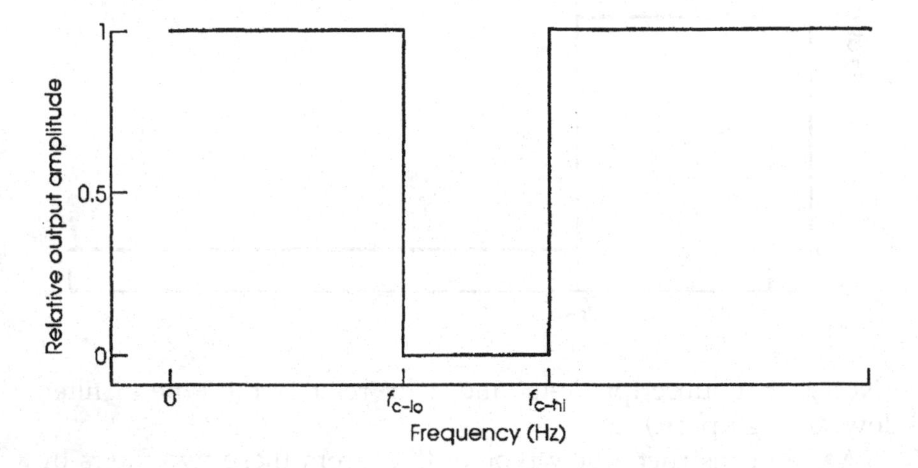

As usual, this is an idealization. A real-life band-stop filter might have an amplitude response more like this:

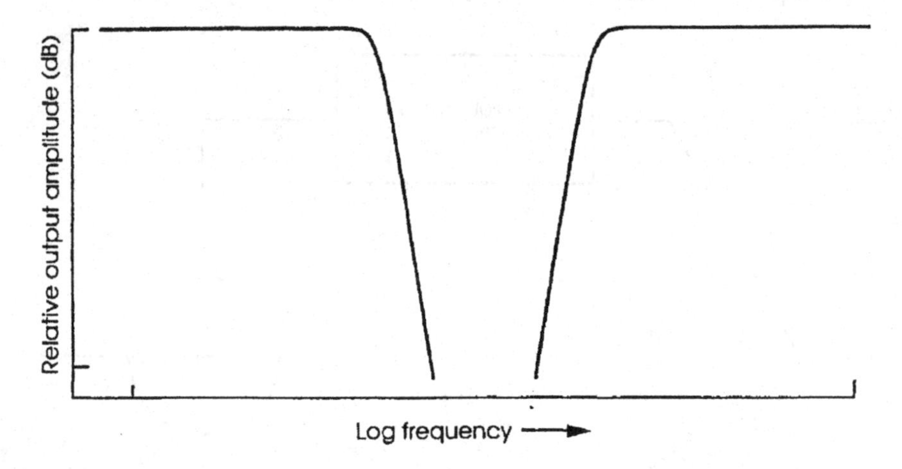

The characterization of such a system would then be specified by the two cutoff frequencies and the two roll-offs.

Our analysis of this system assumed that it was LTI, and, in fact, such a parallel arrangement of two linear systems is always LTI. This results from the fact that the addition operation performed on the outputs of the two LTI systems is itself LTI.

Systems in cascade

Such a parallel combination of systems, although not unusual by any means, is much less commonly encountered than systems connected serially. Here, the output of one system is fed directly to the input of another system; this is often known as a *cascade* of systems, because the signal is passed from one system to another in the same way that water passes from one section, or cascade, of a broken waterfall to another:

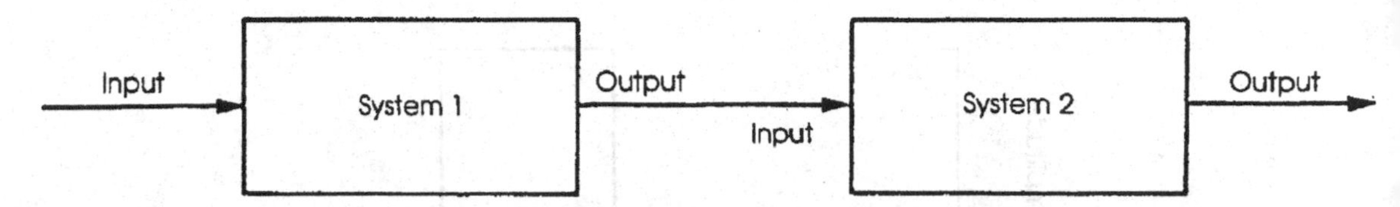

As you can see, the input to the second system is simply the output of the first system. Because the component systems making up the cascade are LTI, it turns out that the cascade as a whole is LTI too.

Band-pass filters

Having established that two LTI systems in cascade can be considered as one larger LTI system, how does one find the overall amplitude response of such a system? Let's take a simple example first, a combination of one ideal high-pass and one ideal low-pass filter. If the cutoff frequency of the high-pass filter is higher than the cutoff frequency of the low-pass filter, it's obvious that nothing will be able to come through the system. Any sinusoid that has a frequency low enough to pass through the low-pass section will be filtered out in the high-pass section, and vice versa. Such a filter might be called no-pass and is of little use! A more interesting situation arises when the cutoff frequency of the high-pass filter is below the cutoff frequency of the low-pass filter:

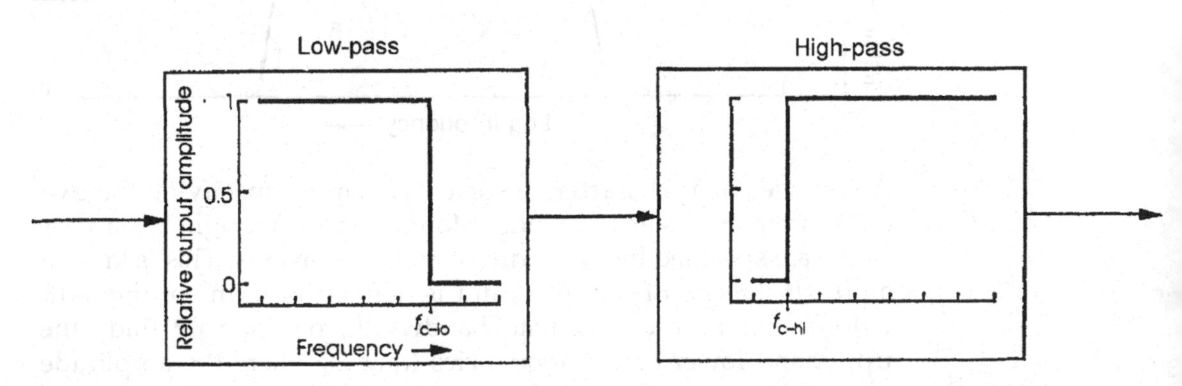

Here a sinusoid can fall into one of three regions, according to its frequency. A sinusoidal input with a frequency above $f_{c\text{-}lo}$ will be filtered out in the first (low-pass) section of the system and thus not appear in the output. A sinusoid with a frequency below $f_{c\text{-}hi}$, although it will pass through the low-pass section, will be filtered out on its passage through the high-pass section. It is only sinusoids with frequencies between $f_{c\text{-}lo}$ and $f_{c\text{-}hi}$ which will get through both systems, and in this case come out at their initial amplitudes. Therefore, the overall amplitude response of the system is:

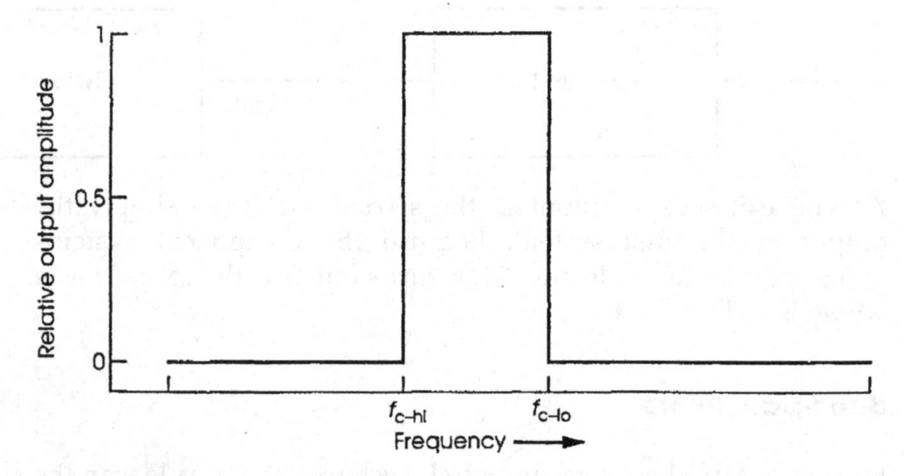

Since all frequencies within a particular band are passed through the system, this is known as a *band-pass* filter.

As usual, real systems never have responses like this, but could be something more like:

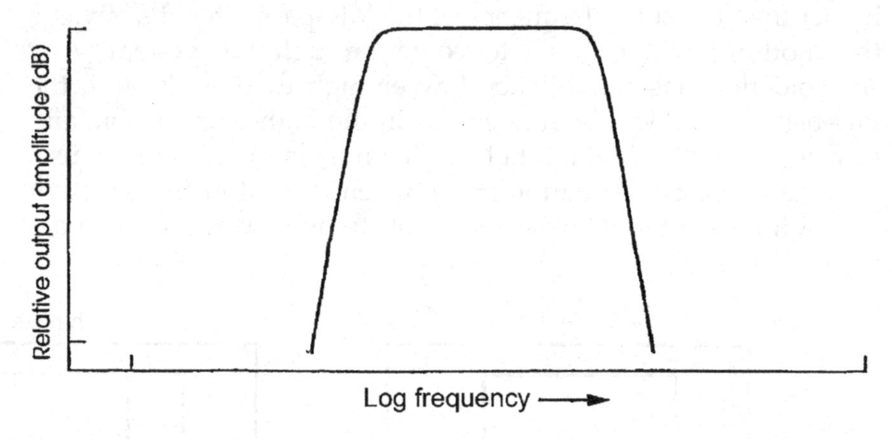

Again, we could characterize such a system by specifying the two cutoff frequencies and roll-offs. More typically, though, band-pass filters are specified by the width of their pass-bands. This is known as the *bandwidth* of the filter, and is defined in terms of the 3-dB cutoffs. In order to calculate bandwidth, one simply finds the upper and lower cutoff frequencies (that is, where the amplitude response has dropped by 3 dB) and calculates the difference between them. So, the bandwidth of this band-pass filter is 150 Hz:

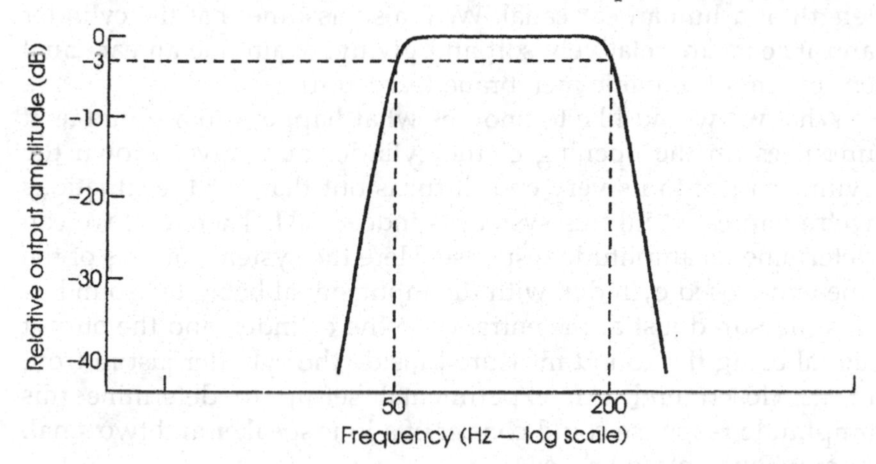

Band-pass responses in simple physical systems

Although band-pass responses *can* be created by the serial combination of a high- and low-pass filter, there are many systems which exhibit such a response directly as a result of their physical characteristics. In these systems, it is more useful to think of them as having a band-pass characteristic inherently, rather than as being the result of the operation of two subsystems.

One particular such system, which as you will see in later chapters can serve as a useful model for, among other things, the ear canal and the vocal tract, is a cylinder closed at one end and open at the other:

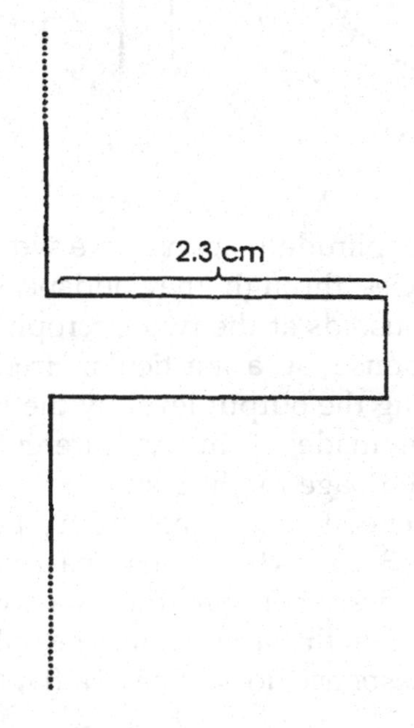

In order to make our results concrete, we'll assume that the cylinder is 2.3 cm (nearly an inch) long, representative of the length of a human ear canal. We'll also assume that the cylinder and its end are relatively soft and giving, again like an ear canal ending in a tympanic membrane (eardrum).

What we would like to know is what happens to a sound as it impinges on the opening of the cylinder and travels down the cylinder right to its very end. It turns out that, for the situations we're interested in, this system is indeed LTI. Therefore, we can determine its amplitude response. Here the system consists of the one-end-closed cylinder, with the input signal being the sound as it is measured just at the entrance to the cylinder, and the output signal being the sound measured inside the cylinder just in front of its closed end. An experimental set-up to determine this amplitude response could consist of a loudspeaker and two small microphones placed thus:

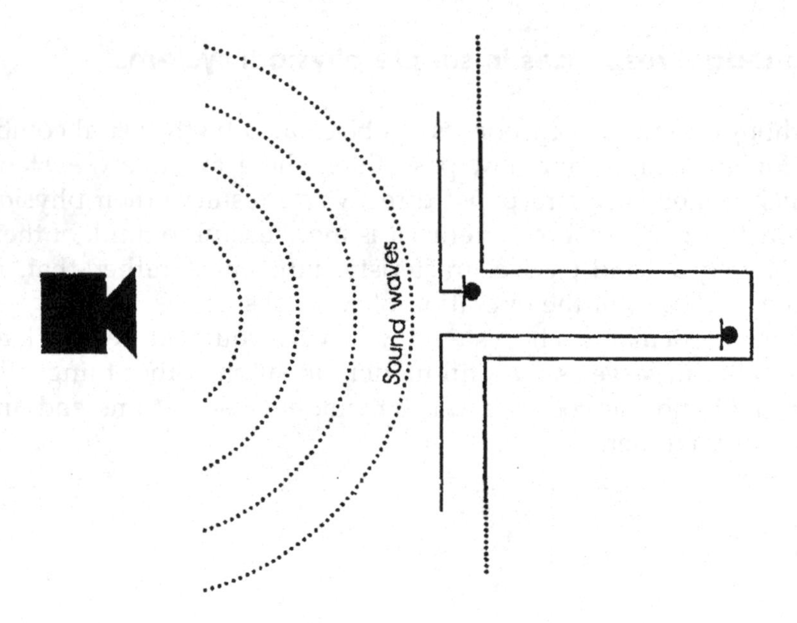

To measure the amplitude response, we would play sinusoids of varying frequencies through the loudspeaker, and note the amplitude of the sinusoids at the two microphones. The value of the amplitude response at a particular frequency would be calculated by dividing the output level by the input level. If such measurements were made at many different frequencies, the curve shown over the page might result.

Note first the general shape of the response. A band of frequencies around 3 to 4 kHz is emphasized, relative to the surrounding frequencies. Therefore, this is a sort of *band-pass* response, of the type seen in the last section. The only real difference is that this band-pass response doesn't have a flat top—it is rounded.

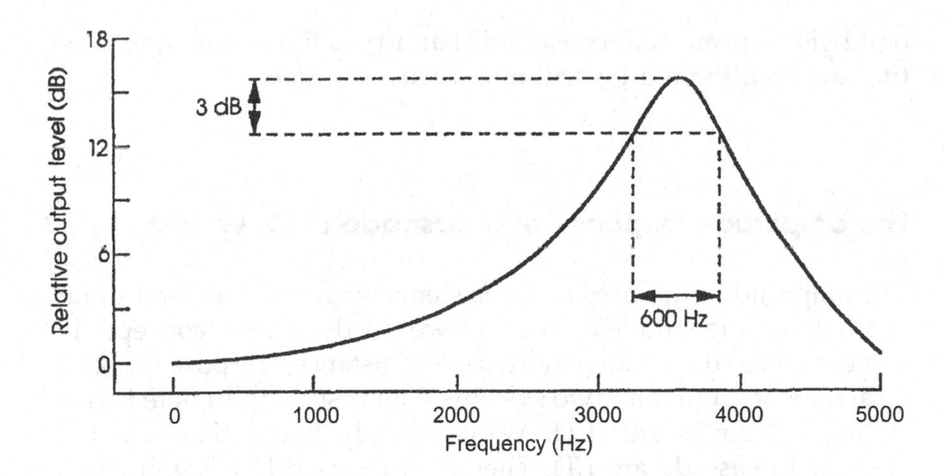

This means that there is one particular frequency at which the system responds best—here about 3260 Hz. Systems which have this sort of response are known as *resonant* systems, and the frequency region over which the response is highest is known as a *resonance*. In other words, a resonance is nothing more than the pass-band of a band-pass system. The single frequency at which the system responds best is known as its *resonant frequency*.

As might be expected from our earlier discussion, a resonance is typically characterized by its 3-dB bandwidth (and resonant frequency). You can see from the figure above that our model ear canal has a bandwidth of 600 Hz.

Often, though, you will see another way of referring to the width of a resonance known as Q, or the quality factor. Q takes into account the relationship of the bandwidth to the resonant frequency by dividing the latter by the former. That is, if 'BW' is the 3-dB bandwidth of the resonance, while f_{cf} is its resonant (or 'centre') frequency, then Q is defined by:

$$Q = \frac{f_{cf}}{BW} \tag{8}$$

For our model ear canal, then, Q has a value of 5.4.

As the bandwidth of the resonance decreases, Q will increase. Therefore, the higher the Q, the sharper, or more selective, the filter. The reason why Q is used is that, generally speaking, it is easier to make narrow-bandwidth filters at higher frequencies than at lower ones. Q gives a measure of bandwidth that is normalized for the frequency region in which the resonance appears, and in some sense corrects for the relative difficulty of making a filter of a particular bandwidth across frequency. Also, filters at different frequencies with the same Q have the same *relative* bandwidth, although the absolute bandwidth will be larger for the higher frequency filter. This constant *proportional* bandwidth means that constant Q band-pass filters will look to be the same width on a logarithmic frequency scale.

The amplitude response of a cascade of LTI systems

The amplitude responses of the systems we have discussed so far are only of the simplest sort, yet we need no new concepts to discuss more complicated systems. For instance, suppose we have a cascade of more than two systems. The first thing to note is that such a system is still LTI. We've already noted that two LTI systems in cascade are LTI. Therefore, the two LTI systems in a cascade may be treated as one grand LTI system. If we cascade this system with another, again the total system is LTI. Thus, three LTI systems in cascade form an LTI system. Continuing this process *ad infinitum*, *any* number of LTI systems in cascade will form an LTI system.

There still remains the problem of determining the amplitude response of the total system from the specification of the individual amplitude responses. Let's deal first with the simplest case, a cascade of two systems. From the original definition of an amplitude response, all we need do is, for each frequency of interest, divide the amplitude of the output signal (from the second system in the cascade) by the level of the input signal (to the first system in the cascade). In terms of the amplitude responses of the two constituent systems, then, we must perform four steps for each frequency value of the amplitude response:

(1) Select a value for the amplitude of the input signal.
(2) Determine the amplitude of the output of the first system in the cascade, using its amplitude response.
(3) Determine the output amplitude of the second system in the cascade to an input signal equal to the amplitude of the output of the first system (now using the amplitude response of the second cascaded system).
(4) Divide the output amplitude of the second system in the cascade by the amplitude of the input to the first system.

As an example, let us go back to the ideal band-pass filter constructed from a high-pass and low-pass filter and work out its amplitude response. This time we'll make things a bit more concrete by setting the cutoff of the low-pass filter to 1 kHz and that of the high-pass filter to 200 Hz, so our system looks like this:

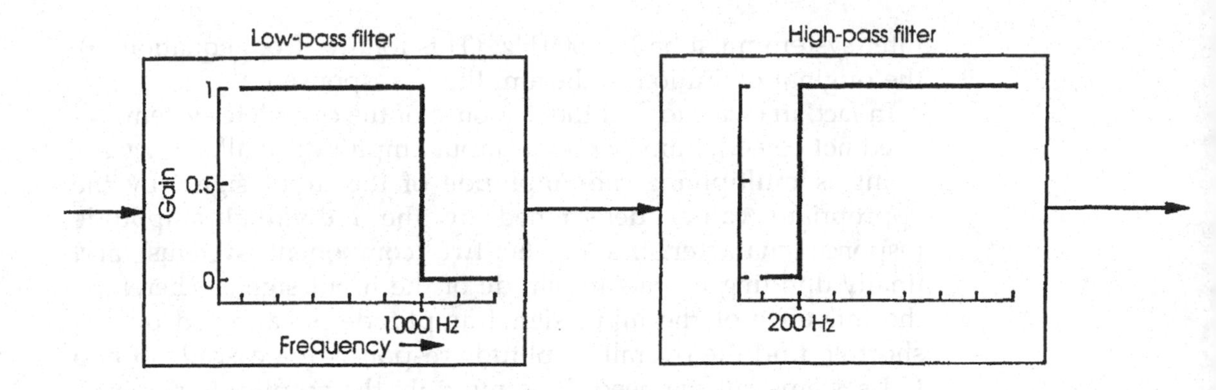

Take an electrical sinusoidal signal of amplitude 2 V at 100 Hz. Let's find out what happens to it as it passes through the first, low-pass, part of the system. First, find the value of the amplitude response at 100 Hz, which is 1, as it is for all frequencies less than 1000 Hz. Now multiply this value by the amplitude of the sinusoid (2 V), to get 2 V. This follows from our definition of the amplitude response—the ratio of the amplitudes of the output of a system to its input:

$$A(f) = \frac{\text{Output}(f)}{\text{Input}(f)} \tag{9}$$

which we have already shown implies that:

$$A(f) \times \text{Input}(f) = \text{Output}(f) \tag{10}$$

Therefore, to determine the output level of a sinusoid, multiply its input level by the value of the amplitude response at that particular frequency.

To find the output from the high-pass filter, we repeat the operation we've just performed, using the appropriate input level (now 2 V) and amplitude response curve (that from the high-pass filter). This results in 0 because the value of the amplitude response for the high-pass filter at 100 Hz is 0 (as it is for all frequencies below 200 Hz). Therefore, there is an output of 0 at 100 Hz. Dividing the amplitude of this output by the amplitude of the input (zero divided by two) gives 0, and so the amplitude response of the overall two-part system has the value 0 at 100 Hz.

Now consider a sinusoid at 500 Hz and 2 V. Again, the value of the amplitude response for the low-pass filter is 1, so the input into the high-pass section is 2 V at 500 Hz. But, in contrast to the result at 100 Hz, the value of the amplitude response for the high-pass filter at 500 Hz is also 1. Multiplying this by the amplitude of the input gives an output of 2 V at 500 Hz. Since 2 V was the amplitude of the initial input to the entire system, the value of the amplitude response of the complete band-pass filter system must be 1 at 500 Hz. (This follows from equation (9), the original definition of the amplitude response.)

In fact, in order to find the response of the complete system, we need not consider any *particular* input amplitude at all. All we are doing is multiplying the amplitude of the input signal by the appropriate values determined by the individual amplitude response characteristics of the two component systems, and finally dividing by the amplitude of the input signal. Therefore, the influence of the input signal amplitude is cancelled out. In short, to find the overall amplitude response of a cascade of two LTI systems, all we need do is multiply the amplitude response curves together. A simple algebraic proof of this can be found in Appendix 6.1 of this chapter.

This result is easily generalized to three or more systems. Two LTI systems in cascade form an LTI system, with the total amplitude response being given by the product of the two individual amplitude responses. This system can then be cascaded with a third, the amplitude response for the total system being given by the product of the third amplitude response with the first two, that is $A_1(f) \times A_2(f) \times A_3(f)$, and so on for a fourth, fifth, etc. system added to the cascade. In short, if a cascade of LTI systems $S_1, S_2, \ldots S_n$ each have an amplitude response of $A_1(f), A_2(f), \ldots A_n(f)$, the amplitude response of the entire system is given simply by $A_1(f) \times A_2(f) \times A_3(f) \times \ldots A_n(f)$.

Another interesting fact can be gleaned from this result. When numbers are multiplied together, it doesn't matter in which order the numbers are multiplied, so for a connection of three systems in cascade:

$$\begin{aligned} A_1(f) \times A_2(f) \times A_3(f) &= A_2(f) \times A_1(f) \times A_3(f) \\ &= A_3(f) \times A_2(f) \times A_1(f) \\ &= A_1(f) \times A_3(f) \times A_2(f) \end{aligned} \tag{11}$$

and so on. Therefore, the overall amplitude response of a cascaded system doesn't depend on the order in which the different subsystems are connected together. So, for example, the band-pass filter just explored could have had the high-pass section first in the chain with the low-pass filter following without affecting the overall amplitude response.

This multiplication of system amplitude responses is only correct, of course, when the y-axis of the amplitude response is on a linear scale. The arithmetic becomes even easier when logarithmic scales are used (dB scales included) since, as we discussed in Chapter 3, taking logarithms turns multiplication into addition. Put in a mathematical form:

$$20 \log[A_1(f) \times A_2(f) \times \cdots \times A_n(f)] = 20 \log A_1(f) + 20 \log A_2(f) + \cdots + 20 \log A_n(f) \tag{12}$$

because in general:

$$k \times \log(ab) = k \times \log a + k \times \log b \tag{13}$$

So, for amplitude response curves on logarithmic scales, the total amplitude response of a cascaded system is calculated by summing the individual amplitude response curves of the component linear systems.

One warning: most of the systems we've described so far have had a value of 1 for the amplitude response across some band of frequencies. This meant that sinusoidal signals in those regions came out at the same amplitude as they went in. We say that the system has a gain of 1, or *unity gain*, for such signals. But, in our classification of response curves as high- or low-pass, the exact value of the gain in the pass-band is irrelevant. When we say a system is low-pass, for instance, all we mean is that sinusoids below the cutoff frequency are passed through the system relatively well compared to signals above the cutoff frequency. One could imagine a low-pass filter with unity gain in the stop-band as long as there was a much bigger gain (an amplitude response value of 100, say) in the pass-band. As in life in general, everything is relative.

Although the gain of the pass-band will not alter the name we give a system, it can still be important if the gain factors in the pass-band of the response are much smaller than 1. This means that even the sinusoids one wants to come through the system are significantly attenuated even if the unwanted ones are attenuated a lot more. Difficulties arise because, as we discussed in Chapter 4, every system has some background noise (even if very low) that will stay pretty constant. If the signal levels get too low and approach the levels of the noise, the signal will be degraded by the noise. One always attempts to get the maximum separation between the levels of the signal and the noise, so that the noise stays small enough (in relation to the signal) so as not to affect the signal.

In times not long past, real-life electrical filters often did give significant losses because they were made up of passive components like resistors, capacitors, and inductors. Such devices do not need to be powered, but because of this they always 'use up' some of the signal. The amount of attenuation a device gives even in the frequency region where transmission is meant to be best is known as the *insertion loss* of the device. Typical real-life filters used to have insertion losses of 6 dB, which is a halving of the amplitude of the signal.

Modern-day electronic filters (consisting of active components such as transistors and op-amps) usually have no insertion loss at all. The energy lost from the signal in the filter is made up for. Of course this power must come from somewhere, and that is why such filters must be plugged into the wall.

The vocal tract as a linear system

Many systems that are of interest to us will have amplitude response curves a bit more complex than those described so far. A good example of such a system is the *vocal tract*, made up of the lips, teeth, tongue and pharynx. For the moment let's consider the velum to be closed, that is, when no air can escape through the nose:

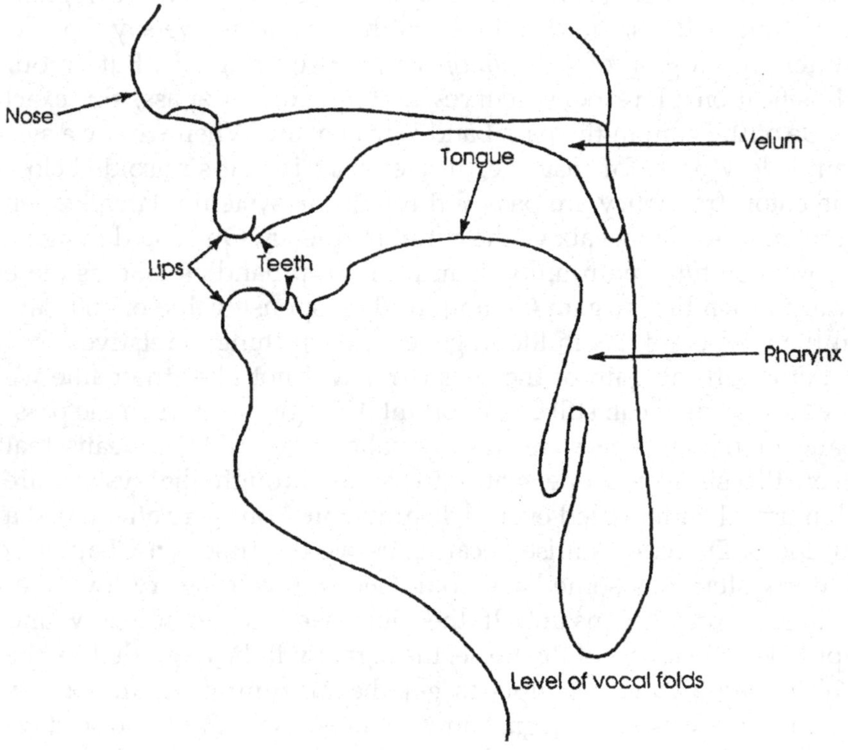

The vocal tract is an acoustic system consisting of a tube of varying width. The bend in it has little effect on its behaviour and so can be ignored. In many instances, the input to the vocal tract system is the sound generated by the vibrating vocal folds. The transmission of the input larynx signal is different for different positions of the articulators. This is another way of saying that the amplitude response for an acoustic tube depends on the shape of the tube. Of course, the sound generated by the vocal folds is never sinusoidal, so we have to wait until we know a bit more to do a complete analysis of the entire vocal-fold/vocal-tract system. We *can* study the effects the vocal tract has on sinusoids, however, by introducing a sinusoidal signal into the vocal tract at the level of the larynx. This is done by pressing a sinusoidal sound source (like a vibrator) against someone's neck and measuring the level of the sound coming out of their mouth as seen schematically at the top of the next page.

Our primary interest here is in what the vocal tract does to sounds passing through it, but there is a complicating factor. The sound induced by the vibrator first has to travel through the tissue of the neck before it gets into the vocal tract. If this transmission is not equal for all frequencies, some aspects of the amplitude response of the overall system will not be attributable to the effects of the vocal tract itself.

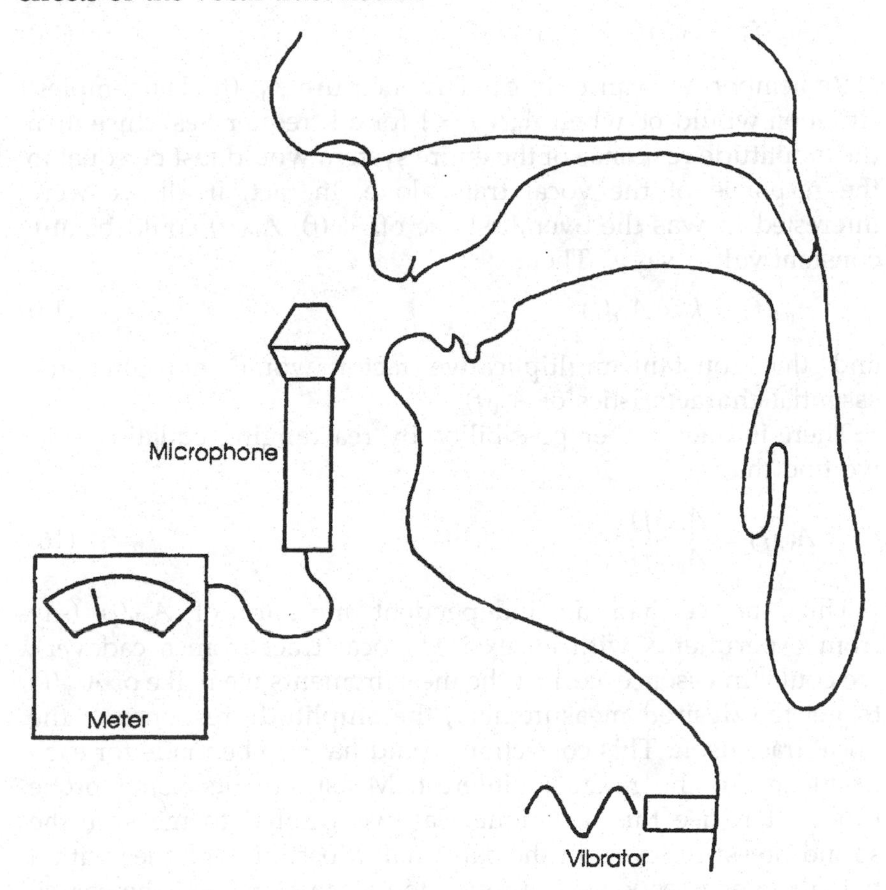

Let's make this more explicit. Here's our system in block diagram form:

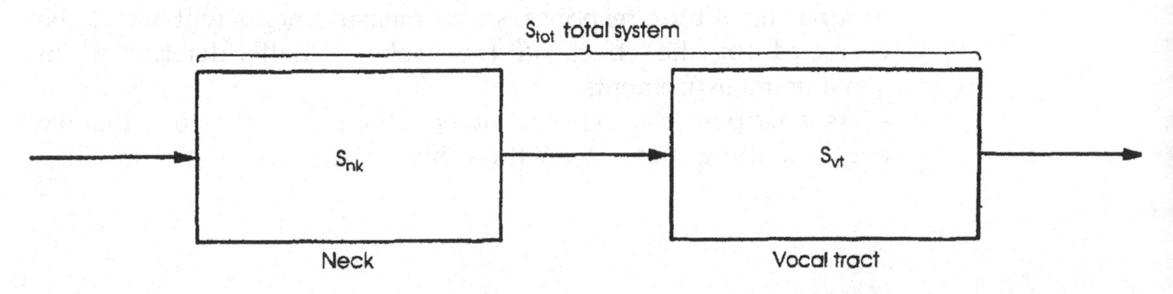

We'll take it for granted that both systems are linear, or near enough so for our purposes (which in fact they are). Secondly, we'll ensure that they are time invariant by having the subject hold a particular vocal tract position during a measurement. In short, both systems are LTI.

Now suppose that systems S_{nk} (the neck) and S_{vt} (the vocal tract) have amplitude responses of $A_{nk}(f)$ and $A_{vt}(f)$. In this experiment, all we can measure is $A_{tot}(f)$, the amplitude response corresponding to the total system, S_{tot}. We have already shown how to calculate the amplitude response at any frequency for two systems serially connected:

$$A_{tot}(f) = A_{nk}(f) \times A_{nk}(f) \tag{14}$$

Remember, we can only directly measure $A_{tot}(f)$. The simplest situation would be when $A_{nk}(f) = 1$ for all frequencies, since then the amplitude response of the entire system would just be equal to the response of the vocal tract alone. In fact, if all we were interested in was the overall shape of $A_{vt}(f)$, $A_{nk}(f)$ could be any constant value, say k. Then:

$$A_{tot}(f) = k \times A_{vt}(f) \tag{15}$$

and this constant multiplicative factor would not alter the essential characteristics of $A_{vt}(f)$.

There is one further possibility. By rearranging equation (14), we find that:

$$A_{vt}(f) = \frac{A_{tot}(f)}{A_{nk}(f)} \tag{16}$$

Thus, if we had an independent measure of $A_{nk}(f)$ (say from experiments with an excised vocal tract from a cadaver), we could, in essence, correct the measurements we make of $A_{tot}(f)$ to get the desired measurement, the amplitude response of the vocal tract itself. This correction would have to be made for each frequency in the range of interest. Measurements using probe tubes often use this technique. Say we wanted to measure the sound pressure levels in the ear canal. Most microphones with a flat frequency response are much too large to fit into the ear, so small probe tubes, which fit over the end of the microphone port, are used instead. However, the probe tube itself will have some non-flat amplitude response, so the measurements will have to be corrected for the effects of the probe, usually determined in separate measurements.

As it happens, the experimenters in the particular study that we are describing used a method involving certain assumptions about the form of $A_{vt}(f)$ to determine $A_{nk}(f)$ from a series of measurements of $A_{tot}(f)$. The details need not concern us here. The only important thing to note is that the effects of the neck have already been corrected for, in the way described above, for the following results.

With the problem of transmission through the neck resolved, we can get back to our measurements. The subject assumes the vocal tract position for a particular vowel, say /i/ as in 'heed', without, of course, actually saying anything, since the input is going to be supplied artificially. Now the vibrating sinusoidal sound is applied to the subject's neck. At the same time, the level of the sound is measured at the subject's mouth. This is done for a large number of single frequencies of vibration. Alternatively, in what amounts to the same thing, the sinusoidal signal is swept slowly in frequency over the desired range and a graphic recorder attached to the microphone continuously records the amplitude of the sinusoidal sound at the subject's mouth. The result is something like this:

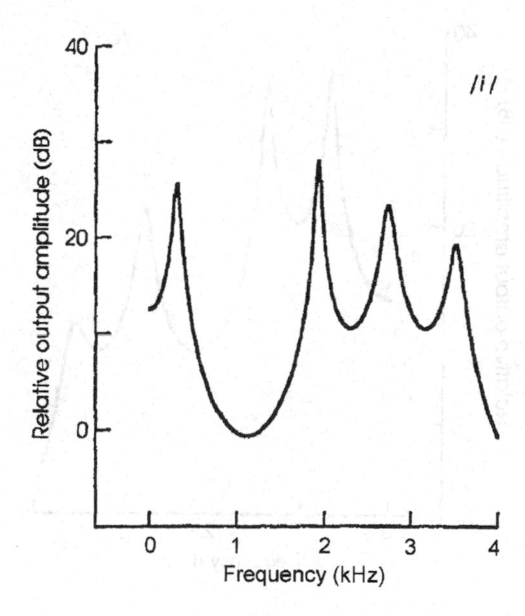

This is a much more complex shape for an amplitude response than any we have previously encountered. Before we try to describe this pattern in simple terms, let's look at the amplitude response curves that are obtained with articulatory positions appropriate for some other vowels. For instance, /ɔ/ as in 'caught':

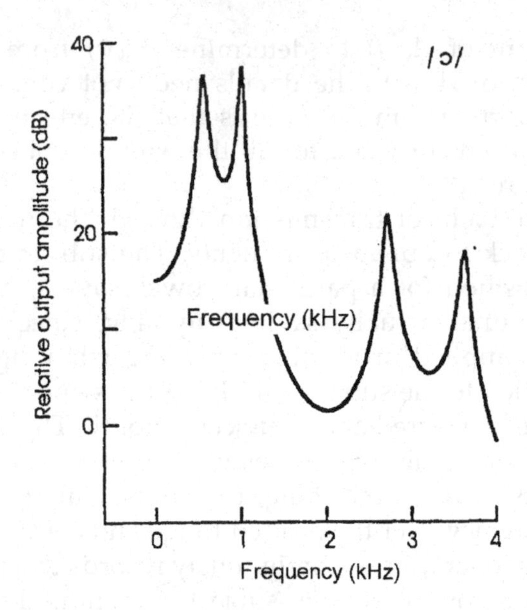

or /æ/ as in 'Sam':

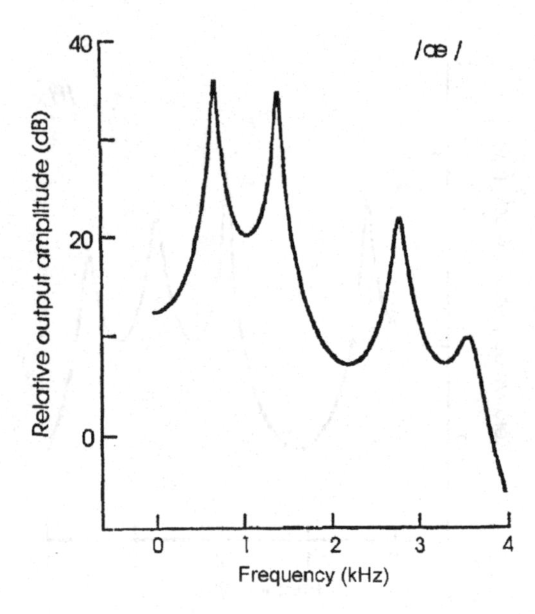

All of these curves are from the same (female) speaker. For a given person, each of these curves is a sort of signature (broadly speaking), for the particular vowel it is associated with.

The crucial feature in all these response curves is that certain frequency regions are enhanced relative to others. In fact, one sees four 'bumps' in the amplitude response curve, and their position varies from vowel to vowel. Each of these is, in fact, just a simple resonance (like a band-pass filter), and thus we see that the amplitude response of the vocal tract can be characterized by a set of resonances. Resonances in the vocal tract have a special name—*formants*—and the position of the three lowest in frequency is known to be a major determinant of our perception of the quality (or identity) of a vowel. Typically, they are numbered from left to right (from low to high frequency) beginning at one, and are abbreviated as F_1, F_2, F_3, etc. So we can say from the response curves that /ɔ/ has a higher F_1 than /i/, but that /i/ has a higher F_2 than /ɔ/.

Just as with resonances, we describe formants with two numbers. Perceptually, the most important aspect of a formant is where its peak falls on the frequency continuum, known as the *formant frequency* (exactly the same concept as the *resonant frequency* described above). For example, the /i/ vowel we just showed has its first four formants at frequencies of 311, 1930, 2740 and 3520 Hz. These are clearly nothing more than the centre frequencies of the four resonances. In many experimental studies (whether involving synthetic or natural speech), only the formant frequencies are given. Sometimes the half-power bandwidths are reported as well, calculated just as they are for resonances. You will almost never see a roll-off specified, for two reasons. Firstly, the effects of various formants are typically mixed together and there is no clear roll-off. For instance, the /ɔ/ shown above has its first two formants so close together that one cannot see the high-frequency roll-off of the first formant. Secondly, and more importantly, the roll-offs of formants that can be clearly seen turn out to be predictable from a simple model. In fact, the entire amplitude response curve in detail is predictable from the formant frequencies and their bandwidths.

Many speech synthesizers work by creating amplitude response curves like these electronically. The simplest way to do this is with a serial combination of simple band-pass filters, each of which has the amplitude response of a simple resonance (seen earlier in the chapter).

A simple speech synthesizer can be constructed by connecting, say, four of these together in cascade:

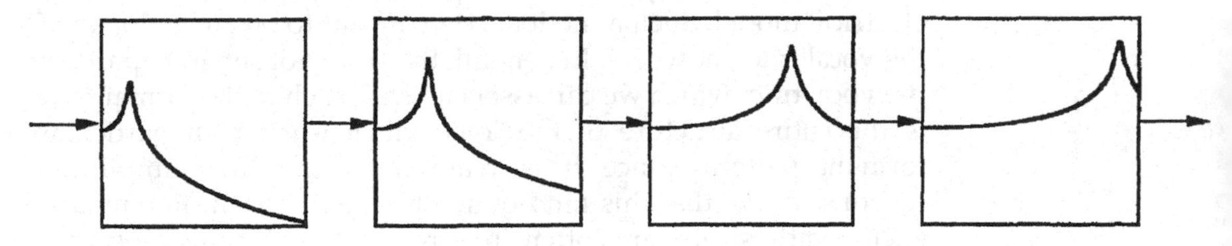

With the centre frequency and bandwidth of each one of these sections adjustable, we can successfully synthesize the major features of the vocal tract amplitude response curves associated with the production of simple steady-state vowels.

We've already shown how to obtain the overall amplitude response of such a system. All we need to do is multiply the individual amplitude response curves by one another. On dB scales, we'd add all the responses together. So, in trying to synthesize an /ɔ/ vowel, we set the formant frequencies at 570, 1030, 2730 and 3630 Hz, and the bandwidths at 80, 55, 90 and 100 Hz, respectively, to obtain:

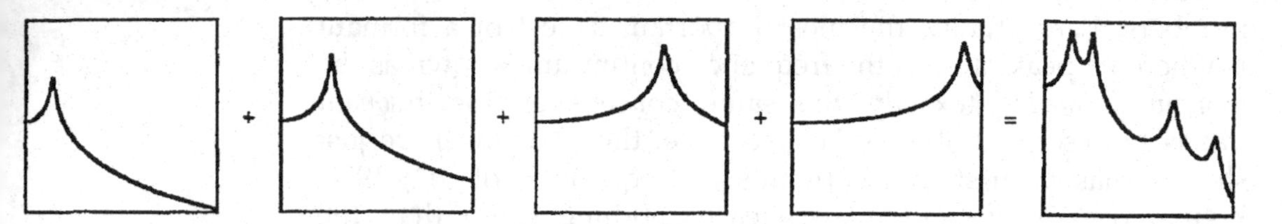

which is very similar to the measured vocal tract response. However, if you carefully compare this version of the amplitude response (synthesized from four resonances only) with the original measurements, you may notice an important difference. The level of the resonances decreases much more across frequency for the synthesized amplitude response. This difference arises because the vocal tract actually has many, many formants extending high into frequency, whose only effect on the audible end of the amplitude response is to act as a high-pass filter. Thus resonance-based speech synthesizers usually include what is known as a *higher pole correction* in the cascade of resonances. This high-pass filtering thus tilts the amplitude response upwards, so that the amplitude response curves obtained from real vocal tracts can be approximated extremely well.

A cascade of resonators is not only a convenient way to construct a part of a speech synthesizer. This series combination of simple band-pass filters is also a model of how the vocal tract works. In this case, however, the model works with electrical signals while the vocal tract works with acoustic ones. But just as an acoustic tube with a varying cross-section (the vocal tract) has four or five varying formants important for speech, so can our electrical model. Such a model is only meant to be an analogue of the vocal tract, however. In general, there are not separate parts of the vocal tract which we can associate with each of the formants. It is the entire structure of the tract which leads to a particular formant pattern. Since it is really only the total amplitude response curve that this kind of synthesizer is meant to emulate, such synthesizers are often referred to as *terminal analogue synthesizers*. This is just another way of saying that, at its final output, the synthesizer gives you something like what the vocal tract does, without mimicking the actual physical processes involved.

Phase responses

Variations in amplitude across frequency for auditory signals are much more perceptually salient than changes in phase, so most attention is concentrated on the amplitude response of systems. Recently, however, there has been increased interest in the perceptual correlates of phase changes, in both normal listeners and the hearing-impaired. Also, phase changes in systems can cause large changes in what a waveform *looks* like (as we shall see in the next chapter), even though there may not be much change in what it *sounds* like. These changes in waveform shape can be crucial in trying to assess relationships between various kinds of waveforms and physiological processes. Therefore, we will discuss phase as well as amplitude, although not nearly in as great detail.

As we've already stressed, sinusoids can only be changed in two ways by LTI systems, in their amplitude and their phase. Just as for amplitude responses, the phase change that a system imparts must be specified for each individual sinusoidal frequency. Such a curve is known, not surprisingly, as a *phase response*. And just as there are two ways of expressing the phase of a sinusoid, in degrees and radians, there are two ways of expressing the phase response. Since these are proportional to one another (as you remember from Chapter 3, there are 2π radians in 360°), the phase response will appear the same regardless of the units used. Such scales will always be linear (never logarithmic), giving the values directly in degrees or radians. In this chapter, to make things simpler, we will always specify phase in degrees.

How, then, do we go about specifying phase responses? Before going into detail, it is as well to note that we are going to define the phase response in a way that makes it easy to use. The properties we found desirable in amplitude responses have analogues in phase responses, and so our development here will closely parallel that already done for amplitude responses. Therefore, we can progress much more quickly, leaving out much of the detailed explanation.

The first thing to decide is the best way to report the relationship between the phases of the input and output sinusoids. We could, of course, simply fix the phases of the input sinusoids at some particular value (90°, say) and report the absolute phases of the outputs. This would, however, restrict the usefulness of the measurement to one particular phase of the input. It would be more convenient if, just as we did for amplitude responses, the phase response were defined in a way that made it easy to predict the phase of the output for all possible phases of the input. A good way to do this is to define the phase response at frequency f, $P(f)$, to be equal to the *difference* between the phases of the input and output sinusoids. In a simple formula:

$$P(f) = O_{\text{phase}}(f) - I_{\text{phase}}(f) \tag{17}$$

where $O_{\text{phase}}(f)$ is the phase of the output and $I_{\text{phase}}(f)$ is the phase of the input, both at frequency f. In effect, the phase of the output is expressed relative to the phase of the input. This avoids the necessity of making an absolute determination of the phases of both the input and the output, which would mean defining an essentially arbitrary 'zero' time.

A little rearrangement of terms shows that, given the phase of an input sinusoid at a particular frequency, f, and the value of $P(f)$, the phase of the sinusoidal output is given by:

$$I_{\text{phase}}(f) + P(f) = O_{\text{phase}}(f) \tag{18}$$

To put this in words: in order to determine the phase of the output sinusoid, simply add the value of the phase response at that frequency to the phase of the input.

Let's take a look at a particular system (System Y) to make things a bit more concrete. Suppose we want to determine the phase response at 250 Hz. Just as was done for the determination of amplitude responses, we enter a 250-Hz sinusoid as the input, and examine the system output:

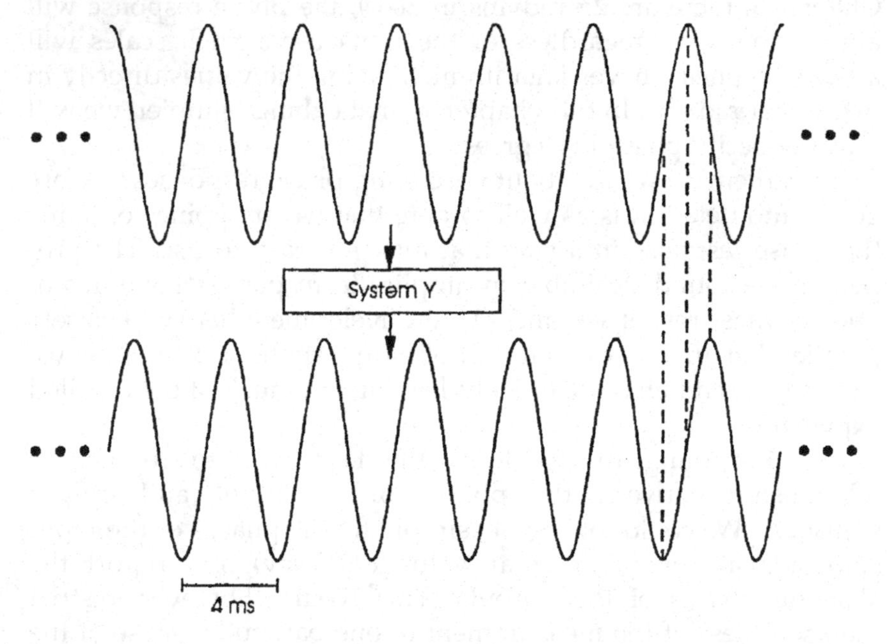

You can see that the input hits its positive peak 1/4 of a cycle before the output. In other words, the output lags the input by 90° (= 360° × 1/4), and so the value of the phase response at 250 Hz is −90°. To determine the phase response at other frequencies, separate measurements would be necessary.

Defining the phase response as the difference between output and input phases, as in equation (17), has another major advantage. Whatever phase input signal is used for the measurement, $P(f)$ will have the same value. A simple proof of this fact is given in Appendix 6.2 of this chapter. Here we only note that the proof takes advantage of two facts: (1) a shift in phase for a sinusoid is completely equivalent to a shift in time, and (2) the system is time invariant.

Linear phase responses

Let's consider the phase response of System Y in more detail. So far, we have only measured it at one frequency, 250 Hz. If we did a series of measurements at other frequencies, we might find the following result:

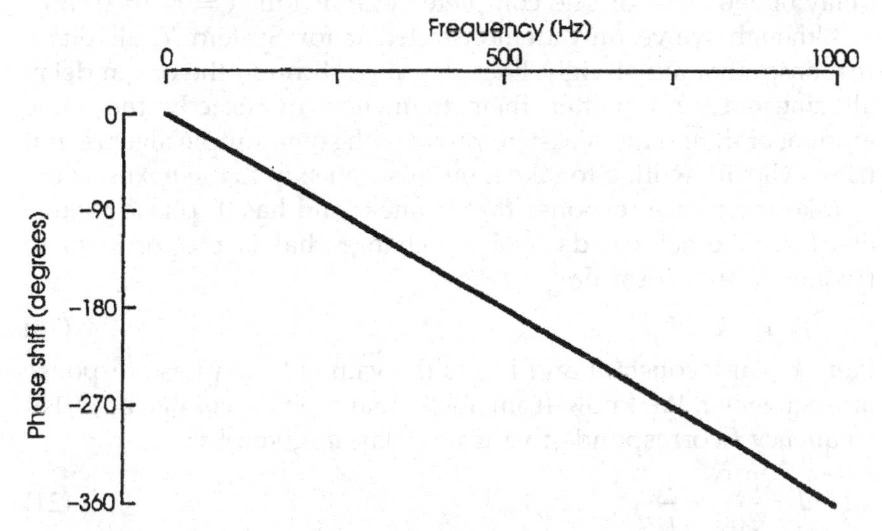

The phase response curve is a straight line passing through the origin. A system with a phase response like this has a special property—every sinusoid passed through it is delayed by exactly the same amount of time. We can show that this is true for this particular system by calculating the delay due to each phase change at a number of frequencies. First let us develop (for sinusoids only, of course) a general formula for calculating the time delay arising from a phase change.

Note first that each complete cycle of a sinusoid contains 360° (no matter what its frequency), and so a phase lag of $\Phi°$ would simply lead to a delay of $\Phi/360$ cycles. To calculate the time this represents, we multiply the number of cycles by the duration of each cycle of the sinusoid, that is, its period. Since the period of a sinusoid is given by the reciprocal of its frequency f (that is, $1/f$), the time delay, d, for a phase lag of $\Phi°$ at frequency f

is given by:

$$d = \frac{\Phi}{360} \times \frac{1}{f} \tag{19}$$

We'll now calculate the value of the time delay for a number of frequencies. Take 250 Hz first. As already noted (and seen from the phase response curve), the value of the phase response is −90 °; that is, the output lags the input by 90°. A phase lag of 90° is equivalent to a time delay of 90/360° or 1/4 of a cycle. The duration of one cycle is 4 ms (= 1/250 Hz), so one-quarter of this period is simply 1 ms.

At 500 Hz, the phase lag is 180° and so the time delay is now ½ (= 180/360°) of a cycle of 2 ms—or 1 ms. At 750 Hz, the phase lag is (270/360°) × (1/750), which is 1 ms yet again. And finally, at 1000 Hz, a sinusoid undergoes a phase change of 360°, which means that it is back to 0° phase again, and has experienced a time delay of 360/360° or one complete cycle of 1 ms (= 1/1000 Hz).

Although we've only demonstrated it for System Y, all phase responses that are straight lines that pass through the origin delay all sinusoids, no matter their frequency, by exactly the same amount of time. This is easy to prove with some simple algebra, but those who are willing to take it on trust can skip to the next section.

Take any phase response that is linear and has 0° phase change at 0 Hz, in other words a phase change that is proportional to frequency. In a formula:

$$P(f) = k \times f \tag{20}$$

Here k is any constant and $P(f)$ is the value of the phase response at frequency f. We know from above that a phase change of $P(f)$ at frequency f corresponds to a time delay, d, given by:

$$d = \frac{P(f)}{360} \times \frac{1}{f} \tag{21}$$

Substituting for $P(f)$ from equation (20):

$$d = \frac{k \times f}{360} \times \frac{1}{f} \tag{22}$$

and cancelling the variables on top and bottom:

$$d = \frac{k}{360} \tag{23}$$

This expression doesn't depend on f, so the time delay is a constant determined by the slope of the phase response, k. In System Y, for example, since there is a phase change of 360° at 1000 Hz, k is equal to 0.36 (= 360/1000) and so the constant time delay is 0.36/360 = 0.001 s or 1 ms.

To summarize, a system with a linear phase response will delay all sinusoids by the same amount of time. Therefore, the phase relationship between different sinusoids at the input of the system will be preserved at the output. It is not enough that the phase response be a straight line—it must also pass through the origin.

Wrapped and unwrapped phase curves

You may have noted that the phase response for System Y has a value of −360° at 1 kHz. Recall from Chapter 3 that sinusoids that differ by exactly 360° are identical—this results, of course, from the periodicity of sinusoids and there being exactly 360° in one complete cycle. Any particular phase difference could have 360° added to or subtracted from it, and would be equivalent. So, for example, a phase difference of 0° would be equivalent to −360°, 0° or 360°. And there is no reason why any multiple of 360° couldn't be added on—each 360° is equal to one complete cycle. So 0° could be expressed as 360°, 720°, 1080° or more. This fact comes to the fore when determining phase responses empirically because a phase difference can only be specified to within a 360° range. So, for example, if an input and output signal are exactly in phase, it may be that the phase difference is actually 360°, 720° or more, or even the negative of these values.

This essential uncertainty about the determination of phase should not trouble you, as it is simply another expression of periodicity. We discuss it at length here, because it can have important effects on the way phase responses look.

Because phase can only be determined within a 360° range, phase response curves will often be restricted to such a range (e.g. −180° to +180°, or 0° to 360°). This can make a phase response curve look very unusual:

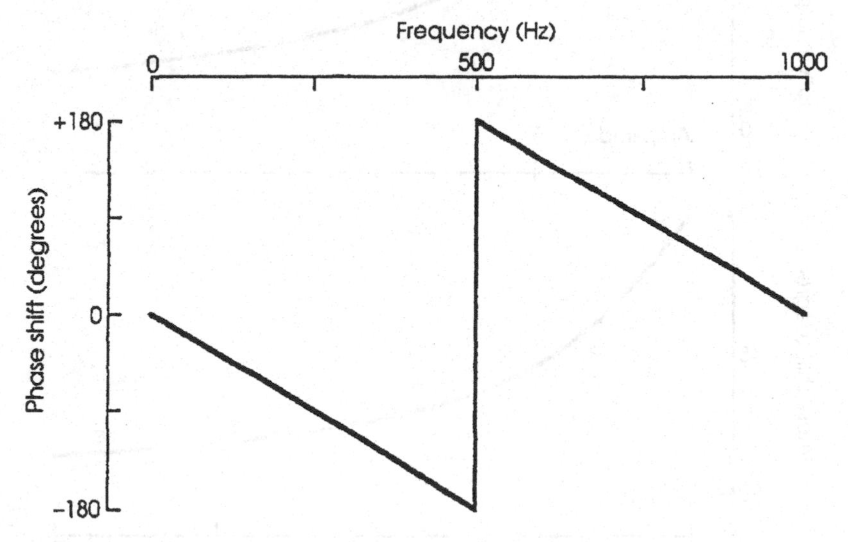

In fact, this is nothing more than the phase response of System Y again, only drawn so as to have phase values between −180° and

+180°. You can see that, as the phase goes increasingly negative from 0°, it shoots suddenly upward from −180° to+180° at 500 Hz. *This does not indicate any discontinuity in the response of the system near 500 Hz.* It results simply from restricting the phase to between −180° and 180°. If the range had been defined differently (e.g. between −360° and 0°), the curve could have 'jumped' at a different frequency. The 'normal looking' phase response can be obtained from the discontinuous one by subtracting 360° from phase values for frequencies greater than 500 Hz. We call this process, of going from a discontinuous phase response to one that is smooth (but has a range greater than 360°), *phase unwrapping*. A number of signal-processing techniques require it.

To summarize, phase response curves often look discontinuous because phase can only be determined uniquely within a 360° range. Phase response curves for the same system that look different from one another can be transformed by adding or subtracting multiples of 360° from any particular value.

Other phase responses

We've chosen a rather simple form for the phase response of System Y, and one that is hardly typical. Here is an amplitude and phase response of a commonly occurring system, a very simple low-pass filter:

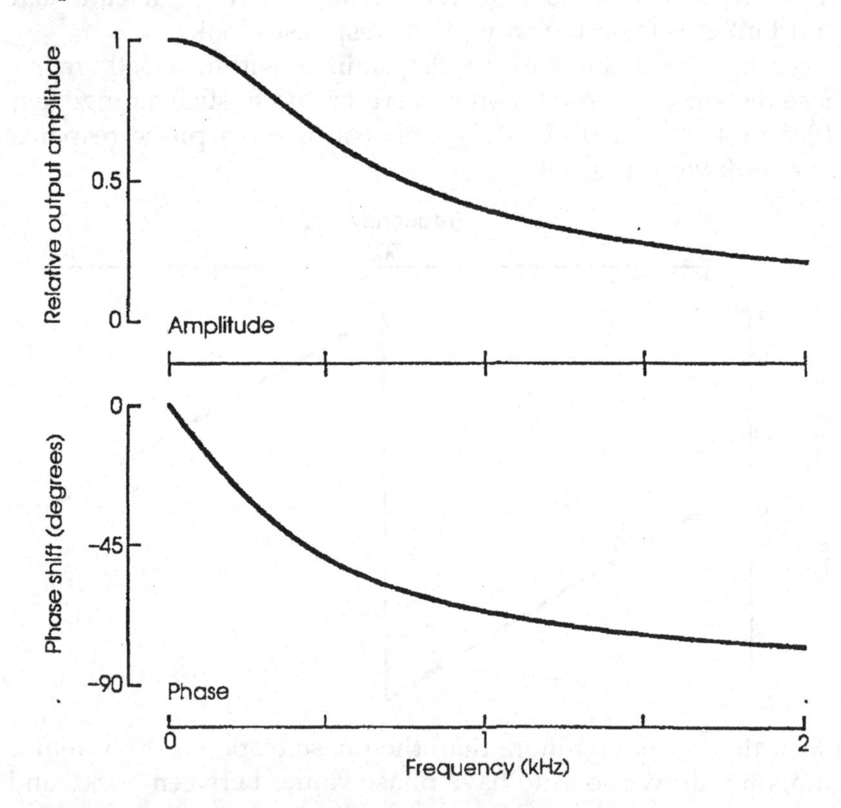

Note how the phase changes rather steeply as frequency increases from 0 Hz, but then levels off to an asymptotic value of −90°. The phase response associated with a simple resonance is interesting, in that the phase changes relatively rapidly across the frequencies in the filter pass-band, and slowly otherwise:

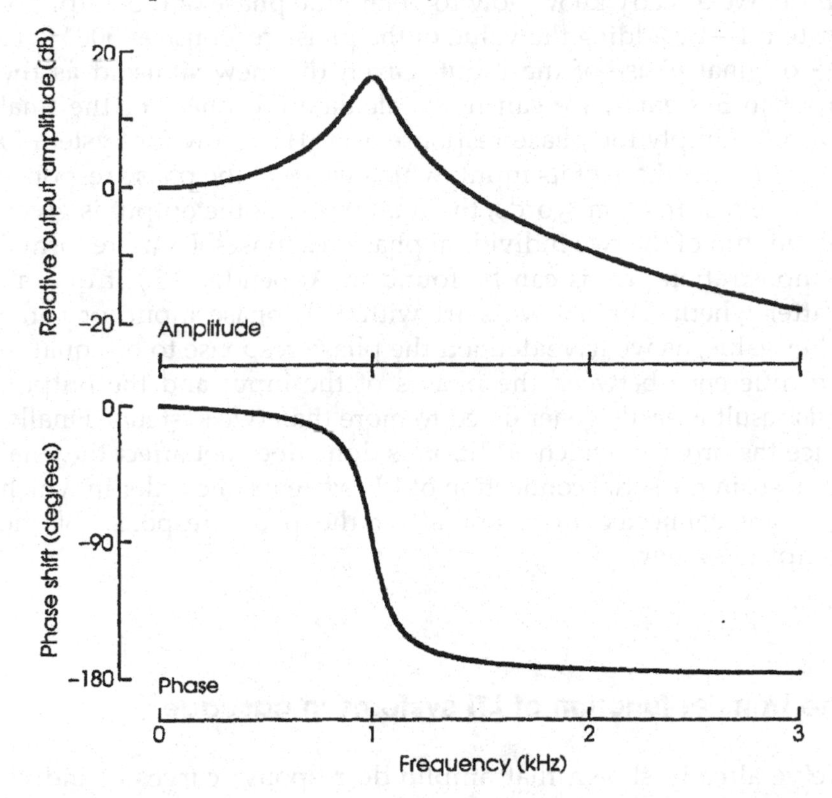

Although linear phase responses are often highly desirable, most practical systems do not have them. On the whole, systems are designed with primary regard to their amplitude response, and the phase is left to do what it will, especially in systems that have sound outputs.

The phase response of two LTI systems in cascade

Although we've only dealt with single systems so far, it is often the case that we connect up systems in cascade, feeding the output of one into the next, and so on. Just as we did for amplitude responses, we'd like a way to work out the phase response of the entire cascade from the phase response of the individual systems.

Let's keep things simple (at least at first!) by only considering two systems in cascade (at top, overleaf). Intuitively speaking, it should be fairly easy to see how to obtain the phase response of

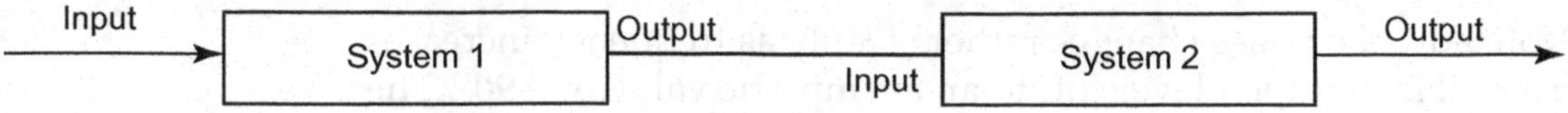

Input
System 1
Output
Input
System 2
Output

the entire system. Take a particular sinusoid, say at 500 Hz and 0° phase. We already know how to obtain the phase of the output of System 1—by adding the value of the phase response at 500 Hz to the original phase of the input. Given this new sinusoid as the input to System 2, we can now calculate the phase of the final output—simply the phase response at 500 Hz (now for System 2) added to the phase of its input, which was just the phase response of System 1. In other words, the final phase of the output is given by the sum of the two individual phase responses. (A more formal demonstration of this can be found in Appendix 6.3.) It doesn't matter whether or not we start with a 0° phase input, or some other value, as we have defined the phase response to be equal to the difference between the phases of the input and the output. This result is easily generalized to more than two systems. Finally, since the order in which addition is done does not affect the final sum, so, in the serial connection of LTI systems, the order in which they are connected does not affect the phase response of the complete system.

The transfer function of LTI systems in cascade

We've already shown that amplitude response curves of individual systems in a cascade are multiplied by (on linear amplitude scales) or added to (on logarithmic amplitude scales—like dB) one another to obtain the final amplitude response. The order of connection thus does not affect the amplitude response of the entire system.

Since LTI systems are completely specified by their amplitude and phase response (jointly known as the transfer function), we have just shown that a serial cascade of systems can be hooked up in any order and the same result will obtain. Beware! This is only true in general for LTI systems and it is not hard to give examples of non-LTI systems where the order in which the systems are connected *does* matter.

For example, let S_1 be a non-LTI system that takes the reciprocal of the input. Let S_2 be an amplifier that multiplies all signals by 5. Let's pass the number 1 through the two systems, starting with S_1. The output of S_1 will be 1, and entering this into S_2 gives a final output of 5. But if we enter 1 into S_2 first, to get 5, and then enter this into S_1 the final result is 1/5, not the same as 5. Only for LTI systems is the order of processing always irrelevant.

Exercises

1. Below you will find the results of a series of measurements in which sinusoids of various frequencies, but always at an amplitude of 2 V, were put through six different systems. To make your task easier, the output levels are expressed in dB V re 2 V. Plot each of the amplitude response curves (comparing linear and logarithmic frequency scales), determine what filter shape they represent, and calculate band widths, filter cutoffs, *Q*, and roll-offs as appropriate. Any roll-offs should be calculated in dB per octave. For systems A, D and E, convert the dB values back into volts and plot. You will probably find it easiest to do these exercises using a spreadsheet program like Excel.

Frequency (Hz)	Amplitude of output sinusoid from each system (dB re 2 V)					
	A	B	C	D	E	F
100	0.1	0	−87.4	0	−57.7	−68.4
200	0.3	−0.1	−27.2	−3	−39.6	−32.2
300	0.6	−3	8	−36.4	−29.1	−11.3
400	1.2	−12.7	30	−21.3	−21.7	0.7
500	1.9	−22.2	33	−3	−16.3	3.6
600	2.9	−30.1	33	−0.1	−12.5	3.7
700	4.2	−36.8	33	0	−10	3.1
800	6	−42.6	33	0	−8.6	0.7
900	8.5	−47.7	33	0	−7.9	−3.2
1000	11.9	−52.3	33	0	−7.5	−8
1100	14.9	−56.4	33	0	−7.3	−12.8
1200	11.9	−60.2	33	0	−7.2	−17.3
1300	7.6	−63.7	33	0	−7.1	−21.4
1400	4.1	−66.9	33	0	−7.1	−25.3
1500	1.4	−69.9	33	0	−7	−28.9
1600	−0.8	−72.7	33	0	−7	−32.2
1700	−2.6	−75.3	33	0	−7	−35.4
1800	−4.3	−77.8	33	0	−7	−38.4
1900	−5.7	−80.2	33	0	−7	−41.2
2000	−7	−82.4	33	0	−7	−43.9

2. Sketch the frequency response of the individual systems, and the system made up of a cascade of two ideal filters, one high-pass with a cutoff frequency of 150 Hz, and one low-pass with a cutoff frequency of 1000 Hz. What kind of filter results?

3. Earlier in the chapter, we discussed the determination of the amplitude response of a cylinder in a sound field. Do we need to

worry about the amplitude response of the loudspeaker used? Give your reason(s).

4. As mentioned above, probe microphones are often used to measure sound pressure levels in small cavities. Unfortunately, the probe itself has its own non-flat amplitude response which must be taken into account. Suppose that the amplitude response of the probe+microphone system is measured to be:

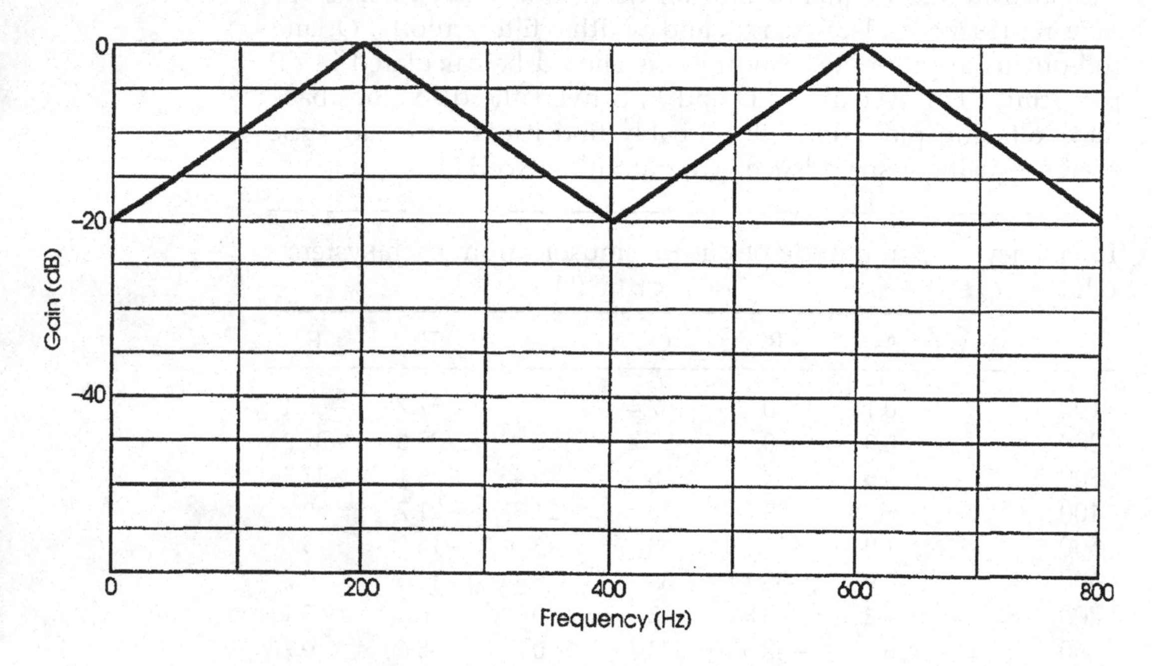

Using this probe microphone, an acoustic system is found to have the following amplitude response:

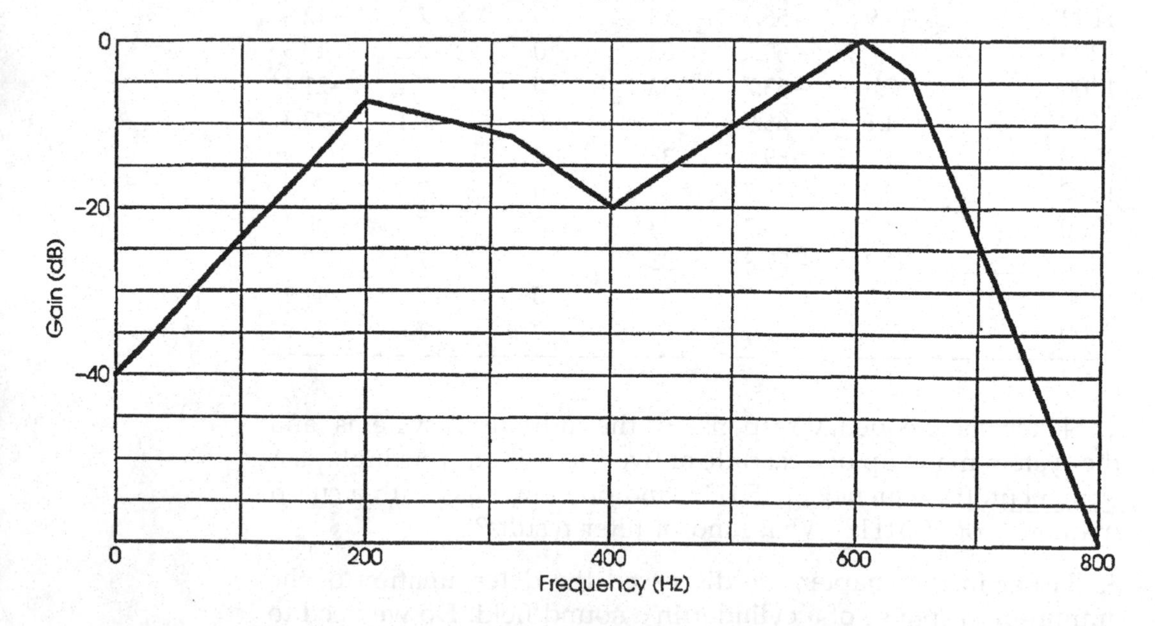

Correct this measurement using the known properties of the probe to obtain the true amplitude response of the measured system.

Similarly, the phase shift of the probe is given by:

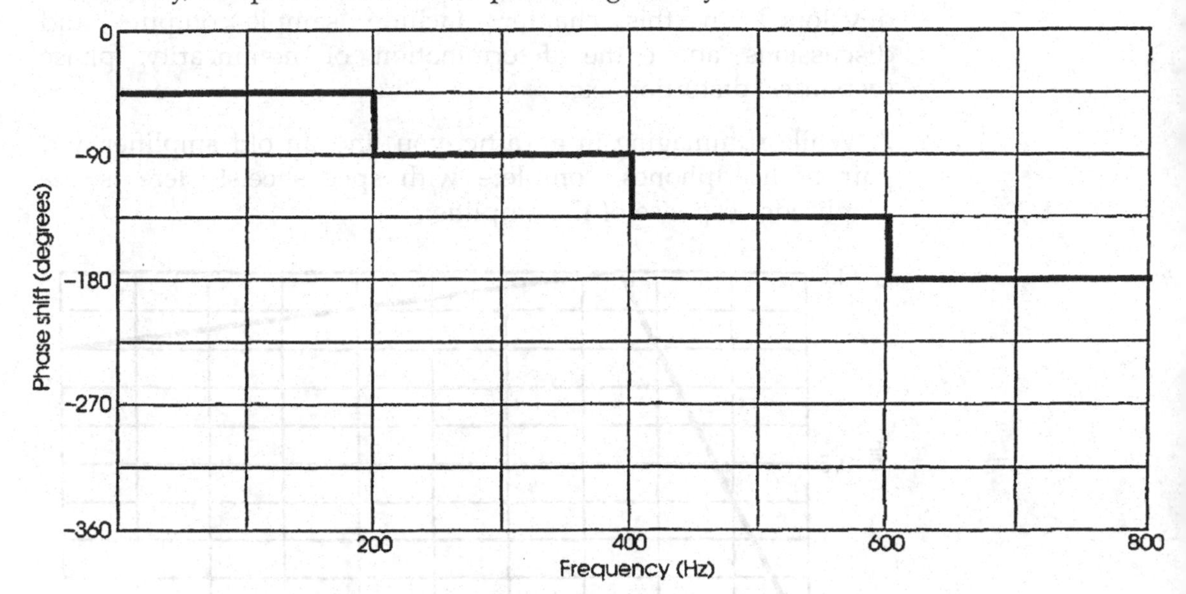

Correct the following obtained phase response to obtain the true phase response of the system:

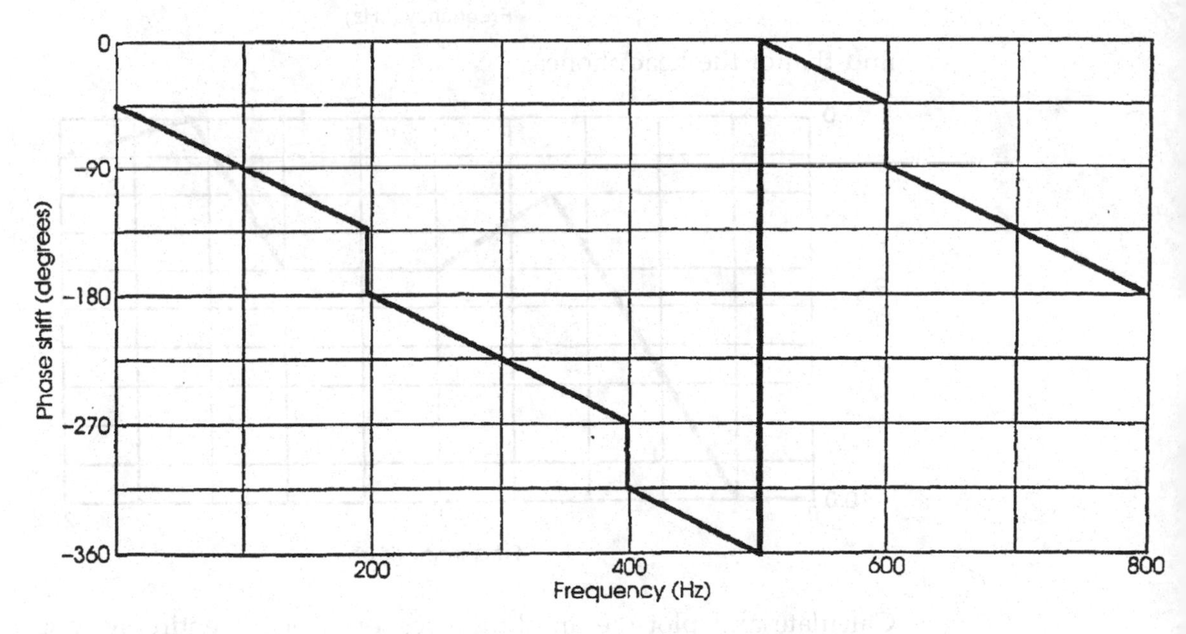

Before doing the correction, unwrap the phase in this curve.

5. Consider a cascade of two systems, one of which amplifies the input signal by a factor of 2, and one of which squares the signal

amplitude. Show that the order in which these two systems are arranged in the cascade affects the output.

6. Obtain the specifications of a hearing-aid test-box from the manufacturers and write an essay about it using the concepts developed in this chapter. Include sample outputs and discussions about the determination of nonlinearity, phase responses, distortion, etc.

7. While rummaging in an attic, you find an old amplifier and pair of headphones, complete with spec sheets! Here is the amplitude response of the amplifier:

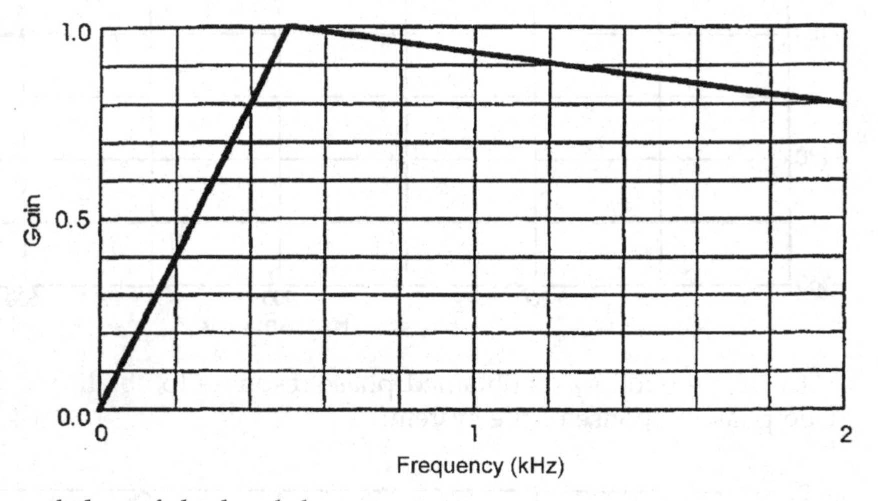

and that of the headphones:

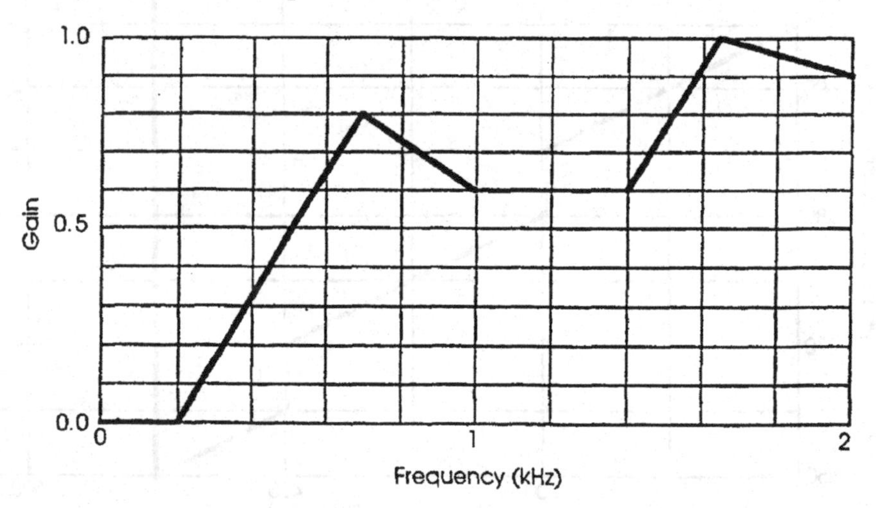

Calculate and plot the amplitude response of the entire system 'amplifier + headphones'. If you wanted to improve the operation of this system, and could only buy one new piece of equipment, what would it be and why? Similarly, here is the amplifier phase response:

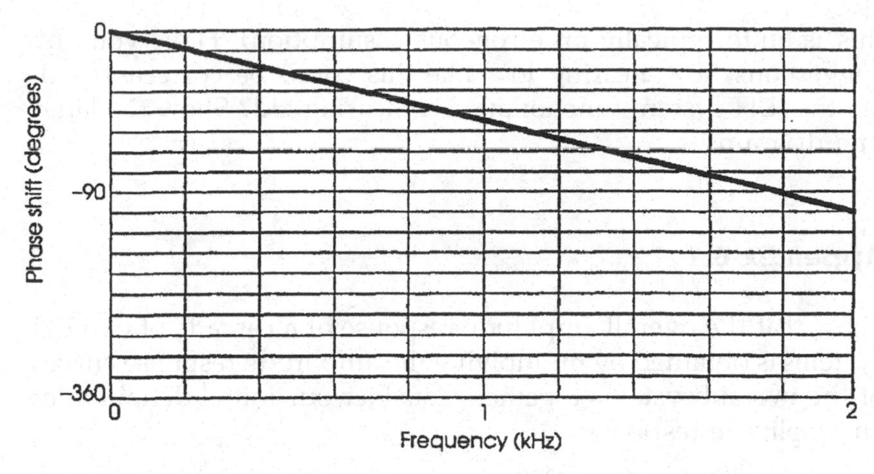

and the headphone phase response:

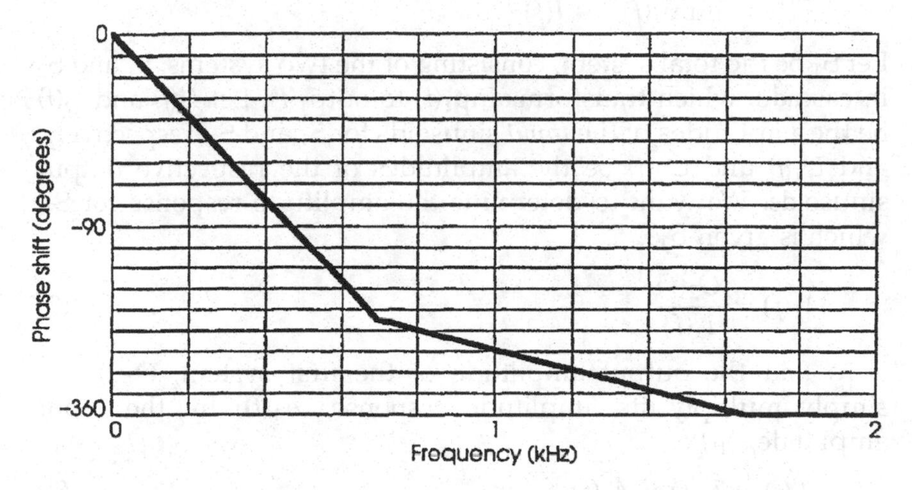

Calculate and plot the phase response of the cascaded system in both a wrapped and an unwrapped form.

8. Show that the half-power point of a filter is 3 dB down.

9. Get a sheet of specifications for at least two tape recorders and two amplifiers. Compare their amplitude responses.

10. Sketch realistic characteristics for the following filters:

(a) A filter which is flat from 20 Hz up to 3 kHz, then drops at 2 dB/octave for an octave and then rises at 12 dB/octave for an octave.

(b) A low-pass filter which is 3 dB down at 100 Hz, 8 dB down at 2 kHz and 20 dB down at 8 kHz.

11. Voiceless sounds have energy at high frequencies. A commonly encountered hearing disorder involves raised thresholds for hearing high frequencies, with hearing at low frequencies remaining normal. Depict this in a diagram assuming that this hearing loss resulted in a change in an LTI system (even though

this is quite typically an erroneous assumption). Have you any suggestions how hearing loss like this could be corrected with another LTI system functioning as a hearing aid? Show the latter in a diagram.

Appendix 6.1

Show that the overall amplitude response of a cascade of two LTI systems is obtained by multiplying the amplitude response curves of the two subsystems together. Consider our initial definition of an amplitude response:

$$A(f) = \frac{\text{Output}(f)}{\text{Input}(f)} = \frac{O(f)}{I(f)} \tag{1}$$

Let S_T be the total system, consisting of the two systems, S_1 and S_2 in cascade, which transforms $\text{inp}_1(t)$ to $\text{outp}_1(t)$. Let $I_1(f)$ and $I_2(f)$ be the amplitudes of the *input* sinusoids for S_1 and S_2, respectively, and $O_1(f)$ and $O_2(f)$ be the amplitudes of the respective output sinusoids. We want to determine the amplitude response for S_T, which is given by:

$$A_T(f) = \frac{O_2(f)}{I_1(f)} \tag{2}$$

To find the output amplitude of the first system, $O_1(f)$, we simply multiply its amplitude response, $A_1(f)$, by the input amplitude, $I_1(f)$:

$$O_1(f) = I_1(f) \times A_1(f) \tag{3}$$

Similarly, to find the output amplitude of the second system, we multiply its input amplitude by its amplitude response:

$$O_2(f) = I_2(f) \times A_2(f) \tag{4}$$

In this cascade of systems, $I_2(f) = O_1(f)$, so we can substitute it into the above equation to obtain:

$$O_2(f) = O_1(f) \times A_2(f) \tag{5}$$

and substituting for $O_1(f)$ from equation (3):

$$O_2(f) = I_1(f) \times A_1(f) \times A_2(f) \tag{6}$$

We can now return to our original definition of the transfer function—equation (2)—and substitute in it for $O_2(f)$, given by equation (6):

$$A_T(f) = \frac{O_2(f)}{I_1(f)} \tag{7}$$

$$= \frac{I_2(f) \times A_1(f) \times A_2(f)}{I_1(f)} \tag{8}$$

$$= A_1(f) \times A_2(f) \tag{9}$$

To summarize, the total amplitude response curve for two LTT systems in cascade is given by the product of the individual amplitude responses. This result is easily generalized to three or more systems.

Appendix 6.2

Show that the value of the phase response for a system does not depend upon the phase of the input sinusoid.

Assume that the phase response for a particular frequency, f_1, measured with a sinusoidal input at some particular phase, $I_{\text{phase}}(f_1)$, has the value b.

$$P(f_1) = O_{\text{phase}}(f_1) - I_{\text{phase}}(f_1) = b \tag{1}$$

Suppose we now enter into the system another sinusoid of frequency f_t, but with a different phase. The phase of this new input sinusoid can be expressed as:

$$I_{\text{phase}}(f_1) - \Phi \tag{2}$$

where Φ is the constant phase difference between the new and original inputs. We can think of this new input sinusoid as a time-shifted version of the original, since, as we showed in Chapter 3, a phase change in a sinusoid is completely equivalent to a shift in time.

It is a simple matter to calculate the time shift this corresponds to. Each complete cycle of a sinusoid contains 360° (no matter what its frequency), and so a phase lag of $\Phi°$ would lead to a delay of $\Phi/360$ cycles. To calculate the time this represents, we multiply the number of cycles by the duration of each cycle of the sinusoid, that is, its period. Since the period of the sinusoid is given by the reciprocal of its frequency (here, $1/f_1$), the time shift corresponding to a phase shift of $\Phi°$ is given by:

$$\frac{\Phi}{360} \times \frac{1}{f_1} = \frac{\Phi}{360 \times f_1} \tag{3}$$

Therefore, if the original input sinusoid is symbolized as inp(t), the new, phase-shifted, input sinusoid can be written as inp($t-\Phi/[360 \times f_1]$) as the latter is simply an appropriately time-shifted version of the former (see Chapter 4).

We can now use the property of time invariance to determine the output to the phase-shifted sinusoid. Recall from Chapter 4 what time invariance means. If:

$$\text{inp}(t) \rightarrow \text{outp}(t) \tag{4}$$

then:

$$\text{inp}(t-d) \rightarrow \text{outp}(t-d) \tag{5}$$

where d is some time shift. In the particular situation we face here:

$$\text{inp}\left(\frac{t-\Phi}{360 \times f_1}\right) \rightarrow \text{outp}\left(\frac{t-\Phi}{360 \times f_1}\right) \tag{6}$$

Now outp($t-\Phi/[360 \times f_1]$) is simply a time-shifted version of outp(t). As they are both sinusoids, we can consider them to be shifted in phase rather than time. Clearly, outp($t-\Phi/[360 \times f_1]$) is just outp(t) with a phase lag of $\Phi°$. Therefore, if outp(t) had a phase of $O_{\text{phase}}(f_1)$—as we initially assumed—then outp($t-\Phi/[360 \times f_1]$) has a phase of $O_{\text{phase}}(f_1)-\Phi$.

To summarize, if an input sinusoid of phase $I_{\text{phase}}(f_1)$ leads to an output sinusoid of phase $O_{\text{phase}}(f_1)$, then an input of phase $I_{\text{phase}}(f_1)-\Phi$ will lead to an output of phase $O_{\text{phase}}(f_1)-\Phi$.

We can now calculate the value of the phase response at frequency f_1 using the phase-shifted input:

$$P(f_1) = [O_{\text{phase}}(f_1) - \Phi] - [I_{\text{phase}}(f_1) - \Phi] \tag{7}$$

$$= O_{\text{phase}}(f_1) - \Phi - I_{\text{phase}}(f_1) + \Phi \tag{8}$$

$$= O_{\text{phase}}(f_1) - I_{\text{phase}}(f_1) \tag{9}$$

$$= b \tag{10}$$

as we assumed originally in equation (1).

In short, the value of the phase response is independent of the particular phase of the input used to measure it.

Appendix 6.3

Determine the phase response of a system, $P_T(/)$, made up of two systems in cascade. First, by the definition of a phase response:

$$P_T(f_1) = O_{2\text{-phase}}(f) - I_{1\text{-phase}}(f) \tag{1}$$

where $O_{2\text{-phase}}(f)$ is the phase of the output of System 2 and $I_{1\text{-phase}}(f)$ is the phase of the input to System 1.

We can easily determine the phase of the outputs of each system in terms of the phase of the inputs:

$$O_{1\text{-phase}}(f) = I_{1\text{-phase}}(f) + P_1(f) \tag{2}$$

$$O_{2\text{-phase}}(f) = I_{2\text{-phase}}(f) + P_2(f) \tag{3}$$

where $P_1(f)$ and $P_2(f)$ are the phase responses of Systems 1 and 2, respectively.

In this cascaded system, however, the input to System 2 is identical to the output of System 1, so that:

$$O_{1\text{-phase}}(f) = I_{2\text{-phase}}(f) \tag{4}$$

which can be substituted into equation (3) to obtain:

$$O_{2\text{-phase}}(f) = O_{1\text{-phase}}(f) + P_1(f) \tag{5}$$

Then, substituting into this equation from equation (2) leads to:

$$O_{2\text{-phase}}(f) = I_{1\text{-phase}}(f) + P_1(f) + P_2(f) \tag{6}$$

Finally, we can use this result to substitute back into equation (1):

$$P_T(f) = I_{1\text{-phase}}(f) + P_1(f) + P_2(f) - I_{1\text{-phase}}(f) \tag{7}$$

$$= P_1(f) + P_2(f) \tag{8}$$

To summarize in words, the phase response of a system consisting of two LTI systems in cascade is given by the sum of the two individual phase responses.

CHAPTER 7

The Frequency Characterization of Signals

Recall that our goal is to characterize efficiently the behaviour of an LTI system in order to predict what it will do to any signal. In Chapter 6 we characterized systems by what they do to the amplitude and phase of sinusoids of different frequencies. If we were only interested in sinusoids, this book would already be finished!

But, obviously, sinusoids are not the only thing we want to know about. We want to predict how complex signals (e.g. speech) will be affected by LTI systems. We claimed rather boldly in Chapter 5 that all signals can be considered to be composed of sinusoids of the appropriate frequencies, amplitudes and phases added together. In this chapter, we're going to show that this is indeed true. There are two sides to this fact; not only can any waveform be represented as a sum of sinusoids, but also, given the sinusoids which constitute a signal, its waveform can be easily calculated. The process of decomposing a waveform into its sinusoidal components is called *Fourier analysis* and that of reconstituting the waveform from the sinusoids is called *Fourier synthesis*, in honour of the French scientist who first proposed that these processes have general applicability.

Let's consider for a moment why these two, inverse, processes are important for establishing what a system does to a complex signal. When presented with a signal we first analyze it into its constituent sinusoids, each characterized by a particular frequency, amplitude and phase. The effect of putting these sinusoids on their own through the system can be determined using the system transfer function (amplitude and phase response) as described in Chapter 6. This would tell us how each of the component sinusoids is affected as it is passed through the LTI system. Now, because the system is linear, it's additive, so it doesn't matter whether or not other sinusoids are presented along with a particular sinusoid (Chapter 4). In a linear system, the output of the system to the sum of two (or more) sinusoids can be predicted from the sum of the outputs to each of the inputs alone.

We could stop there and describe the system's output in terms of what sinusoids are present and their amplitudes and phases. Or we could go one step further and reconstitute the amplitude versus time waveform by adding together the appropriate sinusoids to synthesize the waveform. Here are the steps just described in diagrammatic form:

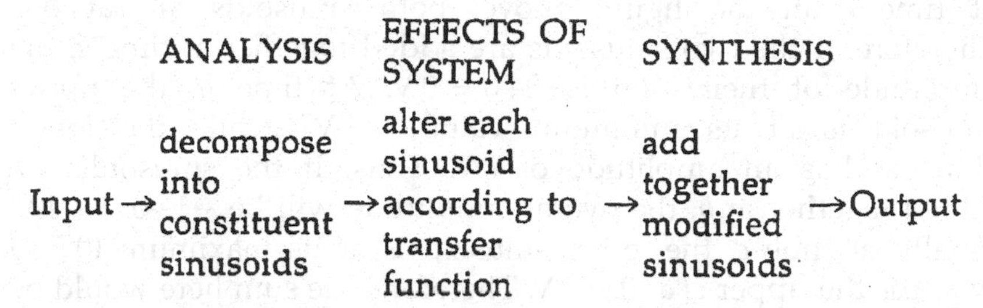

Of the three processes involved, there are two that you don't know much about yet—how to decompose the input signal into its sinusoidal components (that is, perform a Fourier analysis) and how to reconstitute a waveform from its sinusoidal components (synthesize the output signal). In this chapter we're going to describe several important aspects of these two processes.

Adding sinusoids: synthesis

First, what exactly do we mean by 'adding up sinusoids'? Basically, adding up any two waveforms simply involves adding up their respective amplitudes at each point in time, and plotting the result—again as a function of time. Let's make this more concrete by adding together two sinusoids, one with a frequency of 100 Hz and the other of 200 Hz:

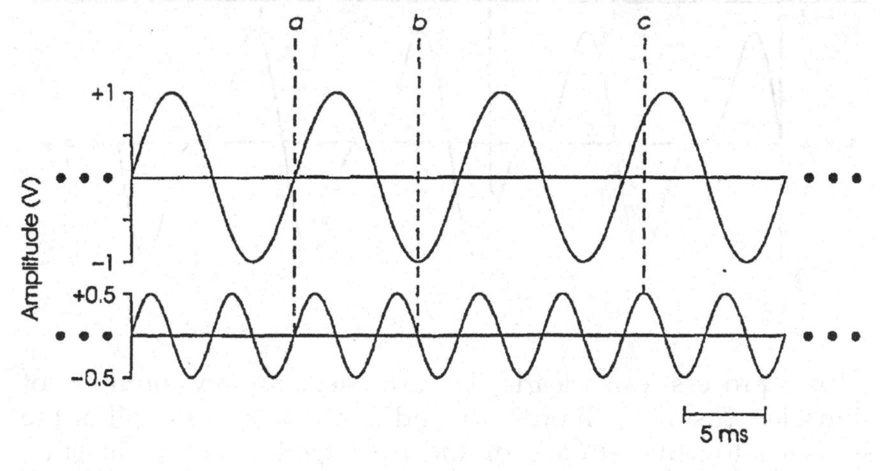

The amplitude of the higher-frequency sinusoid (at the bottom of the figure) is half that of the upper, and both are sine waves—that is, they start from zero and increase. Two sinusoids which match up in time in this way are said to be in *sine phase*.

Since these are time waveforms, the y-axis is amplitude and the x-axis time. At any point in time, we know the amplitude of each of the sinusoids, either from the graph, or from the trigonometric formulae we developed in Chapter 3. For example, at time a in the figure above, both sinusoids are at 0 V. Therefore, if the two sinusoids are added together at time a, the amplitude of their sum is $0+0=0$ V. At time b, the upper sinusoid is at its minimum value (-1 V) while the lower sinusoid has an amplitude of 0 V. Thus, if the sinusoids are added together at b, the overall amplitude will be $-1+0=-1$ V. Finally, at time c, the lower sinusoid is at its maximum (0.5 V) whereas the upper is at 0.707 V. Therefore, the sum here would be 1.207 V. In order to do the complete summation, the two sinusoids have to be added up in this way at every moment in time, obtaining:

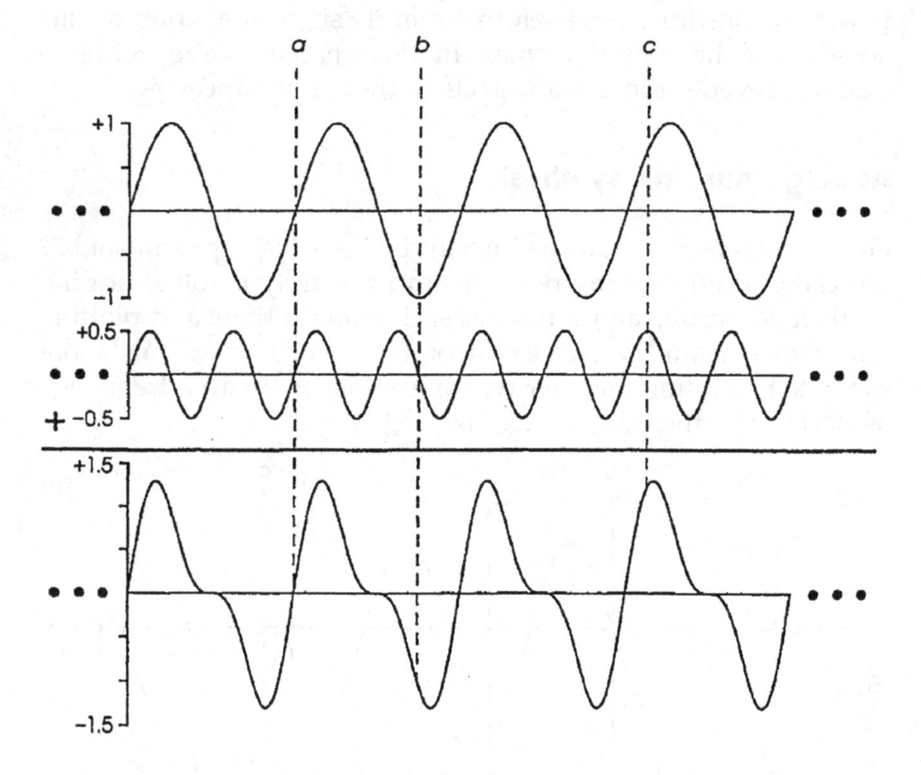

This process can clearly be extended to any number of sinusoids. Just as with ordinary addition, you can add all of the sinusoids together at once, or add two together, obtain the sum, add the third to the sum and so on. Also, again as with ordinary addition, it doesn't matter in what order the sinusoids are added up.

Decomposing periodic waveforms: analysis

Often, though, we aren't told what sinusoids to add together. We're given a particular periodic waveform and have to work out which sinusoids have to be added up to re-create it. This reverse process, *analysis*, is more difficult than *synthesis*, the adding together of sinusoids. What may come as a surprise is not that some particular periodic signal can be made up from sinusoids (as shown above), but that it is possible to do this for *any* signal. *Fourier's theorem* tells us how to do this, and, more importantly, puts constraints on what sinusoids can be used to make up a complex periodic waveform.

Fourier's theorem states that a periodic waveform can only contain sinusoids which are *harmonically related* to the repetition frequency of the original signal (that is, one over its period). 'Harmonically related' means that the frequency of the sinusoid is an integral multiple of some 'basic' or 'fundamental' number (that is, one times it, two times it, three times it etc.). So, if a waveform has a period of 10 ms, its fundamental frequency is 100 Hz (= 1/ 0.01 s) and its harmonically related frequencies are 100 Hz, 200 Hz, 300 Hz etc. Each of these component frequencies is known as a *harmonic* (you'll also see the names *partial, overtone* and *spectral component* used to mean roughly the same thing).

Anything that consists of a sequence of related terms like these harmonics is known as a *series*. Thus, the analysis of a periodic signal into its component sinusoids is referred to as calculating its *Fourier series*. So, to take the example of a periodic waveform which is made up of harmonics of 100 Hz, its Fourier series would be completely specified by the amplitude and phase of the sinusoidal components at 100 Hz, 200 Hz, 300 Hz and so on. Note, too, that for a given periodic waveform, certain terms in the harmonic series may not be present. Thus, you can have a harmonic series in which only odd multiples of 100 Hz are present.

All this may seem pretty abstract, so let's illustrate the important properties of Fourier series with some specific periodic signals.

The Fourier series of a sawtooth waveform

The techniques of calculus are required to obtain the Fourier series of periodic signals from first principles. Not assuming such knowledge on your part, we're going to demonstrate graphically and verbally two basic facts: (1) why the Fourier series can't contain sinusoids with a period more than the period of the complex waveform and (2) why the frequencies of the higher terms (the harmonics) have to be exact multiples of the lowest frequency term.

Here's the periodic signal whose Fourier series we're going to calculate:

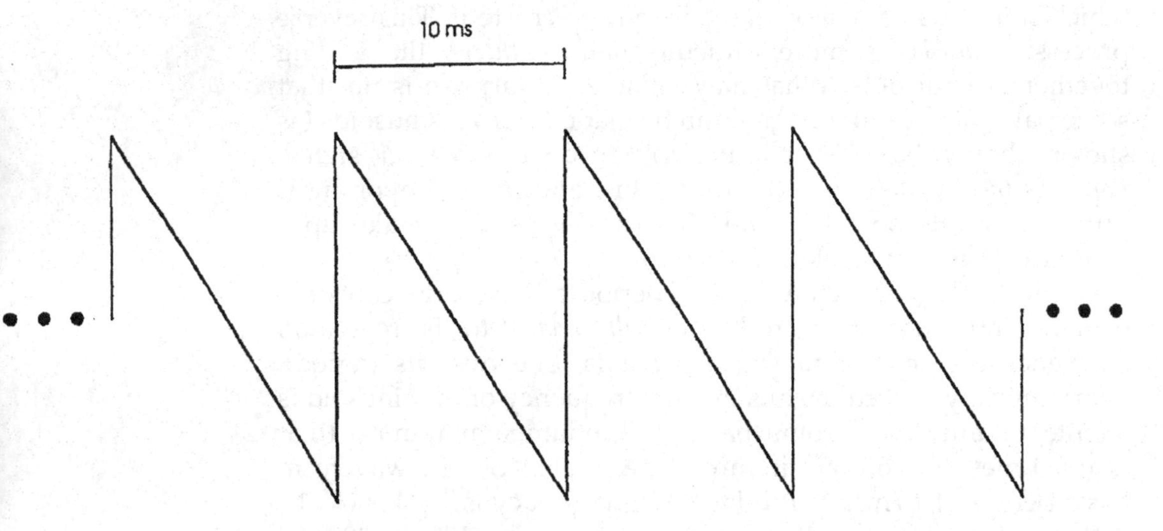

You may recognize this as the sawtooth wave seen in Chapter 3. The sawtooth repeats itself over and over again (that is, after all, what we mean by periodic), and its period is 10 ms.

Here is a sinusoid that repeats at the same rate as the sawtooth, 100 ms:

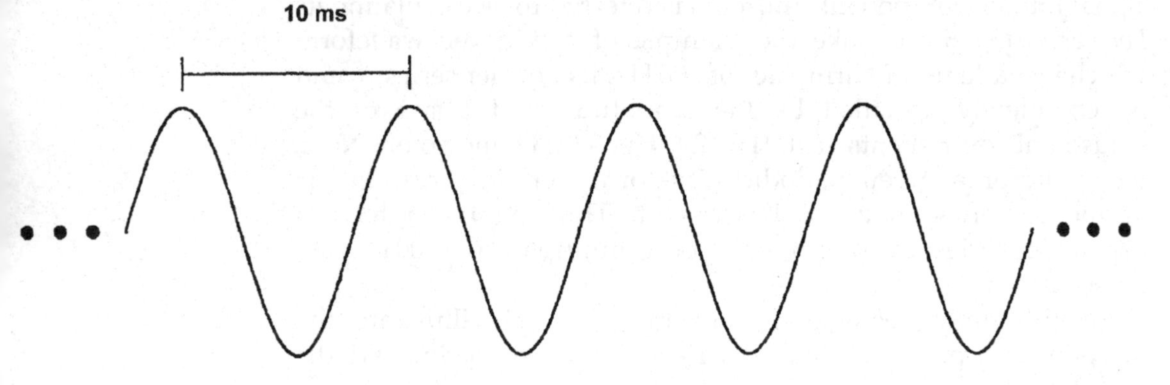

We've set the phase of the sinusoid so that the two signals go up and down together. Superficially, at least, the two waveforms do not look that dissimilar, as you can see when they are superimposed (illustrated on the following page). The sinusoid that we've drawn is a closer fit to the sawtooth than a sinusoid of any other frequency, amplitude and/or phase. It thus looks like a pretty good candidate for the sinusoidal component to start our series. Note that the amplitude of the sinusoid is less than that of the sawtooth. In fact, if the sawtooth has a peak-to-peak amplitude of 1.57 V, this sinusoidal component has a peak-to-peak value

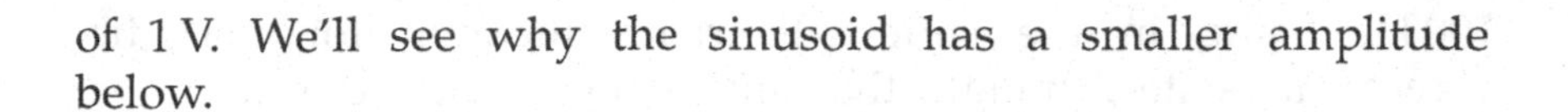

of 1 V. We'll see why the sinusoid has a smaller amplitude below.

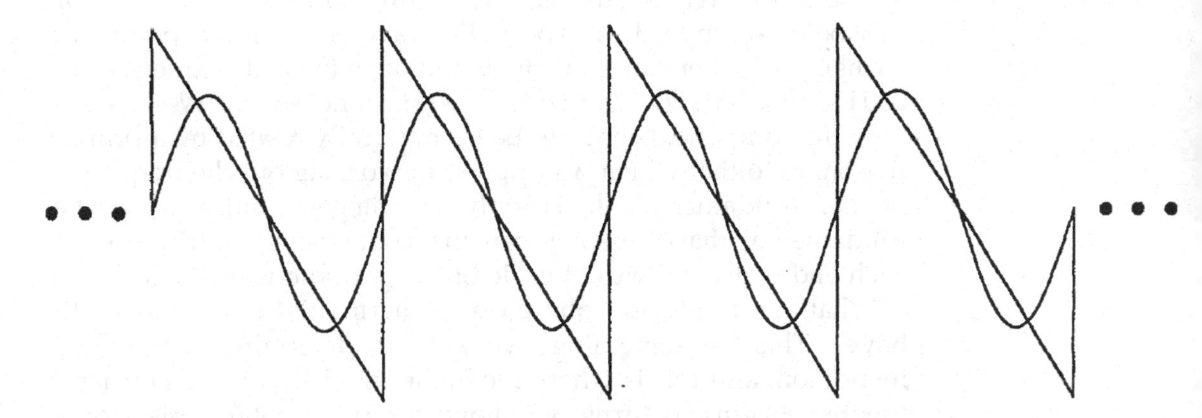

This sinusoid is one term of the Fourier series for this sawtooth. In fact, it is the *first* term (or lowest frequency) of that series. Let's see if we can convince you that this is indeed the lowest possible frequency that can be used to construct the sawtooth—that there cannot be any lower frequency sinusoid that contributes to the shape of the sawtooth wave.

Recall that the sawtooth wave repeats exactly every 10 ms. Therefore, all the constituent sinusoids that give the wave its sawtooth shape must contribute equally within each and every adjacent 10-ms cycle of the signal. Any sinusoid that does not repeat every 10 ms cannot contribute equally to the shape of the wave in every 10-ms cycle. Any sinusoid that repeats with a period longer than 10 ms would make the wave in different 10-ms cycles different in shape. Therefore, 10 ms is the longest period any contributing sinusoid can have. Put the other way, 100 Hz is the lowest frequency sinusoid that can occur in the series because all the component sinusoids must have a period of 10 ms or less.

The sinusoidal component with the same periodicity as the complete waveform is the most important in many commonly encountered periodic waveforms and so it is given a special name—the 'fundamental'. The period of this component is nothing other than the period of the original waveform.

Of course, just because 100 Hz is the lowest frequency that can possibly be present in this sawtooth does not mean that the 100-Hz sinusoid is, in fact, present. However, in this case, the 100-Hz sinusoid looked like such a good candidate from the outset that it still seems reasonable to suppose that it *is* a member of the series.

Our next step, though, is to extend the series to get a better approximation to the sawtooth waveform. There is nothing in what we've said so far that would prevent frequencies higher than

100 Hz being in the series. Fourier's theorem for periodic waveforms states, though, that any higher frequencies must be harmonically related to the first term. So, for this particular sawtooth (or indeed any periodic waveform with a period of 10 ms), the higher components would have to be at frequencies of 200 Hz (2 × 100 Hz), 300 Hz (3 × 100 Hz) and so on. We can see why the components have to be harmonically related by a similar reasoning to that which we applied in working out the frequency of the fundamental. It is only at integer multiples of the fundamental that sinusoids can make the same contribution to each and every individual cycle of the periodic waveform.

What amplitude and phase do the harmonics of our sawtooth have? This is something we cannot determine simply by inspection, and this is where the higher mathematics of Fourier's theorem begins. It turns out, however, that, in the series for a sawtooth, amplitude falls off as the inverse of the harmonic number. So, if the period of the sawtooth is 10 ms, and its peak-to-peak amplitude is 1.57 V, then the amplitude of the fundamental (at 100 Hz) is 1 V peak-to-peak, the second term of the series (at 200 Hz) has an amplitude of 1/2 V, the third (at 300 Hz) 1/3 V and so on. The phase values are even simpler. All the components have to be sine waves (and so are said to be in *sine phase*). Since 0° phase means a cosine, a sine will have a phase of −90°.

Although we cannot demonstrate the calculation of the appropriate amplitude and phases, we can demonstrate empirically that such sinusoids do indeed add up to an approximation of a sawtooth. Here are the first two terms of the series with the correct amplitudes and phases:

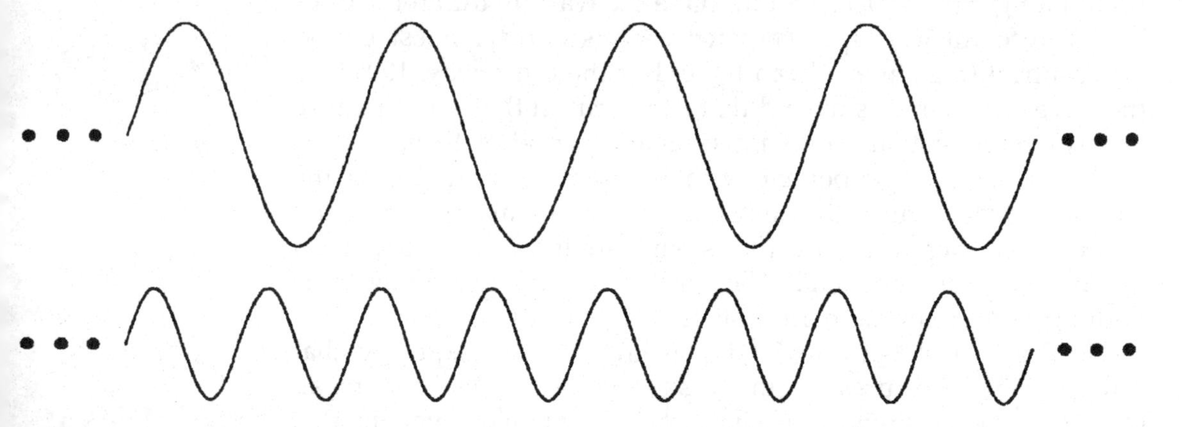

Fourier's theorem states that all the appropriate sinusoids added together would reproduce the original signal, but even these two terms added together give something closer to the sawtooth than the fundamental alone, as seen at the top of the following page.

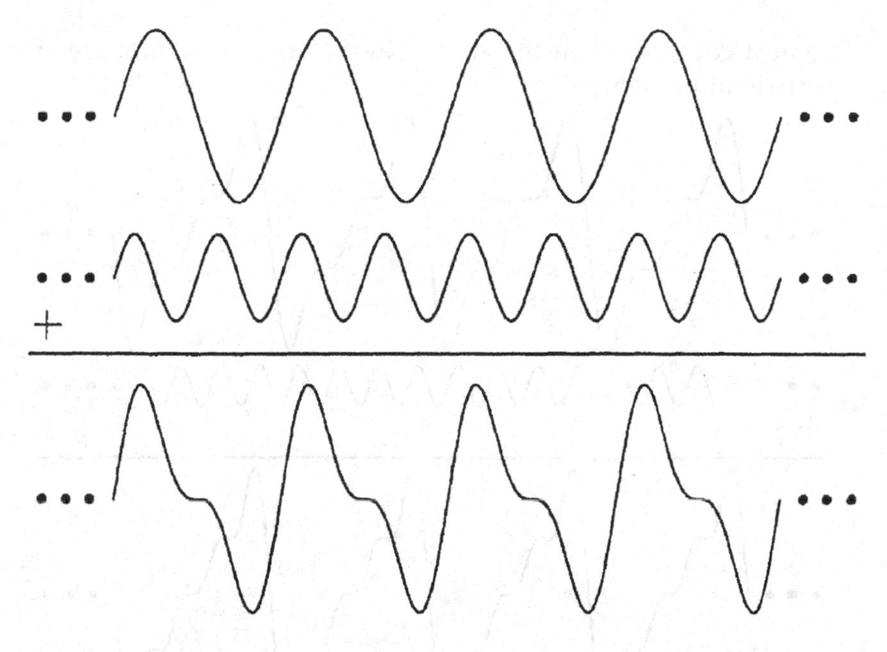

In fact, this is the same waveform we used previously to illustrate the addition of sinusoids. You can see that adding these two terms together has indeed given us a better approximation to the sawtooth. The reason why the amplitude of the sinusoid with the same period as the sawtooth (the fundamental) is not the same amplitude as the sawtooth itself is because the peak height of the sum is determined not only by the fundamental, but also by the higher harmonics.

If we add the next term (300 Hz at 1/3 V), then the approximation is improved further:

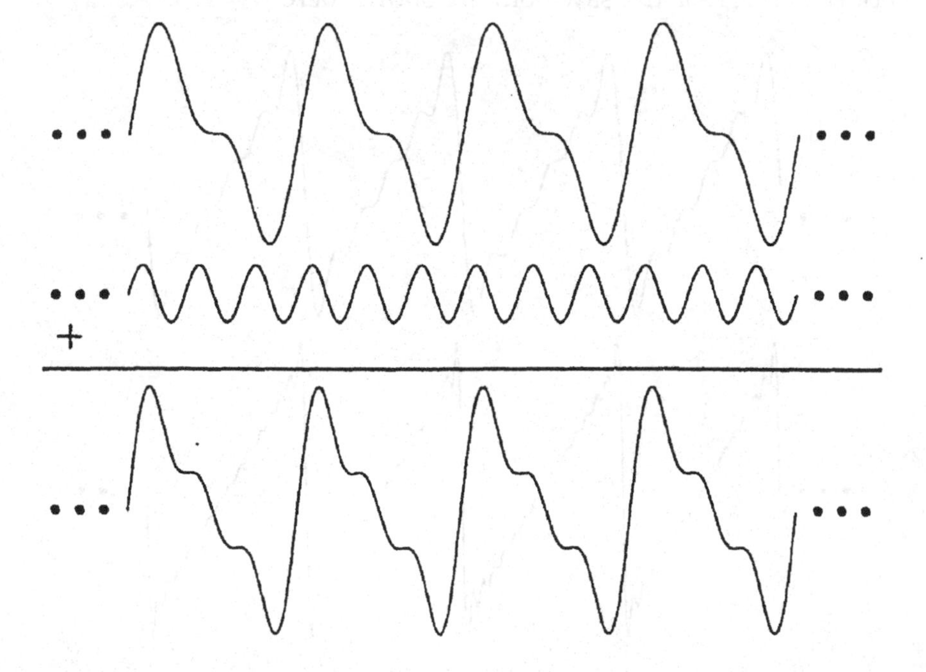

The next component in the series (400 Hz at 1/4 V) improves the approximation again:

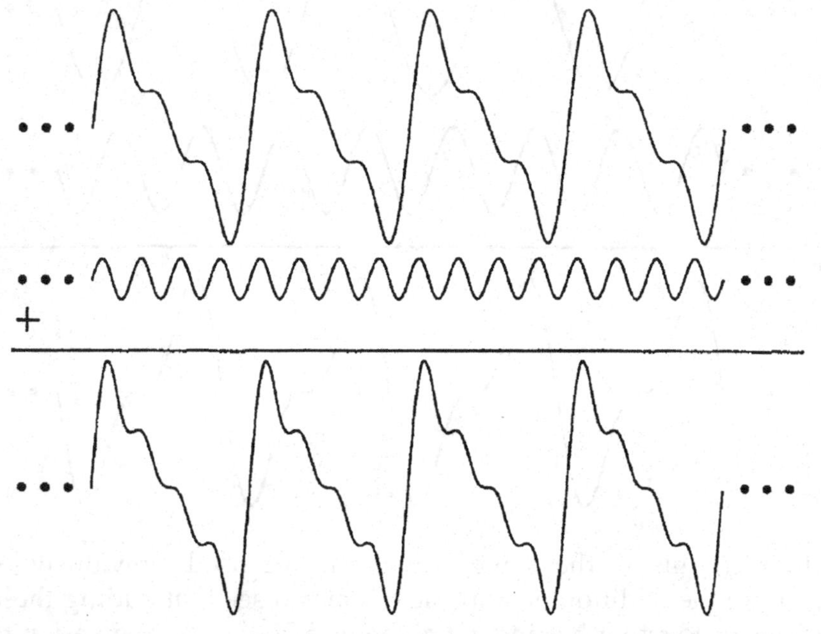

We can keep adding more and more components, each time getting a better and better fit to the sawtooth. To get the sawtooth wave exactly requires the addition of an infinite number of harmonics. But, in practice, we do not have to go to this extreme. Usually a good approximation to the signal can be obtained by adding together a relatively small number of low-order terms. The signals obtained by adding together the first 5 and 15 terms of the Fourier series for the sawtooth are shown here:

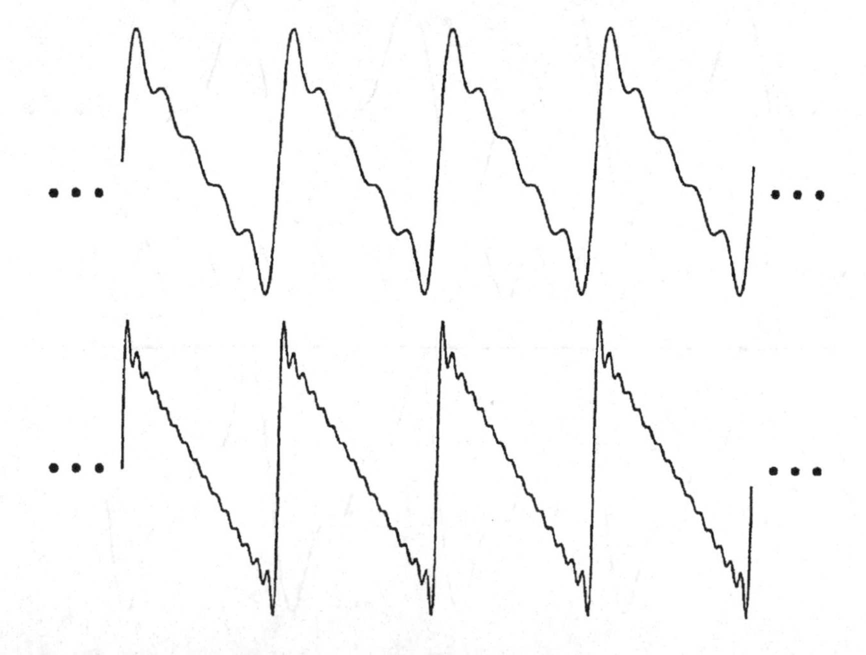

It is clear that the approximation is pretty good even though we've used nowhere near an infinite number of terms! Note, too, that as we go higher in frequency the addition of new components makes relatively less difference. This is often true (although not necessarily so) and results from the fact that the amplitude of the harmonics drops off as their frequency increases.

Although we've told you what the amplitudes and phases of the sinusoidal components are (rather than deriving them), the fact that the results look sensible should make you a bit more confident that we're doing the right thing. This example can be used to illustrate what is meant by analysis and synthesis; analysis when we establish what sinusoids our complex signal contains and synthesis when we show the waveform that results by adding up the components.

Though the series for the sawtooth is infinite, don't get the impression that all periodic waveforms have an infinite number of spectral components (that is, need to be synthesized from an infinite number of sinusoids). Any of the approximations to the sawtooth created from a finite number of terms might be exactly the waveform required.

You may have noticed that the sawtooth synthesized from only the first few harmonics led to a smoother curve than the one synthesized with higher harmonics as well. When waveforms change fast, there is generally some high-frequency content and it's necessary to include the high-frequency harmonics to create fast changes. In the sawtooth wave, for example, the higher harmonics give the sharp corners of the teeth.

Amplitude spectrum of a sawtooth wave

We don't want to have to draw sine waves to represent the frequency content of a signal and so we need a more convenient way of conveying this information. We do this with *spectra*. A *spectrum* is a description of a signal in terms of the sinusoidal components needed to construct it. You can think of a spectrum as a list of the ingredients needed for a recipe to construct any particular waveform. Unlike recipes for cooking, however, the 'ingredients' are always very similar—sinewaves of different kinds!

Just as we did for systems, it is convenient to treat the information about amplitude and phase separately. First we'll consider, for our sawtooth in particular (although the techniques are general), how to represent the *amplitude* of the sinusoidal components that are present—the *amplitude spectrum*.

One way to present the amplitude spectrum of the sawtooth waveform is in the form of a table with two columns—one column consisting of the frequencies of the sinusoidal components present and the other of their amplitudes:

Frequency (Hz)	Amplitude (V)
100	1
200	1/2
300	1/3
400	1/4
500	1/5
600	1/6

This is similar to the table we constructed when considering amplitude response curves in Chapter 6. Just as in that case, however, it is usually more informative to draw a graph. In graphical form, an amplitude spectrum has as its axes amplitude and frequency. Amplitude is drawn on the vertical axis and frequency on the horizontal axis. A vertical line is drawn at frequencies where harmonics are present in the signal. The height of the line is proportional to its amplitude (specified by the vertical axis). The amplitude spectrum of the first 10 components of the sawtooth would look like this:

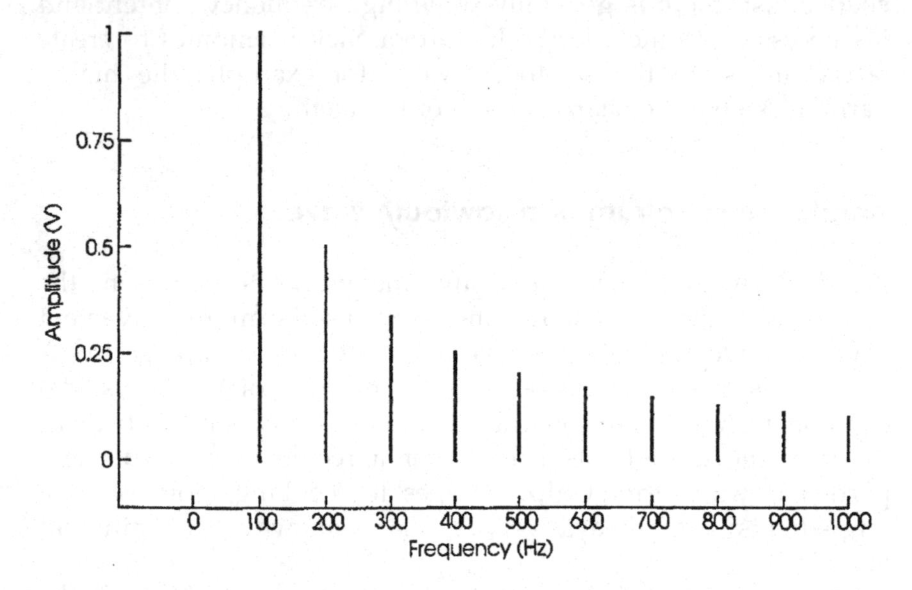

It's easy to see from this graph that the amplitude of the components drops off with increasing frequency. How might this change in amplitude over frequency best be described? One common approach is to draw a smooth curve through the tops of the vertical lines. Such a curve is known as a spectral *envelope* (see overleaf):

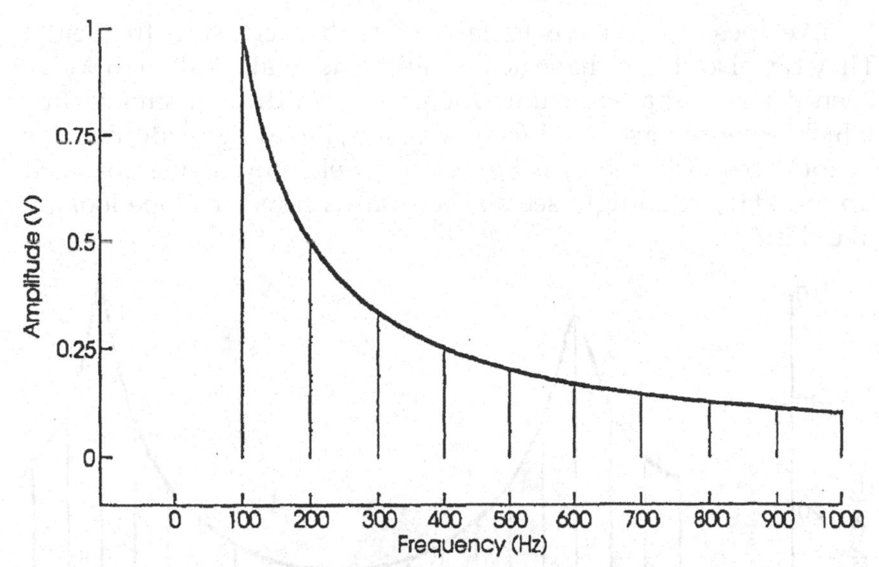

Note that the envelope does not imply that all frequencies (or even all harmonics) are present; only that those that are present lie below the curve.

As the frequencies of the components are doubled (for example, as you go from 100 Hz to 200 Hz, or 200 Hz to 400 Hz), the amplitude drops by half each time. That is, going from 100 Hz to 200 Hz, the amplitude drops from 1 V to 1/2 V, or going from 200 Hz to 400 Hz the amplitude drops from 1/2 V to 1/4 V (1/2 of 1/2 V). As we've noted earlier, a doubling in frequency is an *octave* change and a halving of amplitude is a reduction of 6 dB. The drop-off in this example is, then, 6 dB/octave. It should be clear that if log scales were used on the axes here (log frequency for the x-axis and dB for the y-axis), the envelope would be a straight line sloping downward:

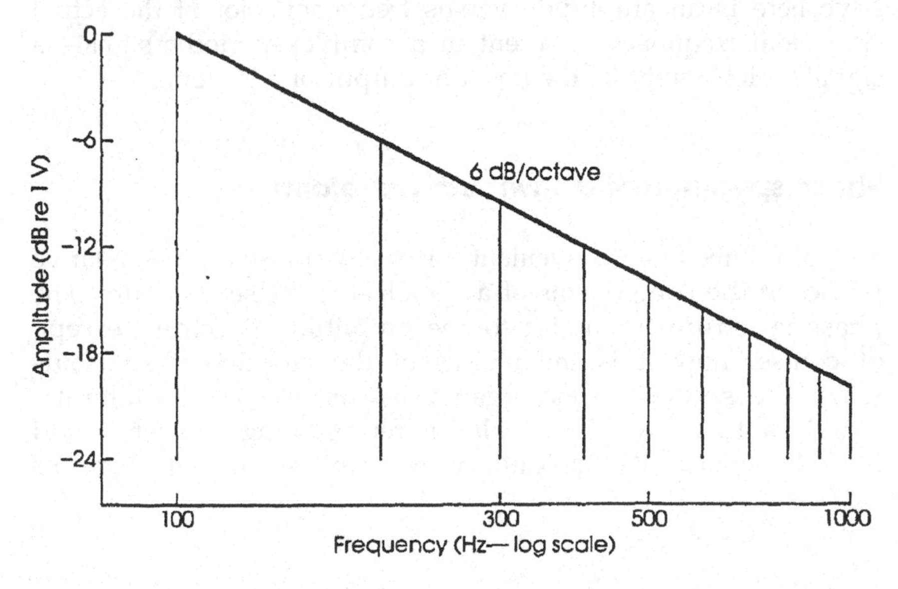

Envelopes don't have to fall off with increasing frequency. They can also rise, or have no definite trend at all. Although we've been discussing a spectrum where you can draw a smooth line which never reverses its direction (that is, always goes down), this is not necessarily the case. So, for a vowel sound being analyzed up to 2 kHz, you might see a spectrum with its envelope looking like this:

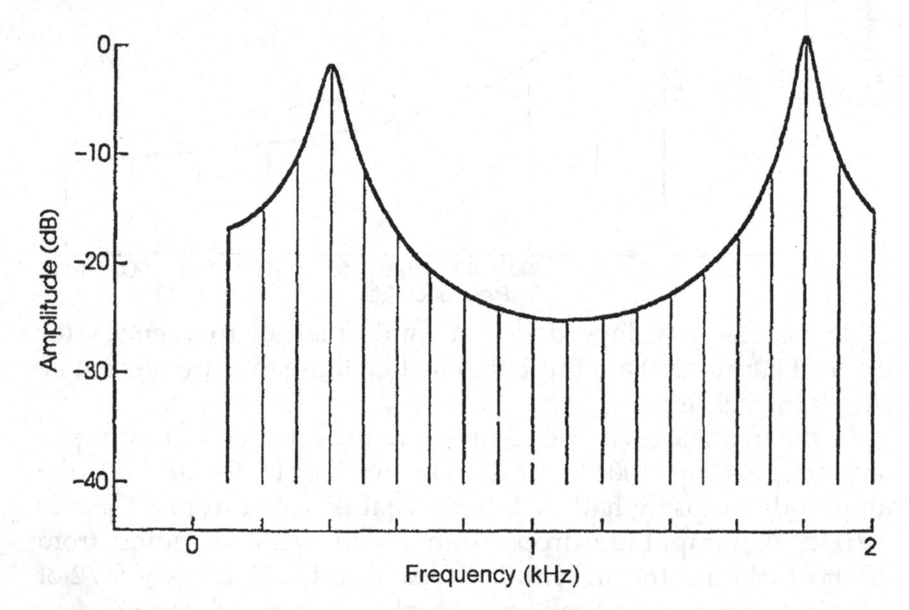

Because they have the same axes, amplitude spectra look something like the amplitude responses we saw in Chapter 6. Remember, though, that they represent different information. Amplitude responses show the ratio between the amplitude of the output and that of the input at particular frequencies. What we have here is an amplitude versus frequency plot of the actual sinusoidal frequencies present in a complex periodic signal—a signal which might be the input or output of a system.

Phase spectrum of a sawtooth waveform

We also need a convenient way to present the relative phases of the components of a signal—the phase spectrum. The phase spectrum is similar to the amplitude spectrum except, of course, now it is the phases of the constituent sinusoids which are specified. Once again, we can present the information in a table, with one column representing frequency and the other phase. We typically measure phase so that 0° means cosine phase. The fundamental component of the sawtooth is a sine function so it will have a phase of –90°. All the other components have the same phase, so the table would look like this:

Frequency (Hz)	Phase
100	–90°
200	–90°
300	–90°
400	–90°

Having all the components in sine phase means that they all cross the *y*-axis moving in an upward direction at the same point in time (at the beginning of each cycle).

The phase spectrum of the sawtooth can, of course, also be presented graphically:

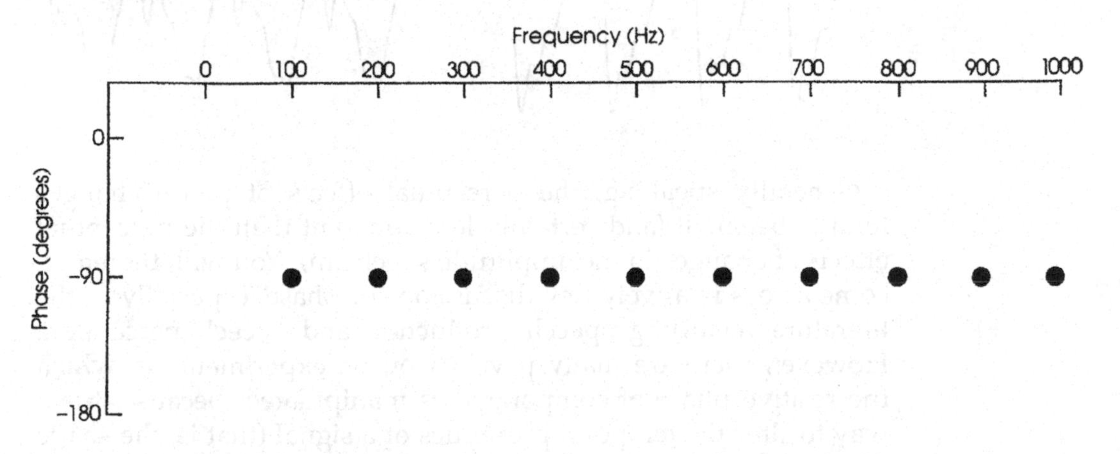

The effect of altering the phase of one component

So far, we've determined the phase and amplitude spectra necessary for a sawtooth. Of course, if we changed the phase or amplitude of any of the constituent sinusoids, the resultant would no longer be the same sawtooth. More specifically, if the phase of one or more of the components is altered, the shape of the wave can alter dramatically. Here, for example, are the constituent sinusoids for two waveforms that look quite different. On the left is our standard sawtooth, summed from the first five components in the series. On the right is shown the sum of sinusoids with the same frequencies and amplitudes, but in which the phase of the fundamental is altered by +90° (changing it to a cosine). Note that the periodicity of the resulting waveform is unchanged:

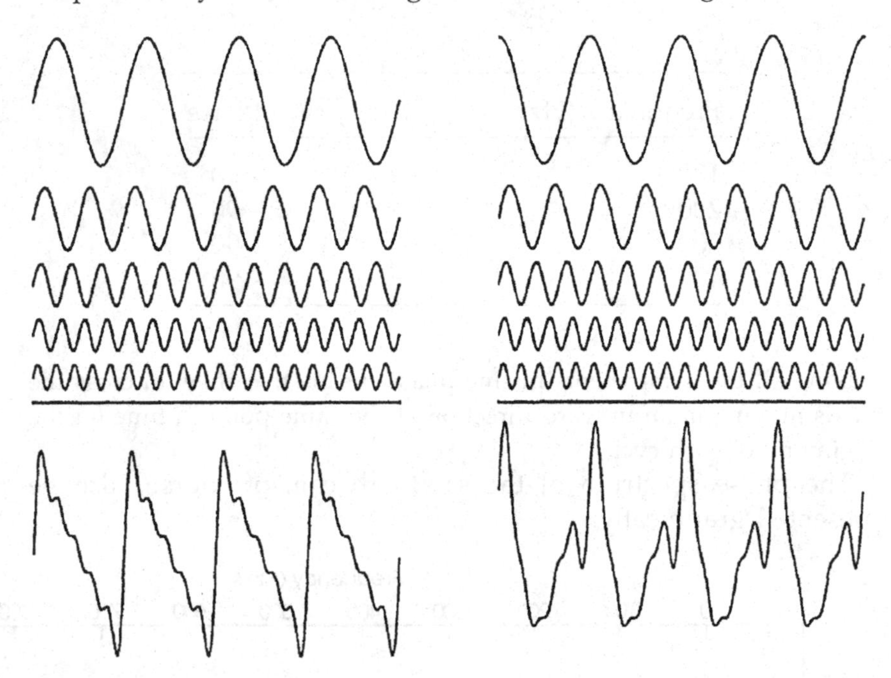

Generally speaking, the perceptual effects of phase changes tend to be small (and certainly less apparent than the perceptual effects of changes in the amplitude spectrum). You will, therefore, come across relatively few discussions of phase, especially in the literature regarding speech production and speech perception. However, there are many psychoacoustic experiments in which the relative phase of components is manipulated, because it is a way to alter the temporal properties of a signal (that is, the shape of the waveform) without altering its amplitude spectrum. Therefore, such phase manipulations can be used to investigate the relative importance of time and place-based processing in the auditory system.

The effect of altering the amplitude of one component

Changing the amplitude of a sinusoidal component will also, of course, change the resulting waveform. Here, on the right of the next figure, is a waveform again based on our five-harmonic sawtooth in which the amplitude of the second harmonic has been altered to 1/4 of what it should be for a sawtooth (at left). Again the shape is altered although the periodicity of 100 Hz remains (see overleaf):

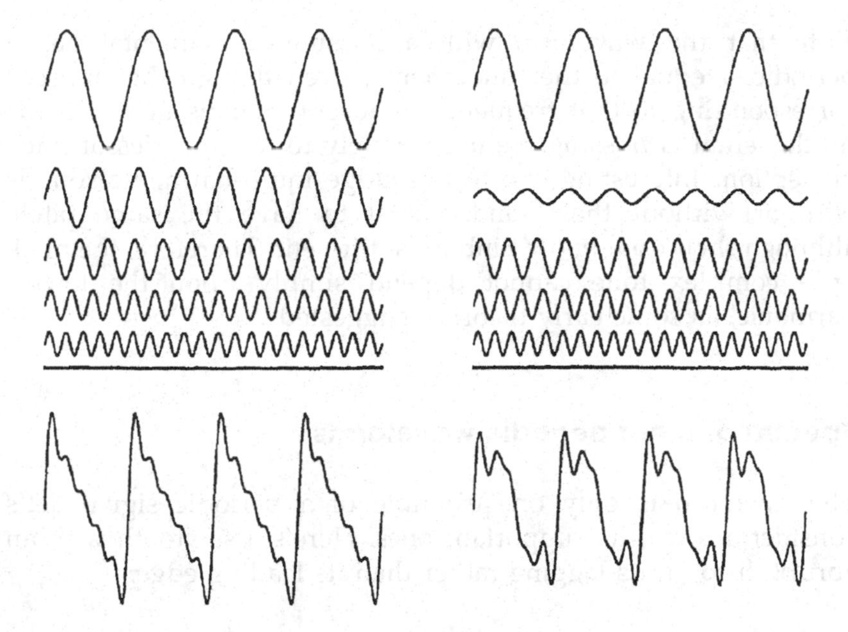

The effect of missing one component out

Besides altering the amplitude and phase of certain of the components, one or more of the constituent sinusoids can be left out altogether (equivalent to setting its amplitude to zero). Any of the harmonics could be omitted, but a particularly important case occurs when the fundamental is left out (in which case the signal is said to have a *missing fundamental*). If the second to fifth harmonics of the sawtooth (with the appropriate amplitudes and phases this time) are taken and added together, the waveform on the bottom right results:

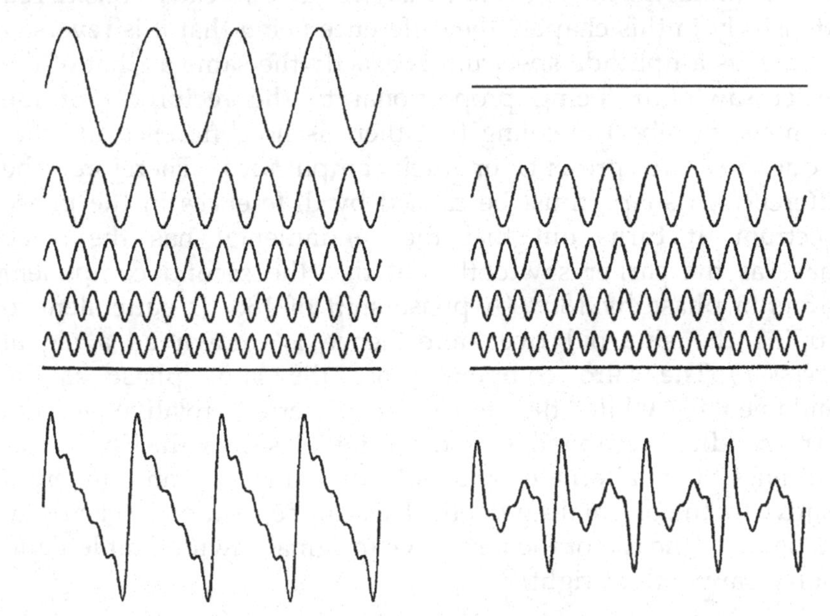

Note that the waveform with a missing fundamental has a periodicity equal to the fundamental even though the sinusoid corresponding to that frequency is absent. Signals in which the fundamental is missing are used widely to test theories of pitch perception. Interestingly, listeners judge multi-harmonic sounds with or without their fundamental to have the same pitch, although their quality (or timbre) is different. Therefore, the pitch of a complex tone cannot depend simply upon the lowest harmonic, as some early theorists suggested.

Spectra of other periodic waveforms

The sawtooth is only one example of a periodic signal: let's consider a few other important ones. Here's a sawtooth with an abrupt drop on its lagging rather than its leading edge:

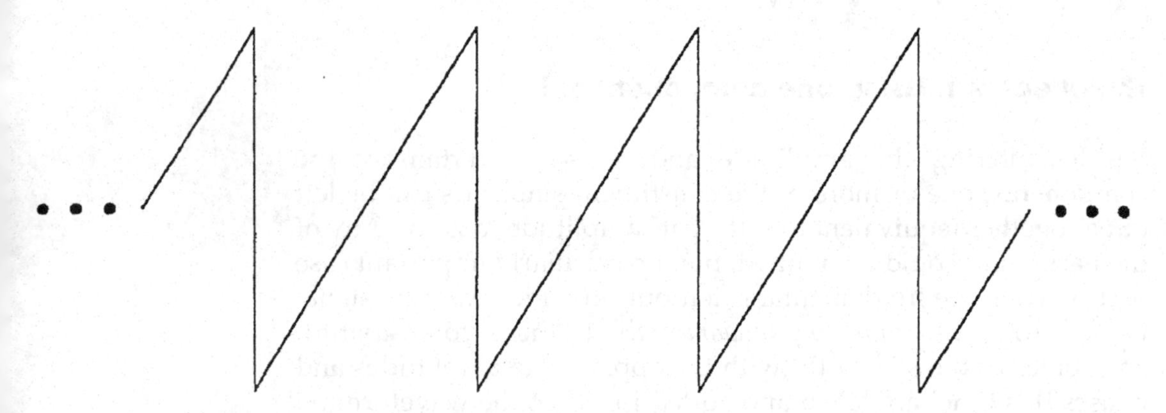

This is similar to the sawtooth that we have already considered extensively in this chapter, the difference being that it is reversed in time. Its amplitude spectrum is exactly the same as that of the earlier sawtooth (being proportional to the reciprocal of the harmonic number) meaning that there is no difference in what frequencies are present, or their amplitudes. Therefore, the difference in shape must be caused by differences in the phase spectrum. It turns out that the fundamental has the same phase as the earlier sawtooth, but that the second component differs in phase by 180°. (A phase shift of 180° is equivalent to flipping the sinusoid over, and hence is known as a signal *inversion*.) The third component has the same phase as the fundamental, while the fourth is inverted relative to the corresponding component in the original sawtooth. The series continues with alternate sinusoids the same as, and inverted relative to, the original sawtooth. This can be seen more clearly in the figure at the top of the next page (original sawtooth at left and altered sawtooth at right):

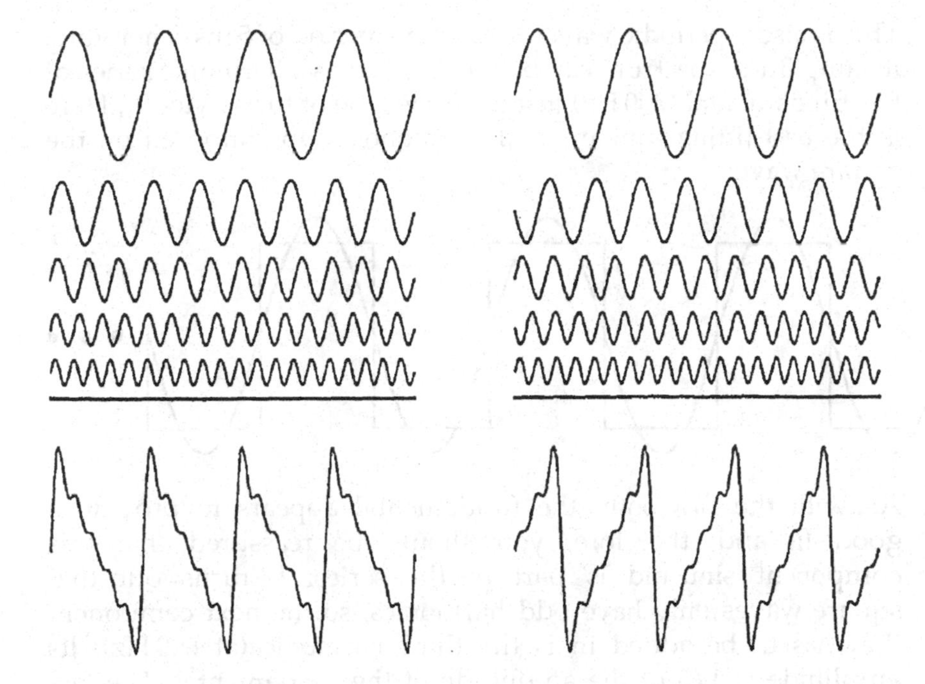

Varying the phases of the component sinusoids changes the shape of the sawtooth, just as we observed above when the phase of one sinusoidal component was varied. Compare the phase spectra of the 'normal' and 'backwards' sawtooth:

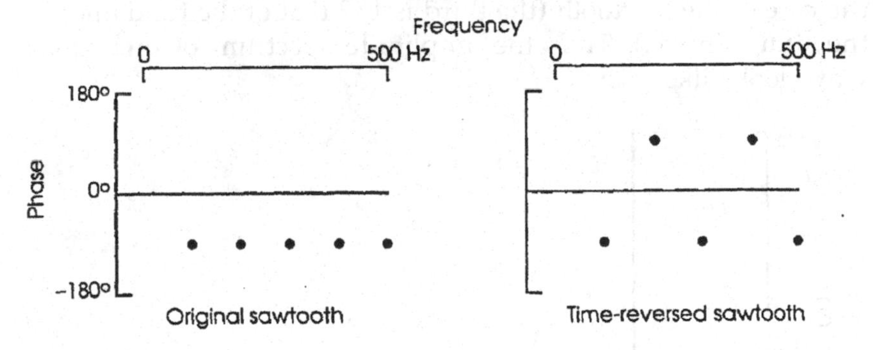

Let's now look at the spectrum of another waveform that you've seen before, a square wave:

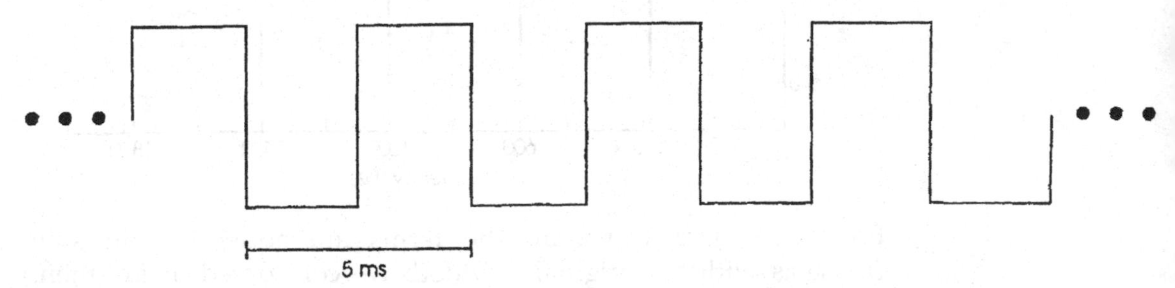

This is also a periodic waveform, with a period of 5 ms. Therefore, its amplitude spectrum can be represented as a harmonic series of the fundamental (200 Hz) just as in the case of the sawtooth. Here is the best-fitting fundamental component superimposed on the square wave:

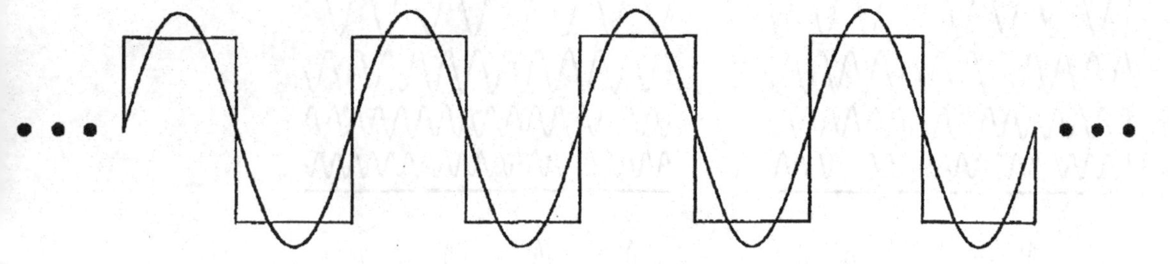

As with the sawtooth, the fundamental appears to provide a good fit and, therefore, you should be reassured that this component sinusoid is part of the series. It turns out that square waves only have odd harmonics, so the next component that has to be added in is the third harmonic (at 600 Hz). Its amplitude is 1/3 of the amplitude of the fundamental. The fact that the second harmonic is absent is one important difference between the square wave and the sawtooth. The absence of even-numbered harmonics continues right up the series (that is, only the first, third, fifth, seventh etc. harmonics occur). The amplitudes of the harmonics when they are present are the same as in the case of the sawtooth (the third is 1/3 that of the fundamental, the fifth, 1/5 etc.). Thus the amplitude spectrum of our square wave looks like this:

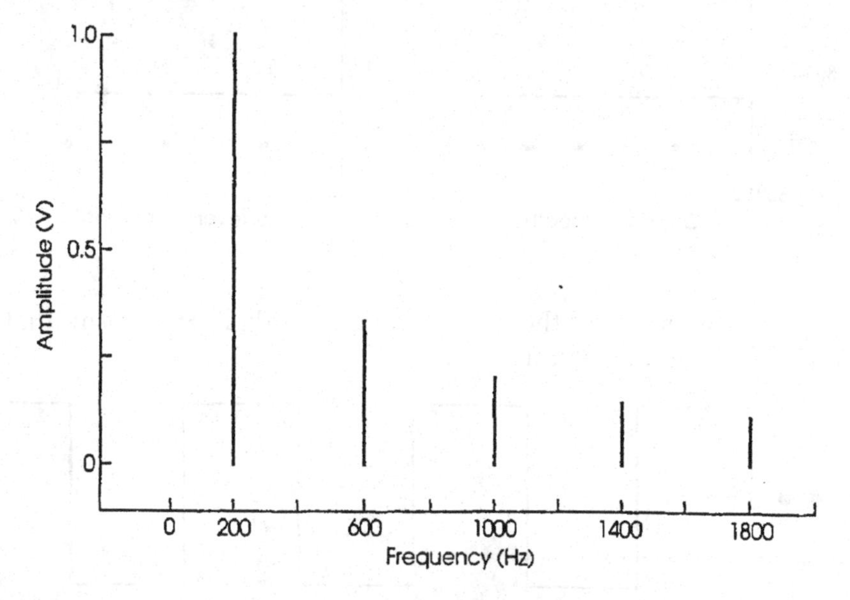

For the square wave, all the terms are added up in sine phase as with the original sawtooth we considered. Here, then, is the sum of the first five components (ninth harmonic) of the square wave:

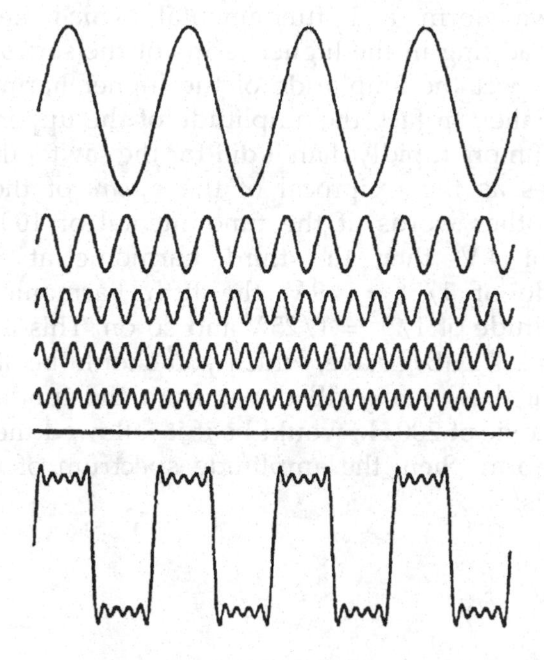

Another example of a waveform that has only odd components in its spectrum is a triangular wave:

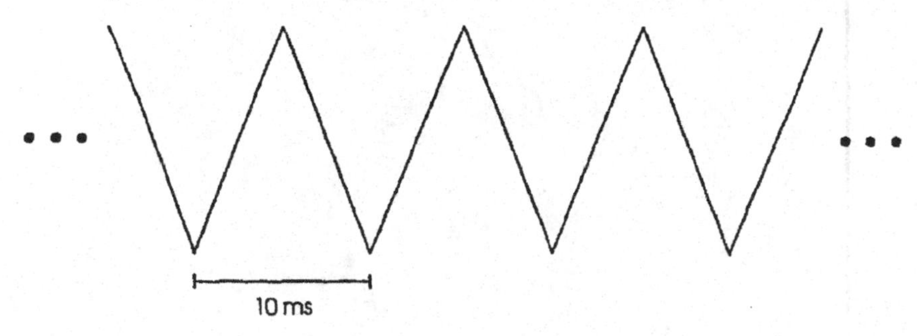

Let's see how well the first odd harmonic, the fundamental, matches the original waveform:

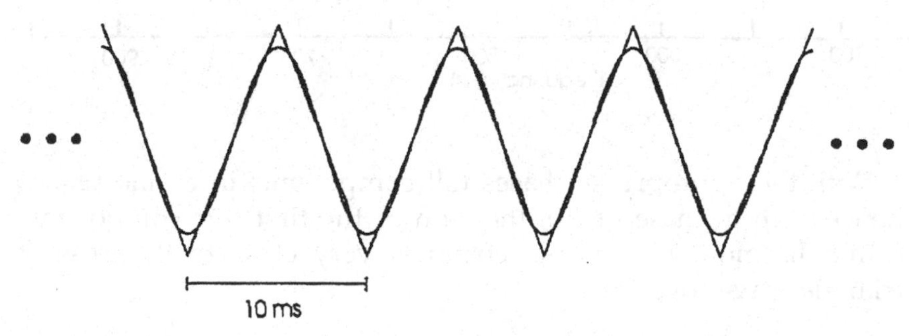

The fact that the fundamental provides a much better fit to the triangular wave than it did to the sawtooth or square wave shows that higher-frequency terms are relatively less important: there is much less difference in shape between the triangular waveform and fundamental which needs to be corrected by adding in the higher terms of the series. Therefore, we would expect the amplitude of the higher harmonics to be relatively smaller. In fact, the amplitude of the upper harmonics does drop off more rapidly than it did for the sawtooths or square wave: it goes as the reciprocal of the *square* of the harmonic number. In other words, if the fundamental of 100 Hz has an amplitude of 1 V, then the third harmonic at 300 Hz has an amplitude of $1/3^2 = 1/9$ V, the fifth harmonic at 500 Hz has an amplitude of $1/5^2 = 1/25$ V and so on. This is equivalent to a drop-off at 12 dB/octave, which you should be able to show by calculating how many dB down from the fundamental the second harmonic at 200 Hz would be if it followed the same rule. In pictorial form, then, the amplitude spectrum of the triangle looks like this:

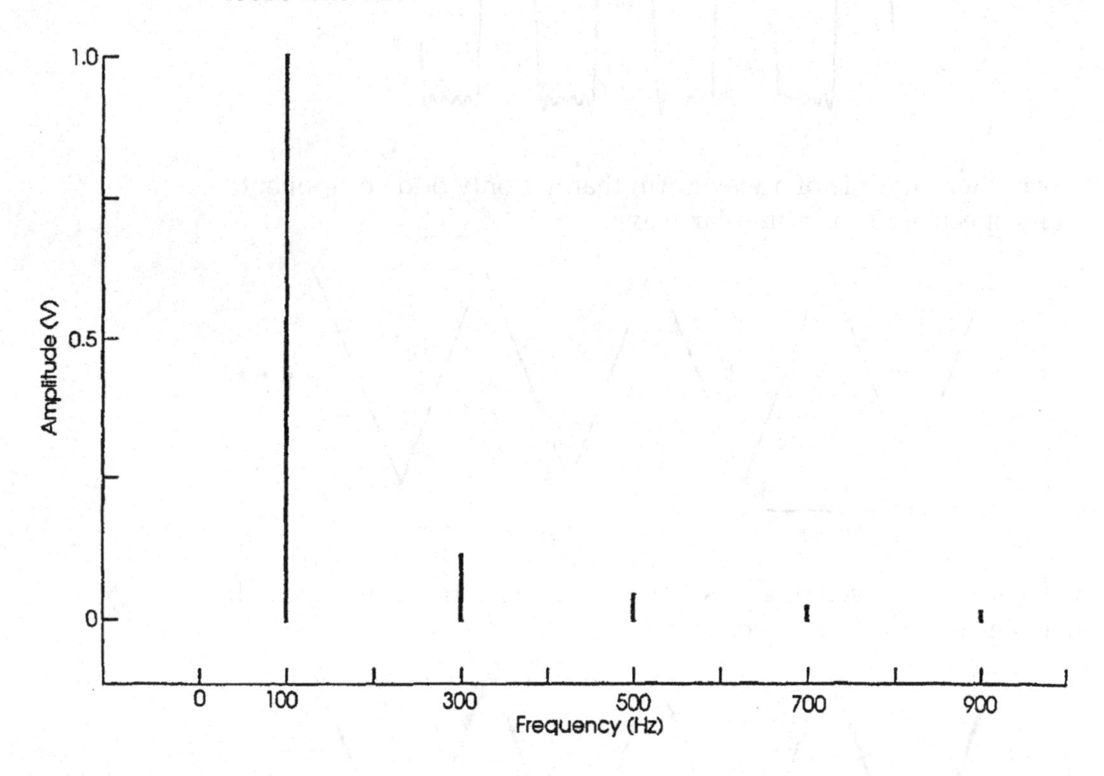

With the appropriate phases (all components in cosine phase, hence with a phase of 0°), the sum of the first five components (ninth harmonic) of the spectrum is very close to the desired triangle wave (overleaf):

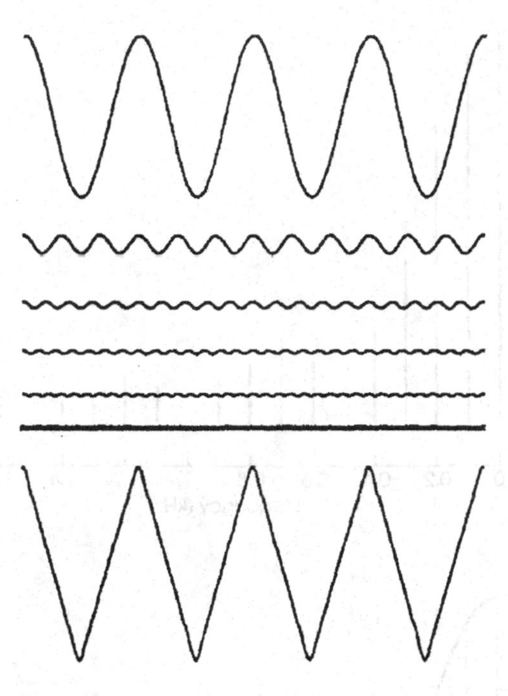

The spectrum of a pulse train

We'll consider one final basic periodic waveform—a pulse train—which will be used to illustrate some important results later on. Here's one with a pulse width of 2 ms and a period of 10 ms:

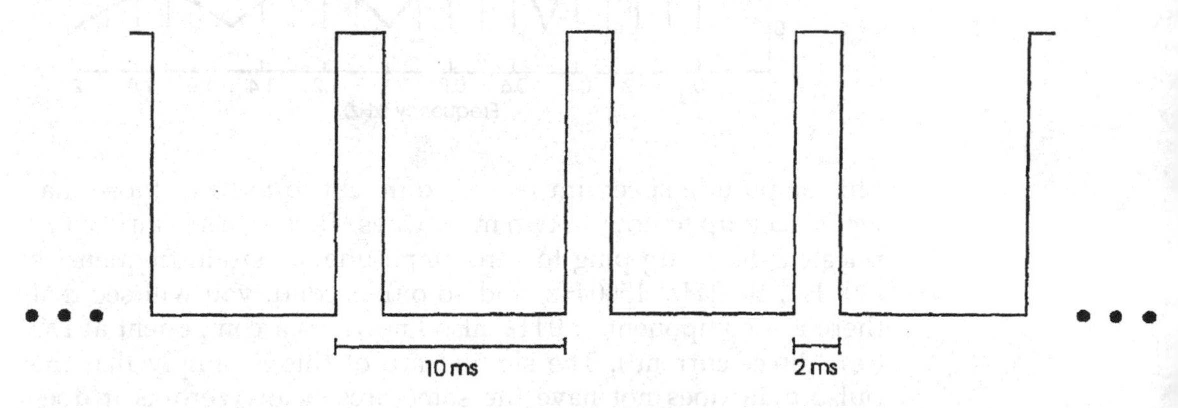

Because the pulses are occurring at a rate of 100 Hz, there can only be energy at 100 Hz, 200 Hz, 300 Hz and so on. It turns out that the amplitude spectrum of this pulse train is as shown at the top of the facing page. Below it is the same spectrum with its spectral envelope drawn in.

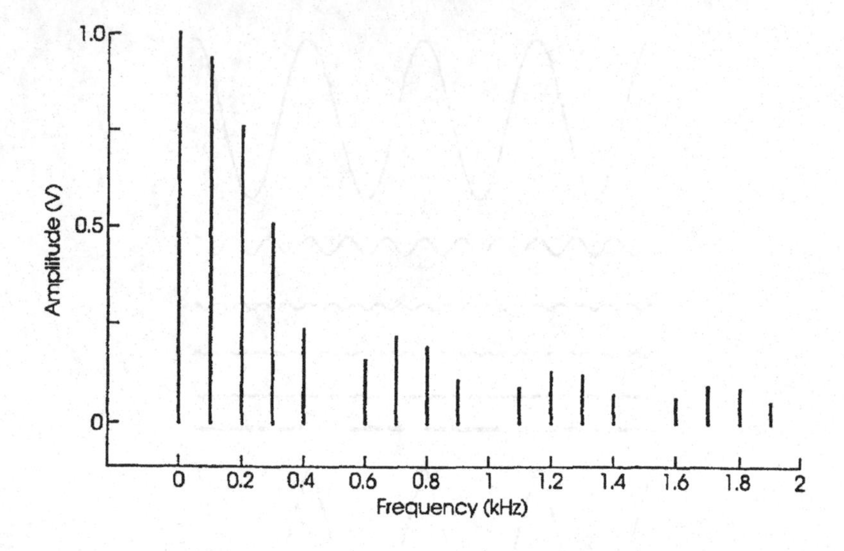

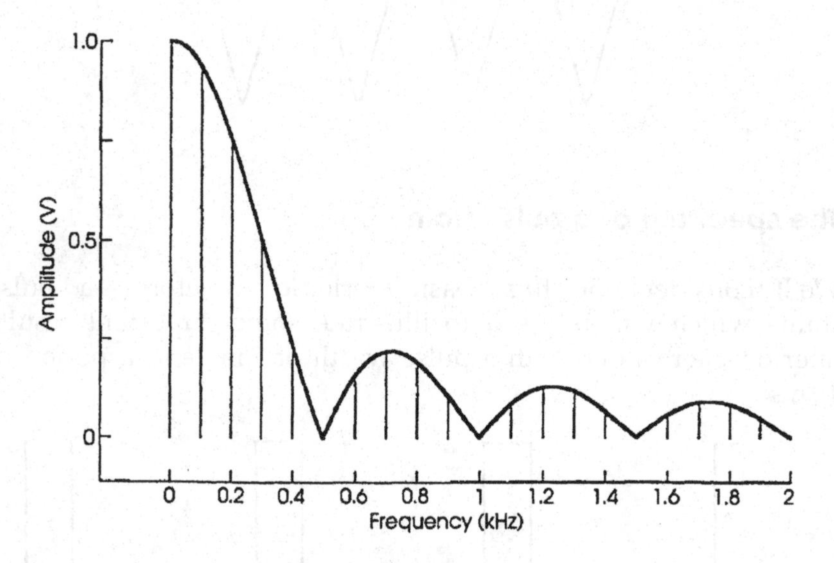

This amplitude spectrum is a bit different to most of those that we've seen up to now, in two main ways. First, it has a fairly complicated shape, dipping to zero amplitude at certain frequencies (500 Hz, 1000 Hz, 1500 Hz, and so on). Second, you will see that there is a component at 0 Hz, also known as a component at DC (for 'direct current'). The significance of this is simply that the pulse train does not have the same area below zero as it does above it. Generally speaking, the DC component indicates the average of the positive and negative bits of the wave over one period. The sawtooth, square wave and triangle were all centred on 0 V, and so had no DC. If the square wave above, for example, were made to start at 0 V and not go negative, it too would have a DC component, but this component would not affect the rest of the spectrum in any way. In some sense, the component at 0 Hz is like any other, but just as a sine wave at 0 Hz can only represent a constant value, so too does the size of the component at 0 Hz represent the vertical placing of the waveform.

The phase spectrum of this pulse train is also somewhat more complicated than those we have previously seen. All components lower in frequency than the first point at which the amplitude spectrum goes to zero (here, 500 Hz) are cosine waves—that is, have a phase of 0°. Above this, the components are inverted (that is, they have a phase of −180°) until the next frequency at which the amplitude spectrum goes to zero (here, 1000 Hz). Continuing upward, the phase flips by 180° again, back to 0° until the next 'zero' in the amplitude spectrum. Thus, the phases flip between 0° and −180°, alternating at each 'zero' in the amplitude spectrum.

Using the appropriate phases, then, the harmonics up to 2 kHz sum to a reasonably good approximation of the original pulse train:

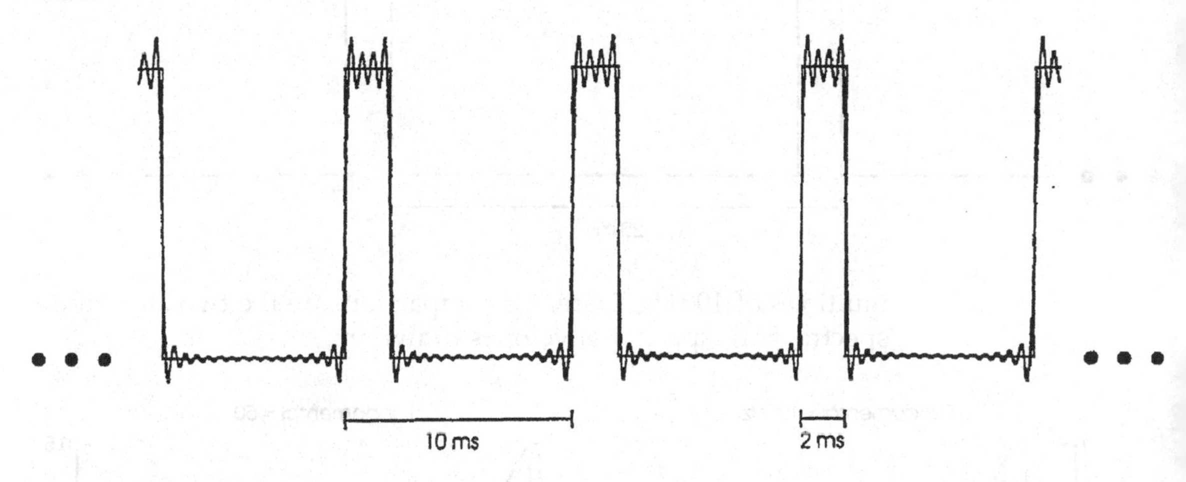

The effect of increasing period with the same duration pulse

Now let's see what happens when the period of the pulse train is lengthened whilst the width of the pulse is kept constant. We'll double the period to 20 ms but leave the pulse 2 ms wide. This longer-period pulse train is at the bottom of the next figure, with the original pulse train above it for comparison (at top, opposite).

It turns out that the phase spectrum remains the same as it was for the shorter-period pulse train (flipping between 0° and −180° at zeroes in the amplitude spectrum), so we'll discuss phase no more. The new wave has a period of 20 ms, which corresponds to a frequency of 50 Hz. Therefore, the amplitude spectrum can only have energy at multiples of 50 Hz, in contrast to the previous pulse train which had components at

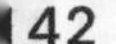

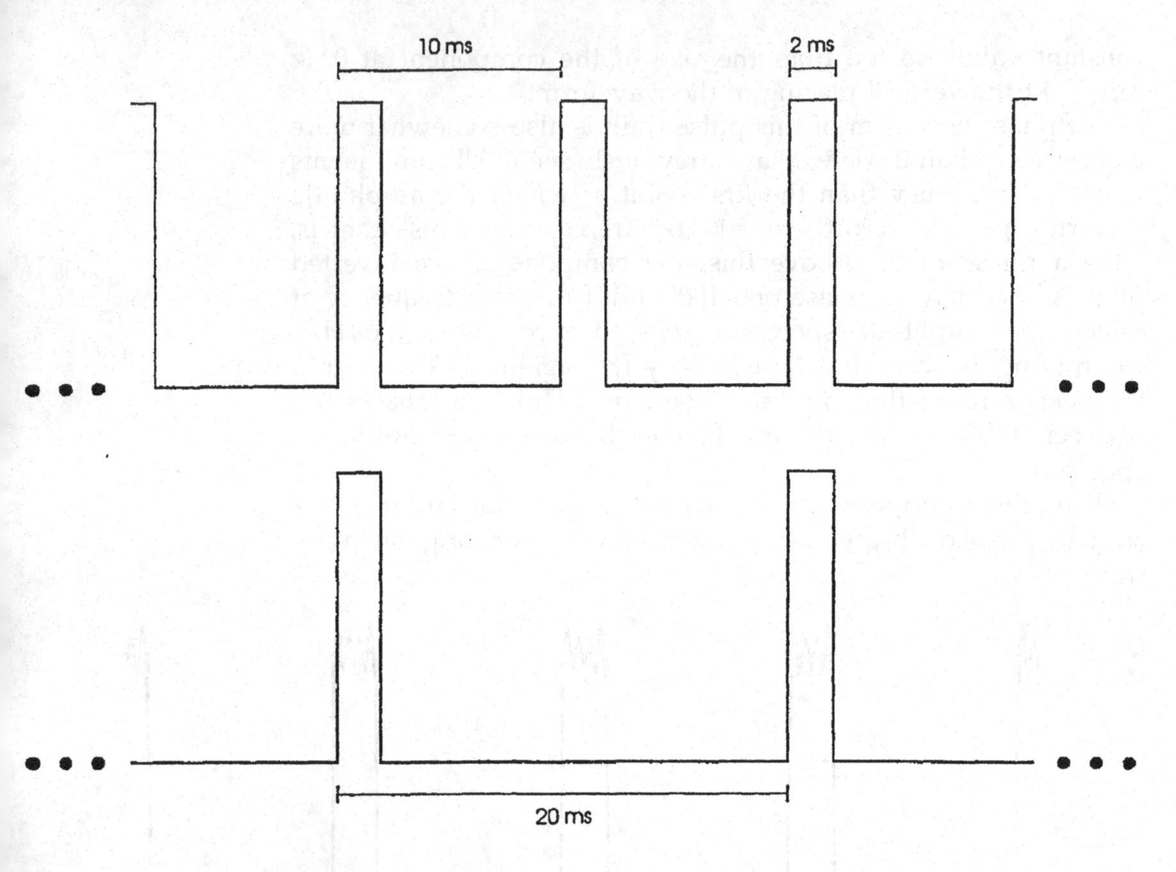

multiples of 100 Hz. Here, for comparison, are the two amplitude spectra, with spectral envelopes drawn in:

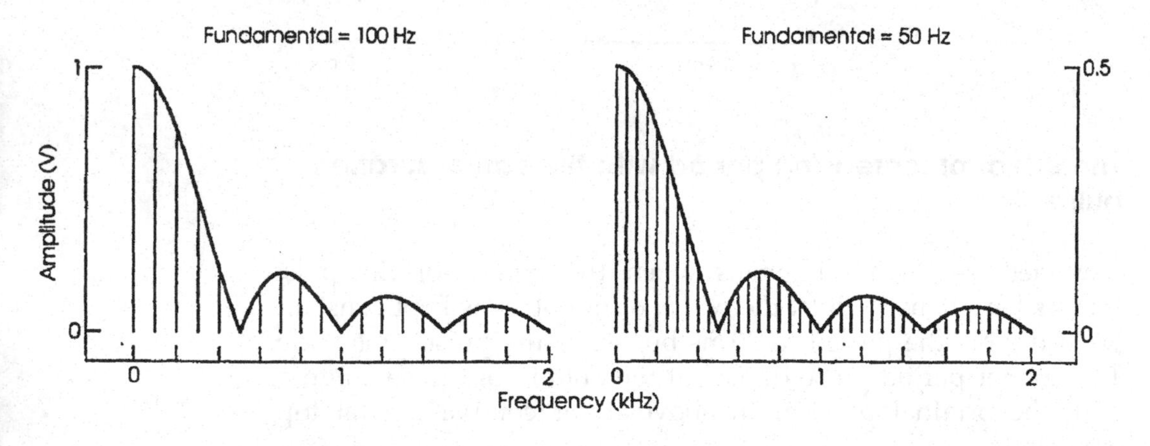

By looking at the scales on the y-axis, you can see that the highest value of the amplitude spectrum of the 50-Hz pulse train is half that of the 100-Hz train. This is not surprising as there is going to be less energy around when the pulse appears half as often. Let's concern ourselves now simply with the shape and harmonic structure of the two spectra, which is easy since we've made the two the same size by appropriate scaling of the y-axis.

In both these amplitude spectra there are dips down to zero. More importantly, the dips occur at the same points in frequency. In fact, the entire shapes of the two spectral envelopes are the same too. This is not surprising if you think about it. The pulses are the same shape (rectangular and 2 ms wide) so you might expect something similar about their spectra. However, although the envelopes are the same, the position of the harmonics are not. The wave with the longer period has more harmonics within any given frequency range, and its fundamental is at a lower frequency than the one with the shorter period.

Let's consider why the pulse train with the longer period has an amplitude spectrum that starts at a lower frequency than the one with the shorter period (ignoring the DC component). The fundamental of a Fourier series is the sinusoidal component with the same period as the original waveform. Thus, as the period of a periodic signal is lengthened, the frequency of its fundamental goes down and the harmonics become closer together. So, in the 100-Hz pulse train, the harmonic series starts at 100 Hz and continues at 200 Hz, 300 Hz and so on. For the 50-Hz pulse train, the harmonic series starts at 50 Hz and continues 100 Hz, 150 Hz and so on. In short, when the frequency of the pulse train is reduced (or, equivalently, its period is increased), the fundamental occurs at a lower frequency and the inter-harmonic spacing is reduced.

Spectra of aperiodic signals

As defined in Chapter 3, an aperiodic signal is one which does not go through repetitive cycles. Another way to look at this is to think of an aperiodic signal as periodic, but with a period that is infinitely long. In other words, no matter how long you waited, this 'periodic' signal would never repeat. Aperiodic signals cannot, then, consist of a series of harmonics based on the periodicity of the signal. Even so, aperiodic signals can be represented by spectra but without spectral components at discrete frequencies. When aperiodic signals are characterized by their frequency content, the process is referred to as calculating a *Fourier transform*. In Chapter 3, we distinguished two sorts of aperiodic signals—transient and random. We'll consider aspects of the Fourier transforms of each in turn.

Fourier transform of a transient signal

Consider taking the Fourier transform of the following single 2-ms pulse:

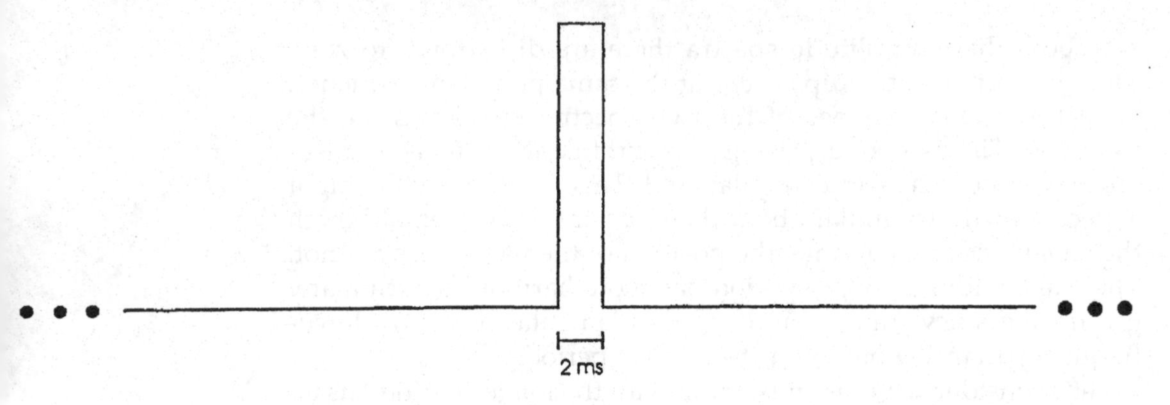

This pulse is identical to one of the pulses in the pulse trains we considered in the previous section. One way to derive the spectrum of this signal is to take the hint above and think of it as a pulse train with an extremely long period—in fact, an infinite one. The progression from a pulse train with a finite period to an infinitely long period goes as follows.

We noted above what happens to the spectrum of a pulse train when its period is doubled from 10 to 20 ms. The fundamental frequency drops by 1/2, the harmonic spacing decreases by the same amount, but the shape of the spectral envelope stays the same. There is, of course, nothing to stop us from doubling the period of the second pulse train from 20 to 40 ms and observing the same effects on the spectrum. The shape of the spectral envelope is unaltered, the fundamental drops further to 25 Hz and the harmonic spacing decreases to 25 Hz:

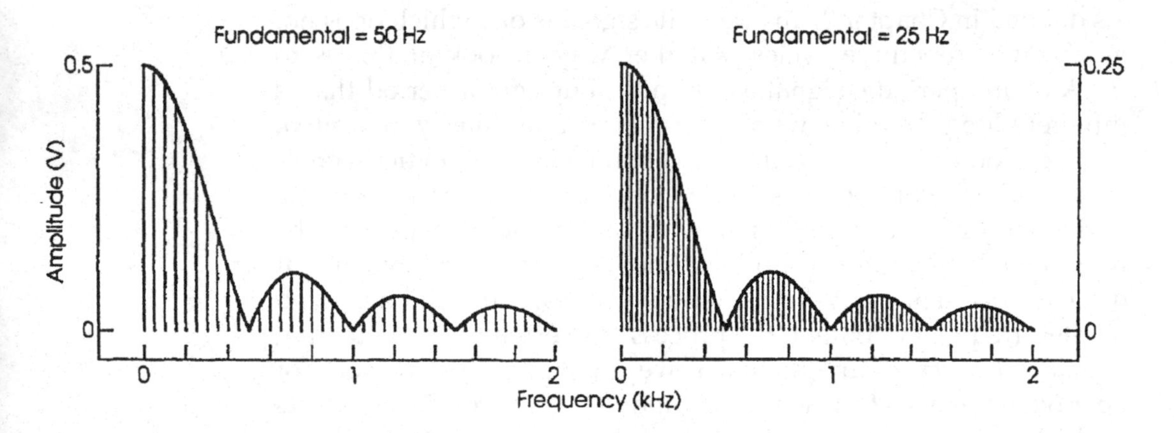

Now, in some sense, the pulse train of period 40 ms is a better approximation to the single pulse than is a pulse train with a period of 10 ms. And there is nothing to stop us getting better and better approximations by repeating this 'period-doubling' process. Each time the period of the pulse train is doubled, the same thing happens to the amplitude spectrum—the fundamental and inter-harmonic spacing drop by 1/2, while the shape of the spectral envelope remains the same.

If we repeated this period doubling enough times, eventually the pulses would be so far apart in time that we could imagine the waveform to consist of a single pulse—the waveform whose spectrum we wish to obtain. As regards the spectrum, every time we double the period, the effects on it are the same—the fundamental falls at lower and lower frequencies and the inter-harmonic spacing decreases. Again, if we repeated this period doubling sufficiently, the fundamental would be so low that we could imagine it to be at 0 Hz on the frequency axis. Similarly, the harmonics would be so close together as to make the spectrum *continuous*. Therefore, the spectrum that corresponds to a single pulse does not have energy at specific frequencies—it has energy spread over *bands* of frequency. The *y*-axis must then represent energy not at a single frequency, but energy summed up within a frequency band 1 Hz wide—this is usually referred to as *spectral density.*

The shape of the spectrum may be obtained by noting that the shape of the spectral envelope doesn't alter with changes in period (because the shape of the pulse doesn't change). Therefore, the spectrum of a single pulse is identical to the spectral envelope of a periodic train of the same pulses:

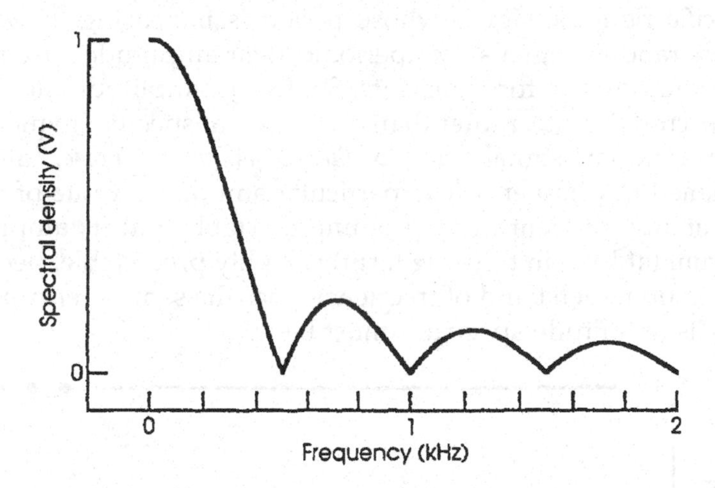

What happens to the phase spectrum? As already noted, the sinusoidal components in the spectrum have frequencies that are effectively continuous. However, as more frequencies have been packed in, the phase of each component hasn't altered. So, the resultant phase spectrum is also continuous over frequency, and has the same values as the components in the pulse train (flipping between 0° and −180° at frequencies where the amplitude spectrum goes to zero):

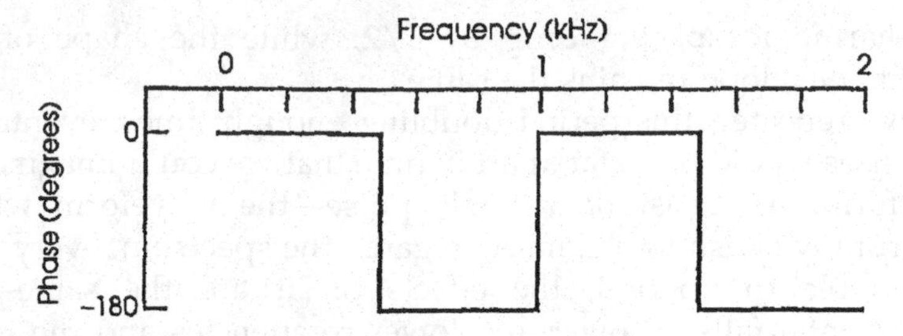

Although we've only demonstrated how to obtain the spectrum of a *particular* transient signal, the same principles can be applied to *any* transient signal. To summarize, the spectra of transient signals are continuous, and obtainable from the spectra of periodic signals created by repeating the transient at appropriate intervals.

Random signals

As noted in Chapter 3, random signals (also known as 'noise') vary in an unpredictable manner over time and typically are considered to be infinitely long. Although it would be possible to explore many properties of 'noisy' signals, we will be concerned exclusively with some aspects of their amplitude spectra.

Because random signals typically don't have a finite duration, we cannot consider them to be approximated better and better by a specific periodic signal whose period is increasing. However, because random signals are aperiodic, their amplitude spectra are continuous, just as for transients. So, too, is amplitude measured as a spectral density, rather than amplitude at specific frequencies.

The random signal *par excellence* is *white noise,* already mentioned in Chapter 3. The particular amplitude value of white noise at any moment in time is unpredictable, but its amplitude spectrum (at least in the long term) is highly predictable. Because, by definition, each band of frequencies has the same energy as any other, its amplitude spectrum must be flat:

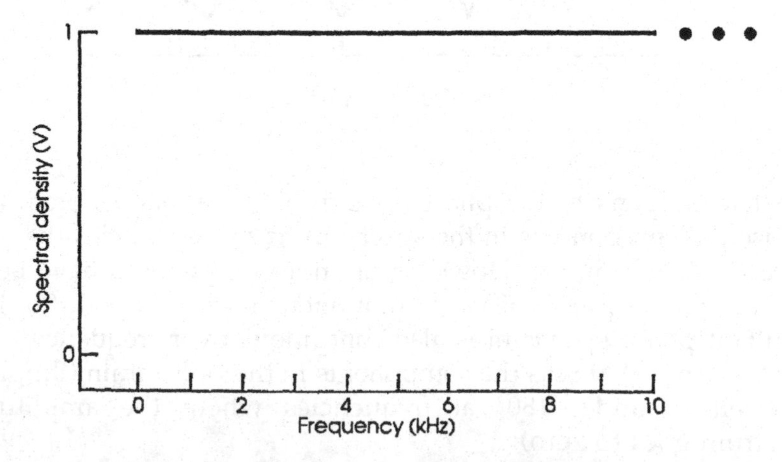

The three dots on the right are to indicate that the spectrum continues out to infinitely high frequencies.

Pink noise is another random signal, used frequently in speech and hearing, which has an amplitude spectrum with a simple pattern across frequency. Pink noise is defined so that each band of frequencies one octave wide contains the same energy. Since an octave covers a wider frequency range as frequency increases, the value of the spectral density must fall across frequency—here from 20 Hz to 5 kHz:

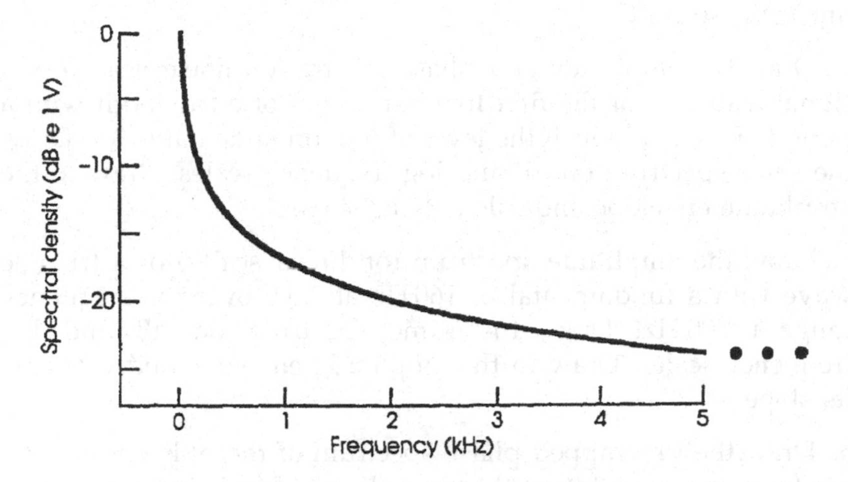

You may be surprised that so-called 'noisy' signals can have such smooth spectra, and rightly so. These spectra represent the frequency content of the signals averaged over an infinite length of time. Any particular stretch of noise analyzed will have an amplitude spectrum that is a 'messy' version of its true spectrum. Here, for example, are spectra obtained from analyzing real-life white noise, in which the averaging time was 80 ms, 400 ms and 8 s, respectively. Note how the spectrum becomes smoother as averaging time increases:

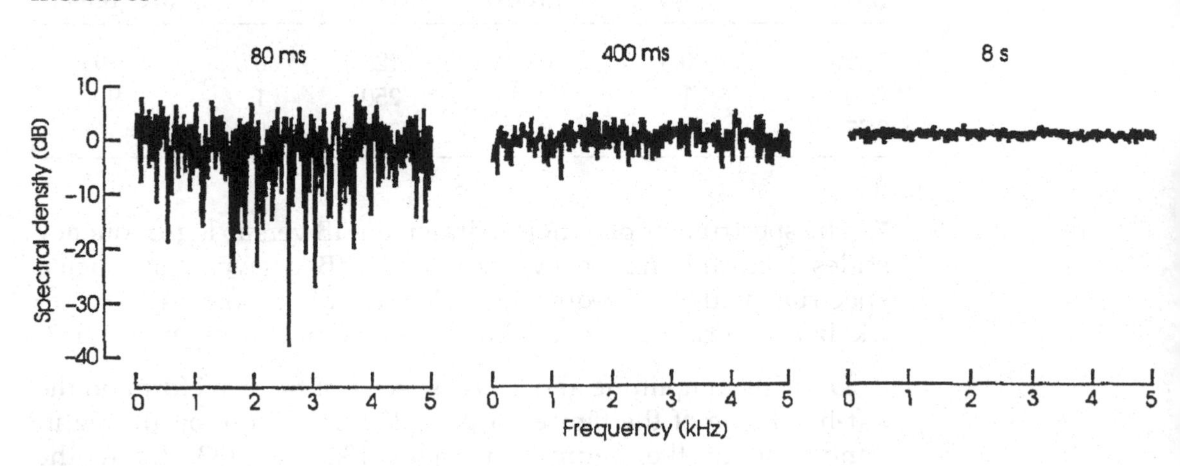

Exercises

1. Draw the amplitude spectrum of a sinusoid of 2 V at 300 Hz.

2. Check that a square wave can be synthesized from odd harmonics that alternate between cosine phase and inverted cosine phase, again by drawing out two cycles of the first three terms and adding them up at various points along their length. Can you relate this phase spectrum to the all sine phase values used in the synthesis of the square wave on page 137, in terms of a linear phase shift?

3. Draw the amplitude and phase spectra (on linear scales) of a signal made up of the first five harmonics of a sawtooth with a period of 4 ms in which the level of the fundamental is 3 V. Draw the same spectrum on dB and log frequency scales. Draw in the amplitude envelope and calculate its slope.

4. Draw the amplitude spectrum (on linear scales) of a triangle wave with a fundamental of 160 Hz at 2.5 V over the frequency range 0–800 Hz. Draw the same spectrum on dB and log frequency scales. Draw in the amplitude envelope and calculate its slope.

5. Draw the unwrapped phase spectrum of the pulse train from the figure on page 146 so that it is always decreasing.

6. (For keen computer users:) Write a simple program using Excel to do Fourier synthesis and display the resulting waveforms. Compare the results you would get for the following spectra, and comment:

	Signal 1			Signal 2	
Frequency (Hz)	Amp. (V)	Phase (degrees)	Frequency (Hz)	Amp. (V)	Phase (degrees)
125	0.5	0	125	0.5	0
250	1	0	250	1	−90
375	0.5	0	375	0.5	0

7. The spectrum of pink noise, drawn on dB versus log-frequency scales, falls off in frequency at a rate of 3 dB/octave. Draw such a spectrum with the frequency scale extending from 31.25 Hz to 2 kHz and clearly marked. Why can't a point be drawn at 0 Hz?

8. Draw the amplitude and phase spectra of the waveform on the left-hand side of the figure on page 132, and those on the right-hand side of the figures on pages 132 and 133. Draw the

amplitude spectra in two ways—using linear scales for both amplitude and frequency, and on dB versus log frequency scales.

9. Sketch the spectral envelope of the following two signals: (a) an envelope rising at 6 dB/octave for one octave starting at 300 Hz, then falling at 2 dB/octave out to 2400 Hz. Draw this on both linear and logarithmic frequency axes. (b) A two-component complex whose second harmonic is 3 dB down at 200 Hz re its level at 100 Hz.

10. A signal contains two sinusoids at equal amplitude, one at a frequency of 1100 Hz and the other at 1200 Hz. What periodicity would this signal have?

CHAPTER 8

Signals Through Systems

We've now seen how to analyze and synthesize signals in the frequency domain, using sinusoids of various frequencies, amplitudes and phases. So, too, have *systems* been described in the frequency domain, in terms of their transmission of sinusoids. Although you will find these concepts on their own useful in understanding much that is written about speech and hearing (and hi-fi stereo gear!), taken together they allow us to do something more—to establish the response of *any* LTI system to *any* signal. This is the final goal we set in Chapter 5.

Two aspects of what we've already discussed allow us to achieve this. First, we know from the previous chapter that all signals can be represented as a (possibly infinite) sum of sinusoids. Second, as was discussed in Chapter 4, the response of an LTI system to a sinusoid at a particular frequency is the same whether that sinusoid occurs on its own, or as part of a complex. Thus, once a signal is represented in terms of its component sinusoids, the response at each component frequency can be ascertained and the output signal synthesized from the outputs to the separate input sinusoids.

Although these ideas apply to both periodic and aperiodic signals, it's easier to see what's happening with periodic sounds and so we'll start with them. First, we'll detail a fairly laborious method which uses only principles that you've already met, and then we'll go on to describe a short-cut.

Here are the six steps involved in the laborious method:

(1) Analyze the signal into its component sinusoids, specifying the frequency, amplitude, and phase of each of the components (in other words, establish the amplitude and phase spectra of the signal, as was described in Chapter 7).
(2) Obtain the amplitude response of the system (Chapter 6).
(3) Obtain the phase response of the system (Chapter 6).
(4) For each sinusoid present in the input, use the amplitude response of the system to ascertain the amplitude of each sinusoidal output (Chapter 6).

(5) Similarly, taking each sinusoidal input component in turn, establish its output phase from the phase of the input and the phase response of the system (Chapter 6).
(6) Now that the amplitude and phases of all the sinusoidal output components are known, the output waveform can be synthesized (using the techniques discussed in Chapter 7).

Putting a periodic signal through a filter

As always, we'll go through a few concrete examples to illustrate these steps in detail. Here is a sawtooth waveform (period = 10 ms) being put through an ideal low-pass filter with a cutoff of 250 Hz:

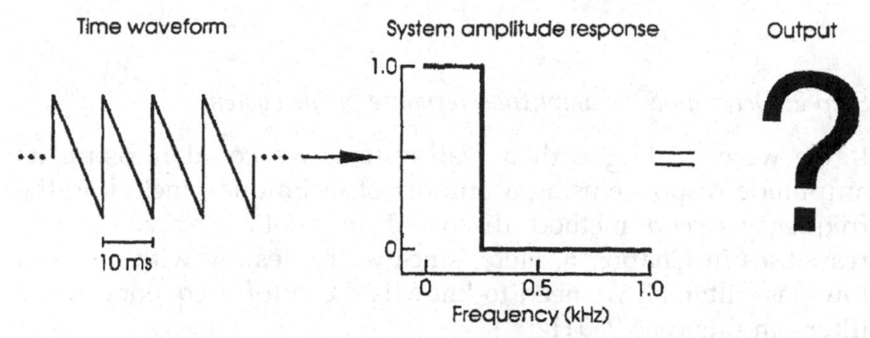

Step 1: Analyze the signal into its component frequencies

The sawtooth is one signal that you've seen often enough now to know its frequency domain representation very well. Its amplitude spectrum is made up of sinusoidal components at integer multiples of the fundamental (here, 100 Hz, 200 Hz, 300 Hz, and so on) which drop off in amplitude as the reciprocal of the harmonic number. So, if the amplitude of the first harmonic is 1 V, that of the second is 1/2 V, the third 1/3 V, etc. Here is the time waveform and amplitude spectrum of our sawtooth (on linear scales):

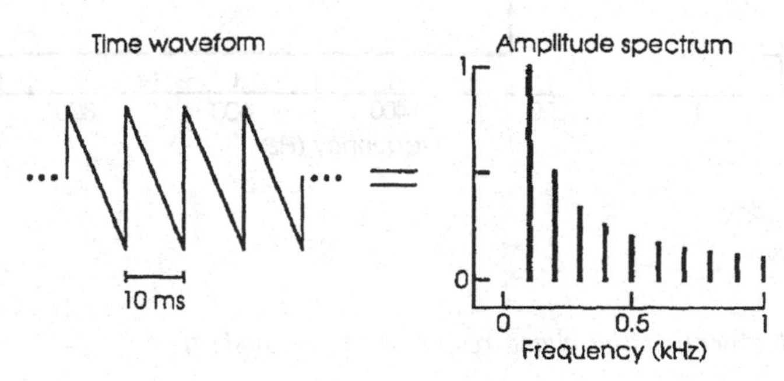

Although you know that this series continues out to infinitely high frequencies, we'll only be concerned with the first 10 harmonics.

You should also recall the phase spectrum of the sawtooth, which, as we showed in Chapter 7, is constant across frequency at −90° (measured relative to the phase of a cosine):

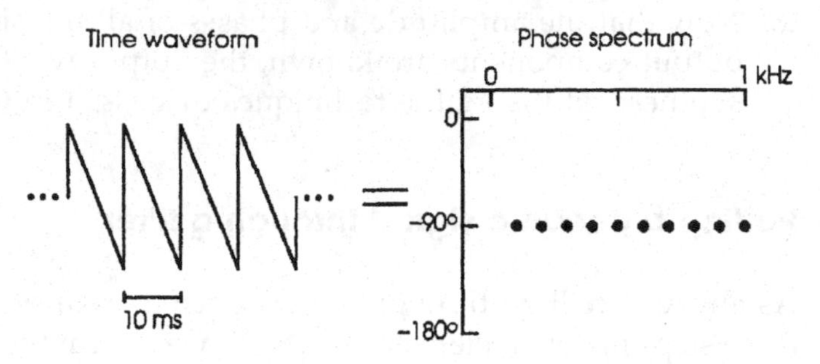

Step 2: Determine the amplitude response of the system

If we were dealing with a real system, we could measure its amplitude response using a number of techniques, including the frequency-sweep method discussed in relation to vocal tract responses in Chapter 6. Here, since we're dealing with an ideal low-pass filter, all we need to know is the cutoff frequency of the filter—in this case 250 Hz:

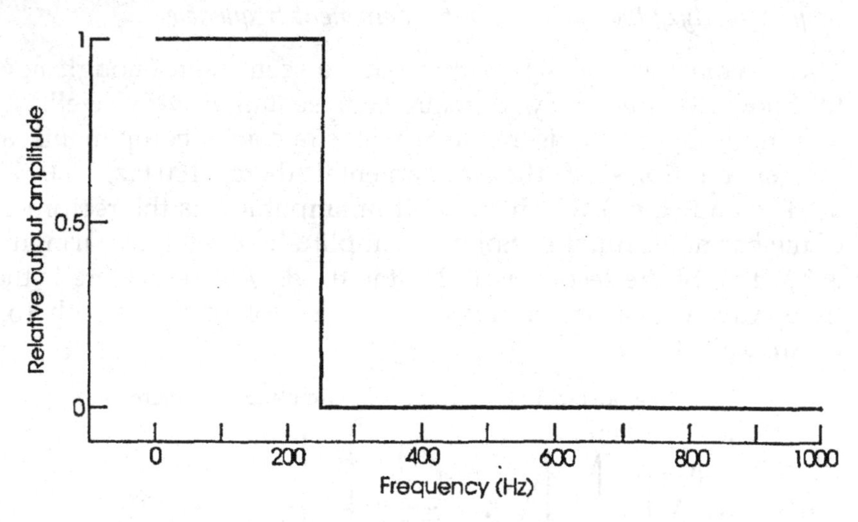

Step 3: Determine the phase response of the system

Again, if we were dealing with a real system, the phase response could be measured empirically as for amplitude responses. As the specification of an ideal low-pass characteristic is concerned solely with the amplitude response, nothing is implied about the phase response. We'll specify one arbitrarily—let's say that the system has no effect on the phases of the inputs:

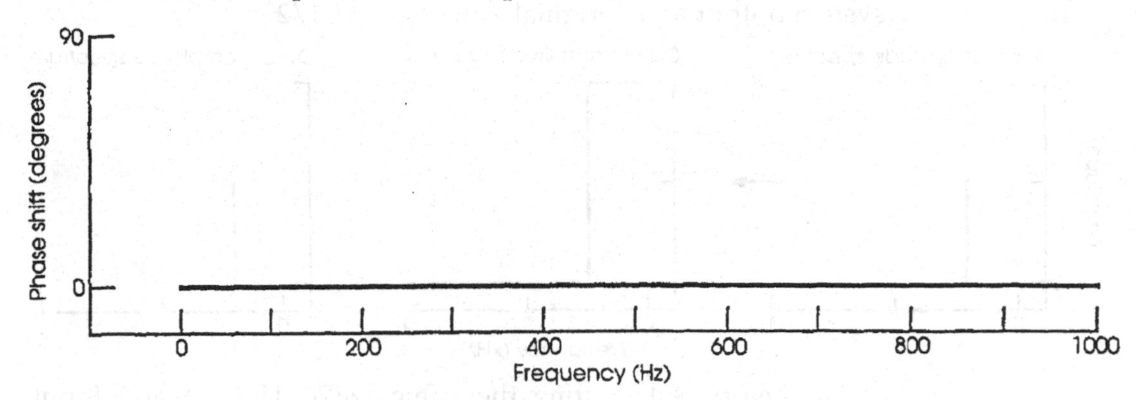

Step 4: Determine the amplitude of each component of the output

For our sawtooth input signal, the lowest frequency sinusoid present is at 100 Hz and has an amplitude of 1 V. The frequency of this sinusoid is below the cutoff frequency of the low-pass filter (250 Hz) and, since the filter is ideal, will be transmitted at its original amplitude. Thus, at the output, there will also be a 100-Hz sinusoid at 1 V. Diagrammatically:

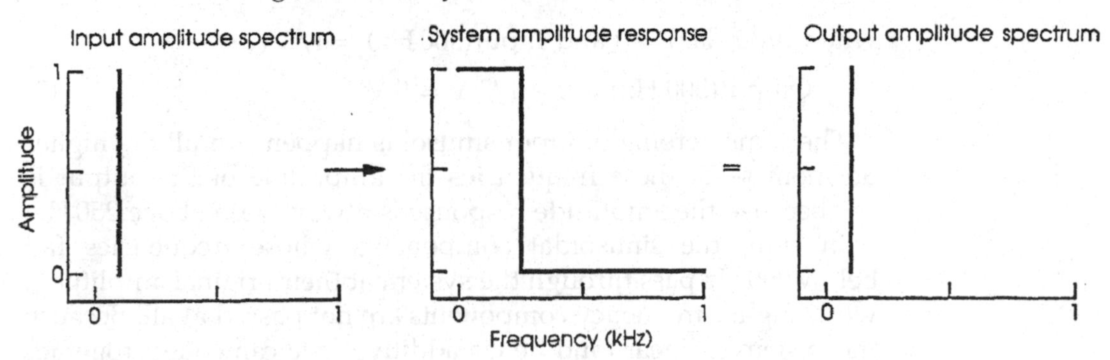

The amplitude of the output can also be obtained directly from this equation:

$$\text{Output}(f) = A(f) \times \text{Input}(f) \tag{1}$$

where $A(f)$ is the value of the amplitude response of the system, Input(f) is the amplitude of the input, and Output(f) is the amplitude of the output sinusoid, all at frequency f. Here, the amplitude of the input is 1 V, and the value of the amplitude response is 1, so:

$$\text{Output}(100\ \text{Hz}) = 1 \times 1\ \text{V} = 1\ \text{V} \tag{2}$$

which shows (again!) that the output is at the same amplitude as the input.

We now take the next component (200 Hz at 1/2 V). This too is in the pass-band of the low-pass filter and so appears at the system output at its original amplitude of 1/2 V:

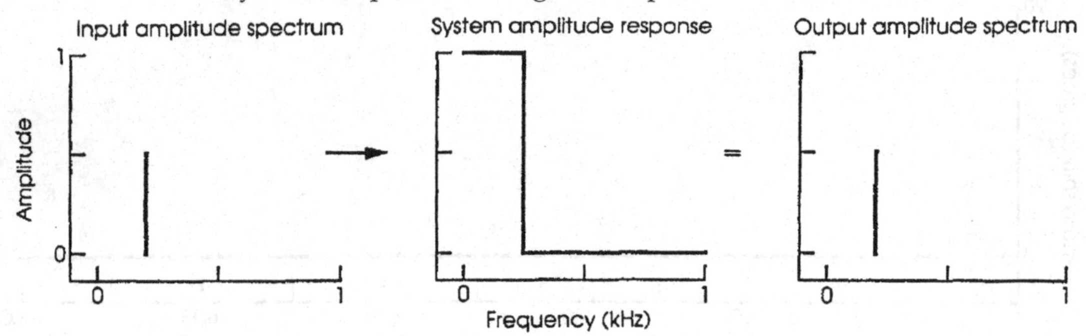

Or, as before, substituting the values $A(200\,\text{Hz}) = 1$ and Input (200 Hz) = 1/2 V into equation (1):

$$\text{Output}(200\ \text{Hz}) = 1 \times 1/2\ \text{V} = 1/2\ \text{V} \tag{3}$$

When the third component of the input is reached, the value of the amplitude response at the corresponding frequency (300 Hz) is zero. This means that the filter will completely block any sinusoid at this frequency, as is best seen from the equation:

$$\text{Output}(f) = A(f) \times \text{Input}(f) \tag{4}$$

when $A(300\,\text{Hz}) = 0$ and Input(300 Hz) = 1/3 V:

$$\text{Output}(300\ \text{Hz}) = 0 \times 1/3\ \text{V} = 0\ \text{V} \tag{5}$$

The same zeroing of input sinusoids happens for all the higher harmonics—at these frequencies the amplitude of the output is 0 V because the amplitude response is always zero above 250 Hz.

In sum, the sinusoidal components whose frequencies fall below 250 Hz pass through the system at their original amplitude, while higher-frequency components are not passed at all. Because the system is linear (and hence additive), we can collect together all the components of the output and put them on a graph (amplitude spectrum of the output).

Thus, in diagrammatic form, the operations we have performed on component amplitudes can be summarized as:

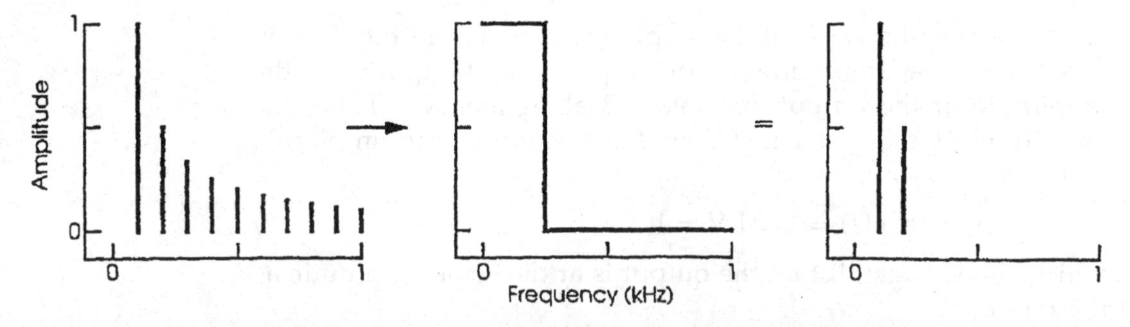

Step 5: Determine the phase of each component of the output

As we noted in Chapter 6, the phase response of a system at a particular frequency is specified as the difference in phase between an input and output sinusoid. In step 1 above we noted that each sinusoidal component of the input signal has a phase of −90° (relative to cosine phase). In Step 3 we decided to give the system a phase response which made no alterations in phase, no matter what frequency sinusoid was considered. Generally speaking, the output phase of the input component at 100 Hz can be calculated from the equation:

$$\text{Output}_{\text{phase}}(f) = \text{Input}_{\text{phase}}(f) + P(f) \tag{6}$$

where $\text{Output}_{\text{phase}}(f)$ is the phase of the output sinusoid, $\text{Input}_{\text{phase}}(f)$ is the phase of the input sinusoid, and $P(f)$ is the phase response of the system (all at frequency f). Here:

$$\text{Input}_{\text{phase}}(100\ \text{Hz}) = -90^\circ \tag{7}$$

while the phase response of the system is always 0° (no alteration to the phase), and so:

$$\text{Output}_{\text{phase}}(100\ \text{Hz}) = -90^\circ + 0^\circ = -90^\circ \tag{8}$$

which is to say that the phase of the output is the same as the phase of the input. This same equation will apply at all other frequencies, so all output phases will be the same as input phases at −90°. In this example, the final result was obvious, so you would have been able to guess the answer straight away without going through all this work. You will, however, need this detailed technique when the system phase response varies across frequency, as we'll show in a later example. As with the amplitude response, we can summarize the input-to-output transformation of all the sinusoidal components on the same graphs (phase spectra of input and output) thus:

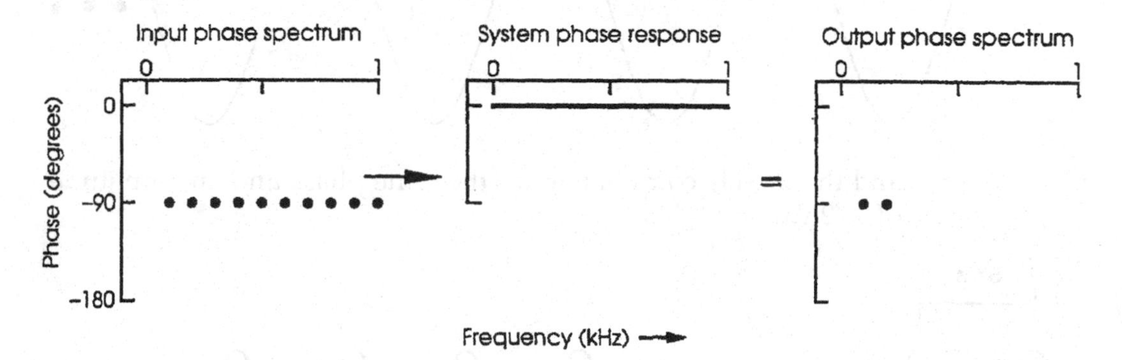

Strictly speaking, we could have drawn points on the output phase spectrum for all the harmonics, rather than just the first two. However, since we already know that only the first two sinusoidal output components are non-zero, we've left out the higher-frequency points for clarity.

Step 6: Synthesize the output wave from its spectrum

We now have all the information necessary (amplitude and phase spectra) for synthesizing the time waveform corresponding to the output signal:

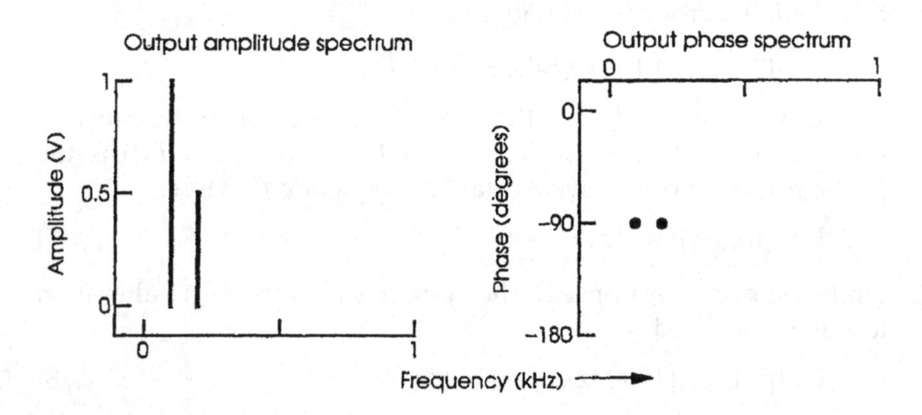

From these graphs, we can construct each component sinusoid of the output. Thus, the 100-Hz component sinusoid has a phase of –90° and is at 1 V:

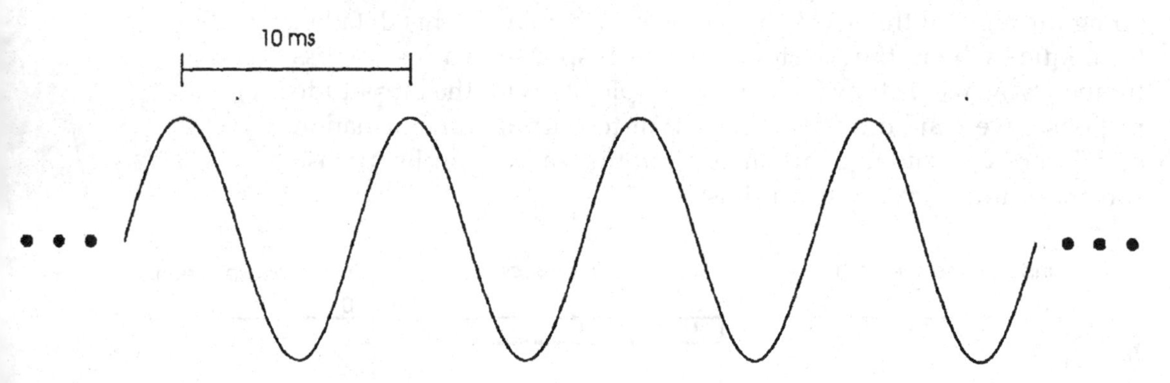

and the 200-Hz component has the same phase and an amplitude of 1/2 V:

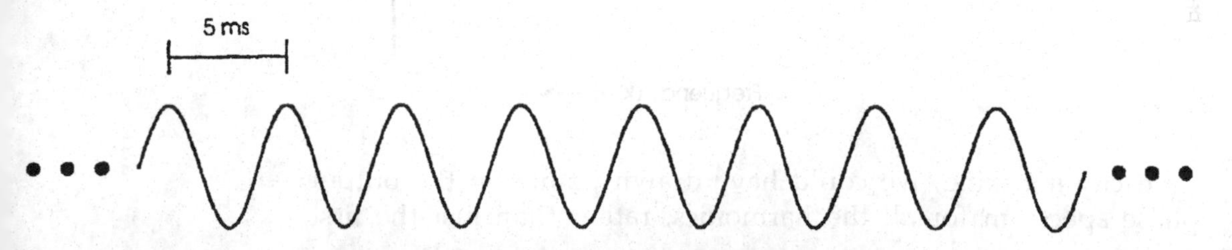

Once the components are plotted as sinusoids, the total output corresponding to their sum can be synthesized by addition just as we did in Chapter 7. Thus we can finally see what the output waveform looks like:

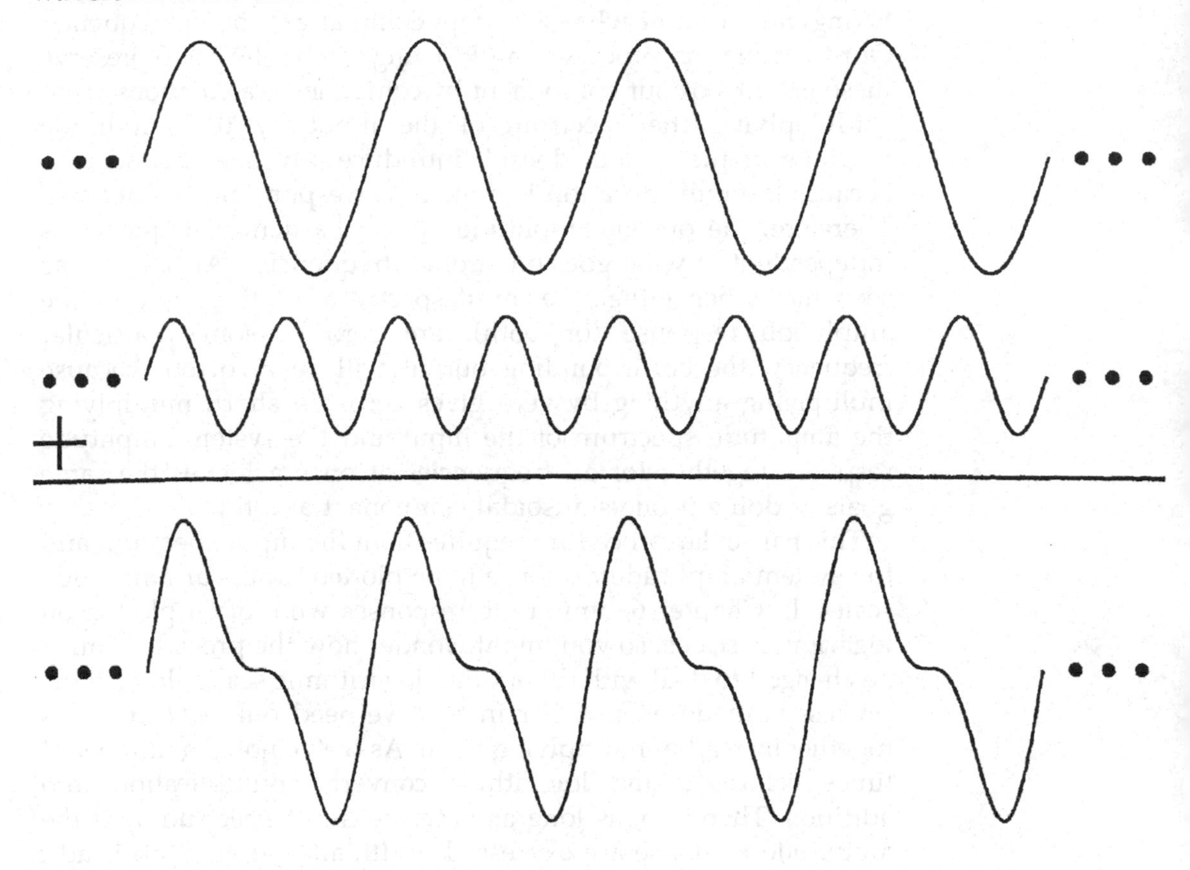

Speeding things up

Doing calculations component by component is simple and easy to follow, but tedious. As we mentioned at the beginning of this chapter, there is an easy way to speed matters up, which is implicit in what we've done already. If amplitude is measured on a linear scale (as has been the case in this chapter so far), the multiplication of the amplitudes of the input sinusoidal components by the appropriate values of the amplitude response can be done all at once rather than one component at a time. Similarly, the phase spectrum of the output can be obtained by adding the phase response of the system to the phases of the input components at corresponding frequencies all at once. Let's look at the calculation of component amplitudes first.

One critical aspect of selecting input components one at a time was the focus on frequency regions where energy was present in the input. Regions with no input energy were ignored. Furthermore, the amplitude of each output component was calculated from the amplitude of an input component, and a value of the amplitude response, at one and the same frequency, without taking into account what was happening at any other frequency. Our 'speeded-up' process of multiplying can be shown to preserve these features of our component-by-component calculations.

Multiplying the spectrum of the input by the amplitude response of the system doesn't introduce any new frequencies, because the multiplication is done at corresponding frequencies. Therefore, the output amplitude at one particular frequency is independent of what goes on at other frequencies. Also, it can be seen that when either the input spectrum, or the value of the amplitude response (or both), are zero at some particular frequency, the corresponding output will be zero too (because multiplying anything by zero gives zero). In short, multiplying the amplitude spectrum of the input and the system amplitude response together for all frequencies at once achieves the same goals as doing it one sinusoidal component at a time.

This particular procedure requires both the input spectrum and the system amplitude response to be plotted on linear amplitude scales. In Chapter 6, amplitude responses were often plotted on logarithmic scales, so you might wonder how the procedure must be changed to deal with dB or other logarithmic scales. In fact, the process is made even easier in that we need only *add* numbers together instead of multiplying them. As we've noted a number of times before, taking logarithms converts multiplication into addition. Therefore, as long as both the input spectrum and the amplitude response are expressed in dB, all you need do is add the two together at corresponding frequencies in order to obtain the amplitude spectrum of the output (again in dB).

In a similar way, the phases of the output components can be obtained by adding together the input phases and the phase response at corresponding frequencies all at once rather than one term at a time. As phases are never expressed as log degrees, there are no added complications (or simplifications!).

Putting a periodic signal through a real low-pass filter

Now that we have a quicker way of determining the response of a system to a signal, we can go on to a couple of more complicated examples. The low-pass filter in the example above is unrealistic in two important respects. Firstly, because its characteristic is ideal, its amplitude response is infinitely steep in going abruptly from the pass-band to the stop-band. Secondly, its phase response is simpler than what would be found in real filters. Nevertheless, the technique that we've developed for determining system outputs is readily applicable to real filters. This can be demonstrated with another low-pass filter, also with a cutoff of 250 Hz, but with a much more realistic amplitude and phase response.

In order to cut down on our work and to give you a better idea of the differences between what happens in an ideal system and what happens in a real system, we'll put the same sawtooth signal through this system. So, from Step 1 (above) the amplitude and phase spectra of the input look like this:

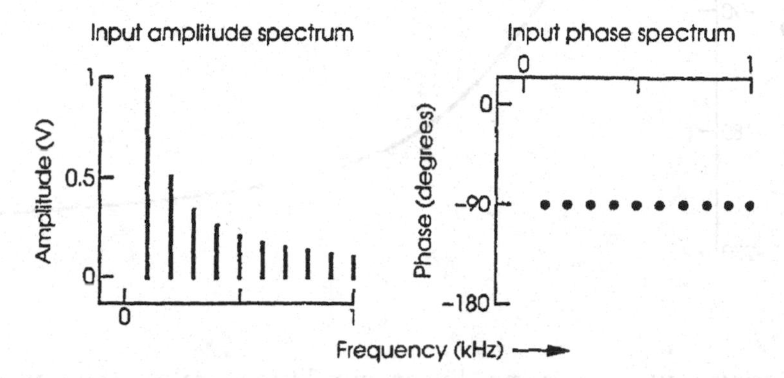

Next, we'll assume that the amplitude and phase responses of the low-pass system have been measured using the frequency-sweep method, obtaining the amplitude response shown here:

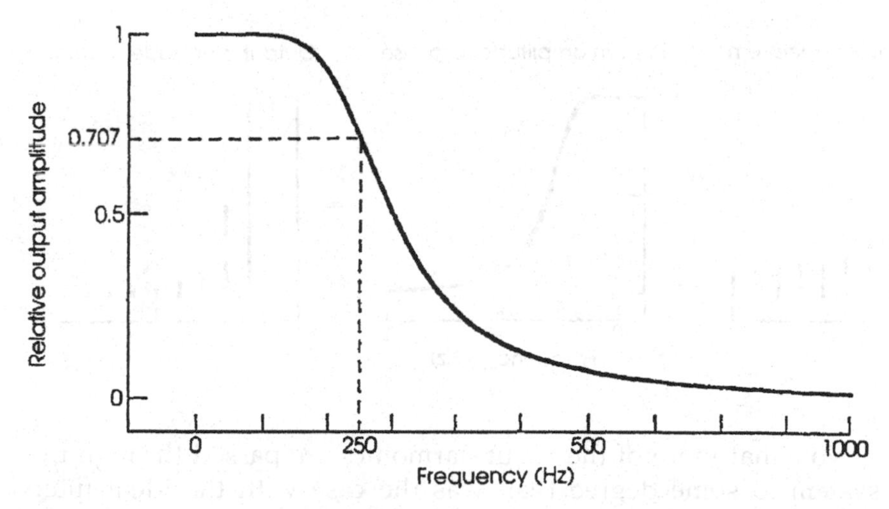

This differs from the ideal response in that the cutoff is not absolute. In the ideal case, a sinusoid at 249 Hz would pass through the filter at full amplitude, while one at 251 Hz would be removed totally. Here, the changeover is more gradual, in that any two input sinusoids that are close together in frequency will pass through the system about equally well. Furthermore, even input sinusoids with frequencies considerably higher than the cutoff will pass through the system somewhat, albeit at reduced amplitudes.

The phase response of the filter is similarly determined (Step 3), and it looks like this:

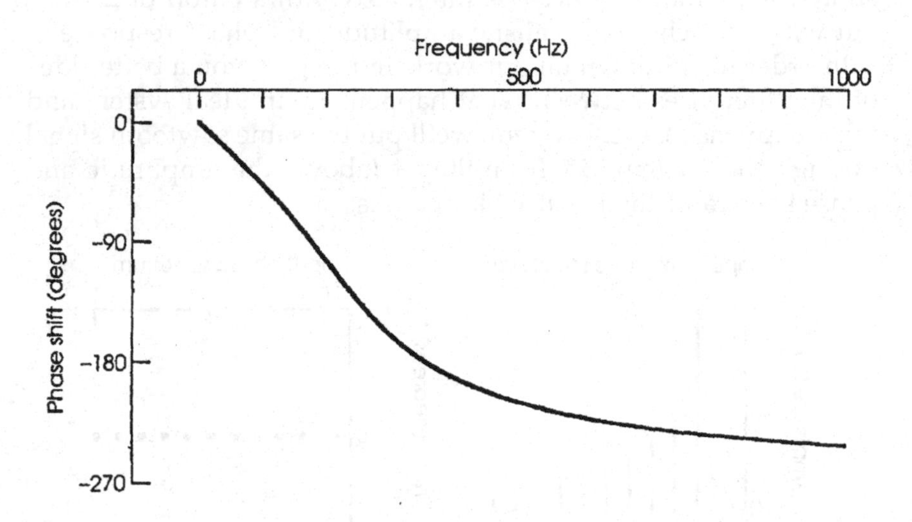

Because we are using linear amplitude scales, the output amplitude spectrum is obtained by multiplying each input component by the appropriate value of the system amplitude response, as summarized pictorially here:

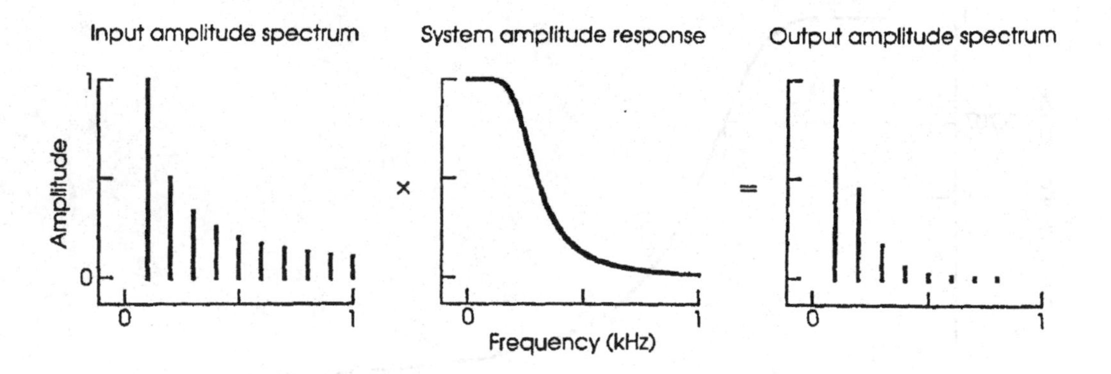

Note that more of the input harmonics are passed through the system to some degree than was the case with the ideal filter. Even so, the amplitude of the harmonics above the eighth has become so small (due to the combined effect of the system increasing its attenuation with frequency, and the progressive reduction in amplitude with frequency for the input spectrum) that they may be safely ignored.

Of course, before we can synthesize the output signal, we need to know the output *phase* spectrum. Unlike our ideal low-pass filter, this more realistic filter alters the phase of all input components:

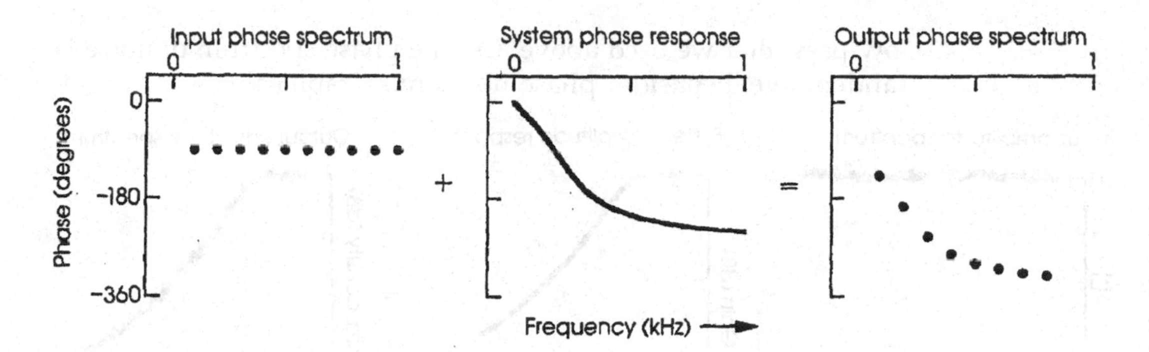

By adding together the eight sinusoidal components with the appropriate frequencies, amplitudes and phases, we obtain the final output waveform:

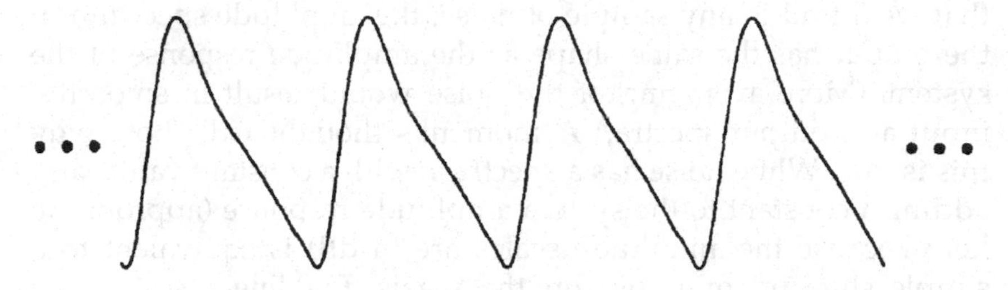

Note how different it looks from the output waveform of the ideal low-pass filter:

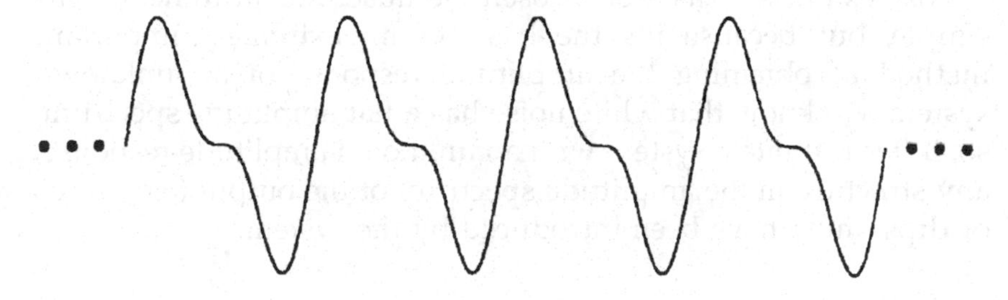

Putting an aperiodic signal through a system

As we've shown in Chapter 7, aperiodic signals have continuous spectra in both amplitude and phase. Even so, the principles illustrated above with a periodic signal apply equally to them. The amplitude spectrum of the input is multiplied by the amplitude response of the system (linear scales) or added (logarithmic scales) to obtain the output spectrum. The phase spectrum of the input is added to the phase response of the system to obtain the output phase spectrum.

A particularly important case arises when white noise is used as an input signal. For example, consider the output amplitude spectrum of white noise passed through the same realistic low-pass filter we used above. (As the phase spectrum of noise is random, we'll consider phase no more.) Graphically:

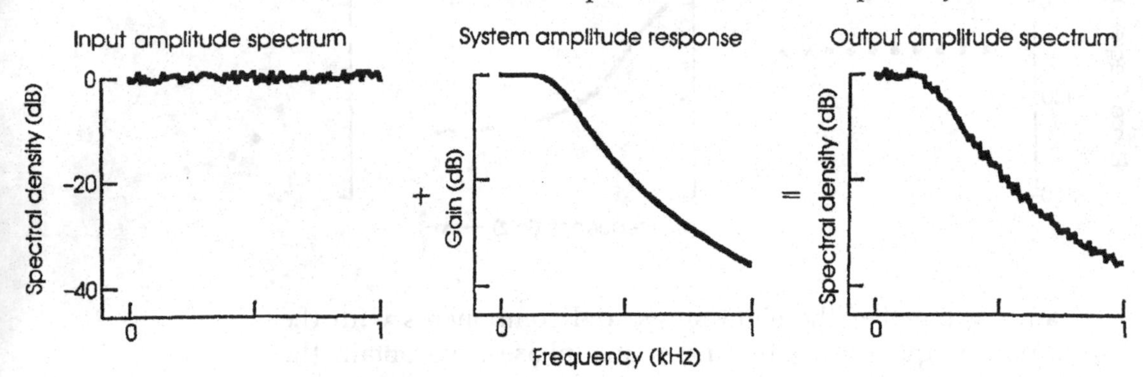

Note that, apart from the irregularities in the frequency spectrum that we'd find in any sample of noise, the amplitude spectrum of the output has the same shape as the amplitude response of the system. (More averaging of the noise would result in smoother input and output spectra.) A moment's thought will show why this is true. White noise has a spectrum with a constant value, and adding a constant to the system amplitude response (appropriate here because the amplitude scales are in dB) is equivalent to a simple shift up or down on the *y*-axis. On linear scales, the amplitude response would only be scaled to be larger or smaller, again without changing its shape.

This example was not chosen because the arithmetic was simple, but because it's the basis of an extremely important method of obtaining the amplitude response of an unknown system. We know that white noise has a flat amplitude spectrum, so, if it is put into a system with an unknown amplitude response, any structure in the amplitude spectrum of the output (e.g. peaks or dips) must have been introduced by the system.

Distortion and the perfect system

Many of the systems we've discussed perform useful transformations of the input signal for various purposes. Quite frequently, however, as for microphones and loudspeakers, we want systems that will reproduce as closely as possible the input signals we put into them. Now that we've explored amplitude and phase responses of systems, and seen the effect they have on a signal, we can define in an unambiguous way what properties a system must have to reproduce a signal perfectly. Of course, this ideal never happens: every system, no matter how good, will introduce changes into the signal, which are known as *distortions*. In everyday language, we say that an audio signal is distorted when it sounds bad to us in one way or another. We can only really talk about distortion, though, in relation to a system that perfectly reproduces a signal.

First, in the perfect system, all sinusoids should pass through the system equally easily. In other words, the amplitude response should be flat—what we call an *all-pass* system. All systems will, of course, have limitations, so it is only necessary for the amplitude response to be flat over some restricted range of frequencies, a range which will be different for different applications. For auditory signals and human listeners, the accepted range is 20 Hz to 20 kHz, the limits of human hearing.

What can go wrong in an amplitude response? One common feature is that certain sinusoidal frequencies are attenuated or amplified relative to others. This is known as *frequency distortion*. Although undesirable in a system designed for accurate reproduction, it is perfectly possible (and often sought) in an LTI system.

Matters can also be not quite right in a phase response. Here the ideal would be zero phase, which would entail no change in the phase of any of the sinusoids put through the system. But we can usually tolerate a time delay in the system, so a linear phase response (Chapter 6) would also be acceptable, since it is equivalent to a constant time delay (the same at all frequencies). Thus, the phase relationships (or relative phases) between the components of the input will be preserved in the output. A system with any other type of phase response is said to have *phase distortion*. The system may still, of course, be LTI.

Other distortions in the proper sense only happen in systems that are not really LTI. For instance, when we put a single sinusoid as input to a system, we often find in the output not only a sinusoid at the input frequency, but also one at twice this frequency added into it. This is known as *harmonic distortion*. Because LTI systems only give a sinusoidal output at the same frequency to a sinusoidal input, a system with harmonic distortion is not LTI. Similarly for *inter-modulation distortion*. Again revealed as a sinusoid of a frequency not in the input, this distortion arises from the interaction of two sinusoids in the input. Quantitative measures of these distortions give an idea how far from an ideal LTI system a particular system really is.

Exercises

1. An input signal consists of the first 10 harmonics of a sawtooth with a fundamental frequency of 200 Hz, and whose fundamental component has a level of 2 Pa. Draw the amplitude spectrum (in dB SPL and using a linear frequency scale) of the output resulting from this signal when put through a filter with the following

amplitude response:

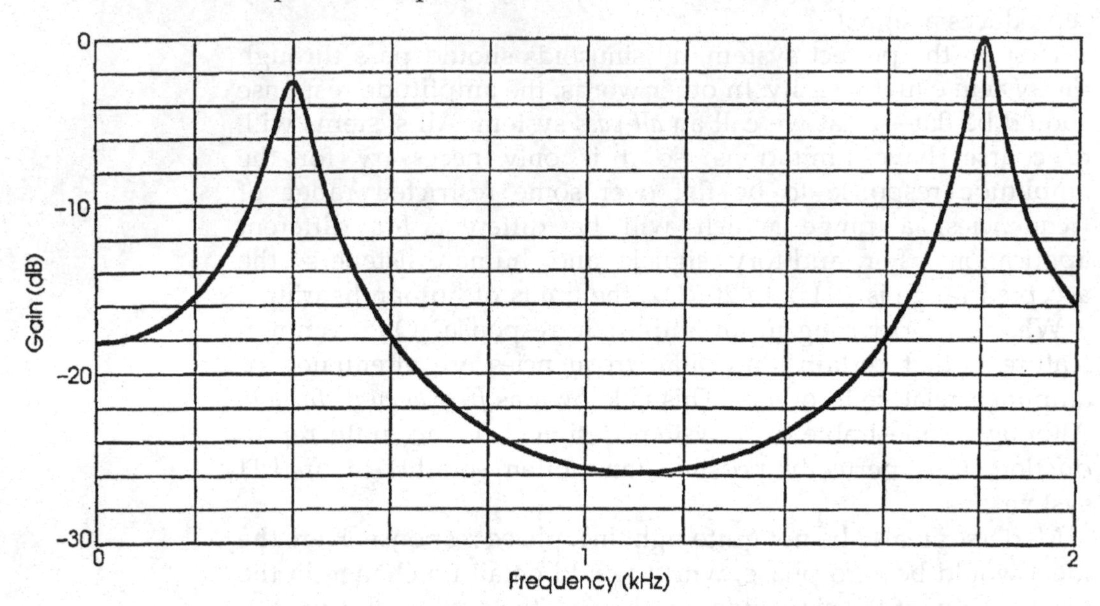

2. *Clipping* is a distortion that is commonly found in amplifiers when the level of the input signal is too large, and results in the limiting of output voltage to a maximum value. Here, for example, is what clipping might do to an input sinusoid:

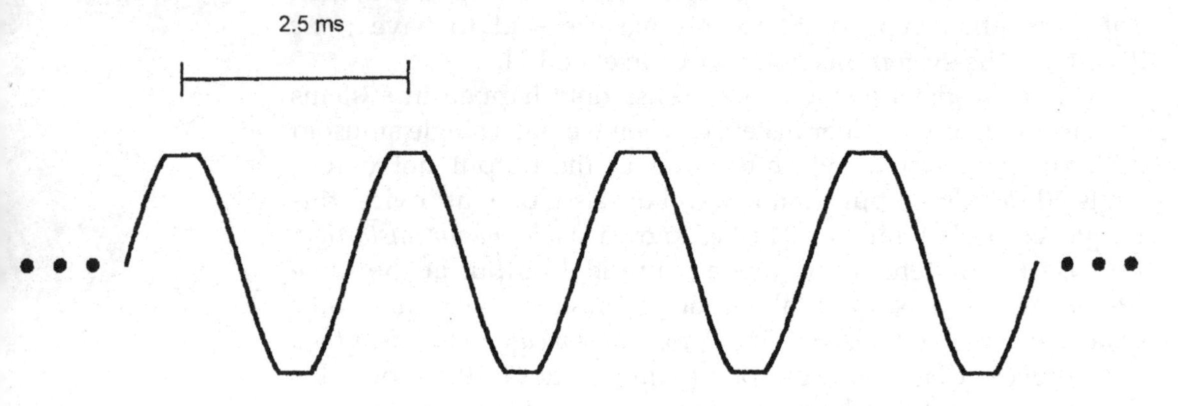

Give the reasons that such a system cannot be considered LTI. (Hint: Compare the forms of the input and output spectra.) What kind of distortion is this?

3. Below is a table giving the input levels of sinusoids (in volts) at a number of frequencies (some repeated), the gain of an LTI system at that frequency (in dB), and the level of the output sinusoid (again in volts). Fill in the missing entries (you may find the short-cuts for converting between simple

ratios and their corresponding dB values at the end of Chapter 3 useful):

Frequency (Hz)	Volts input (V)	Gain (dB)	Volts output (V)
100	2	0	2
200	2	0	2
200	1	0	?
200	?	0	4
400	2	−6	1
400	?	?	2
800	2	20	?
800	?	?	4
900	?	−10	1

4. Draw the amplitude response of a low-pass filter that has a gain of 0 dB from 0 to 450 Hz, and then rolls off at a slope of 12 dB per octave. Suppose you observe an output signal that is a sawtooth waveform of fundamental frequency 100 Hz and whose fundamental component is at a level of 3 V. Draw the amplitude spectrum of the input signal. Use a frequency range of 0–1800 Hz for all your graphs.

5. Suppose the low-pass filter above has a phase response that is 0° from 0 Hz to 450 Hz, and then becomes −90° for higher frequencies. Draw its phase response, and the phase spectrum of the input signal corresponding to the sawtooth output above.

6. A signal consists of the first eight harmonics of a 100-Hz fundamental which rises at 14 dB/octave. This is put through a system, and the amplitude spectrum of the output contains these same harmonics but now they all appear at the same amplitude. Sketch the amplitude response of the system that they were passed through.

7. A square wave with a fundamental frequency of 125 Hz is put through a high-pass filter whose cutoff frequency is 400 Hz and whose roll-off is 6 dB/octave from this point. This is then put through a low-pass filter whose cutoff frequency is 1000 Hz and which has a roll-off of 10 dB/octave. Sketch the output spectrum.

8. Give an example of a situation where it would be undesirable to use a system with a saturating nonlinearity, with reasons justifying your choice.

CHAPTER 9

The Time Characterization of Systems

In Chapter 8, when describing how to predict the output of an LTI system given its frequency response and some input, the result was first expressed as a *spectrum*—in other words, in terms of the frequency content of the output signal. When we wanted to know what the *time waveform* of the output signal would look like (say, on an oscilloscope), the appropriate sinusoidal components were added up (with frequencies, phases and amplitudes given by the output spectrum). We've called this process, of going from a *frequency* representation of a signal (a spectrum) to a *time* representation (a waveform), Fourier *synthesis*. This is in contrast to the inverse process of Fourier *analysis*, where we start with some signal in time, and work out what sinusoids would have to be added up to get the desired waveform. These two ways of describing signals, one based in time (as a waveform with axes of amplitude by time) and one in frequency (as a spectrum with axes of amplitude or phase by frequency), are just two ways of describing the same thing. They are completely equivalent since it's always possible to transform between one description and the other. It's just that one or the other way of describing the signal can be more convenient in particular situations.

When it comes to describing LTI systems, however, we've so far only talked about a frequency domain approach. Here, systems are described by the effect they have on sinusoids—given by their transfer function or frequency response. You can tell that a transfer function is a frequency domain representation by looking at its axes—some measure of amplitude (ratio of output to input) by frequency for the amplitude response, and phase by frequency for the phase response. But just as signals can be described equally well in the time or frequency domain, so too can systems. Just as with signals, sometimes the time domain description of systems is preferable to the frequency-based one. This means, therefore, that the time domain description must contain, perhaps in some disguised form, the amplitude and phase response of the system. Later on, we'll see how to strip away this disguise.

The basic idea behind the time characterization of an LTI system is very similar to that underlying its frequency characterization. An LTI system is completely characterized by its response to sinusoids at all frequencies because any signal can be expressed as a sum of the appropriate sinusoids. The time domain approach rests on the fact that any signal can also be expressed as a sum of very narrow rectangular pulses of the appropriate height and position in time, each of which is known as an *impulse*. Given the response of the system to just one of these impulses—known as the *impulse response*—we can then use the linearity and time invariance of the system to calculate its response to *any* input. The impulse response, thus, holds the same privileged position in the time characterization of systems as the transfer function holds in their frequency characterization.

This may all seem a bit abstract. Let's take a concrete example in order to illustrate the processes of building up a signal out of narrow pulses, and using the system response to only one of these pulses to predict the output for *any* input. Although we'll use a particular system and input signal for the sake of argument, the ideas can be applied to any LTI system and any signal.

What can we learn from a system response to a single pulse?

Let's start by supposing we know the output of a particular electrical system (call it System Z) to a rectangular pulse with a peak amplitude of 3 millivolts (mV) (1 mV is one-thousandth of a volt) and with a duration of 1/3 ms. This is determined either by someone telling us or experimentally:

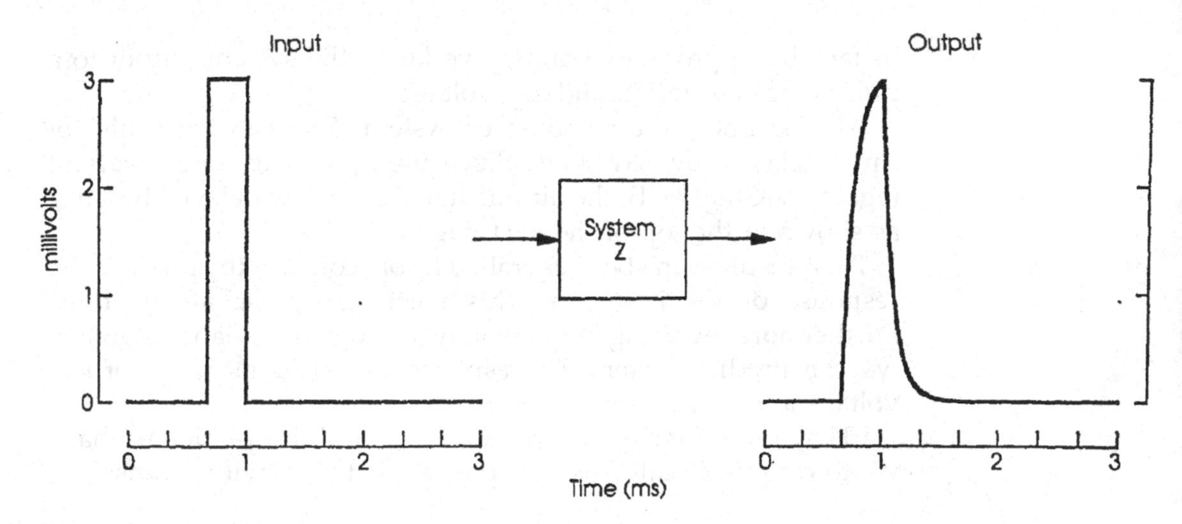

You can see that the response rises quickly and starts to level off a little before dropping suddenly, and then more gradually back to zero.

This doesn't seem to get us very far. As has been stressed before, we need an efficient way of characterizing the system response so that we can predict its output for *any* input signal. We don't want to have to make a catalog of every signal we might put into the system along with its corresponding output. But the situation is not as bad as it looks. Because the system is LTI, we can predict its response to other input signals from the response to this single 1/3-ms pulse.

We know, for instance, what System Z will give out for a pulse of the same duration but with an amplitude of 1 mV Because the system is linear (and by definition homogeneous—see Chapter 4) and we've multiplied the input signal by a constant factor of 1/3, all we need to do is multiply the output signal by the same factor of 1/3:

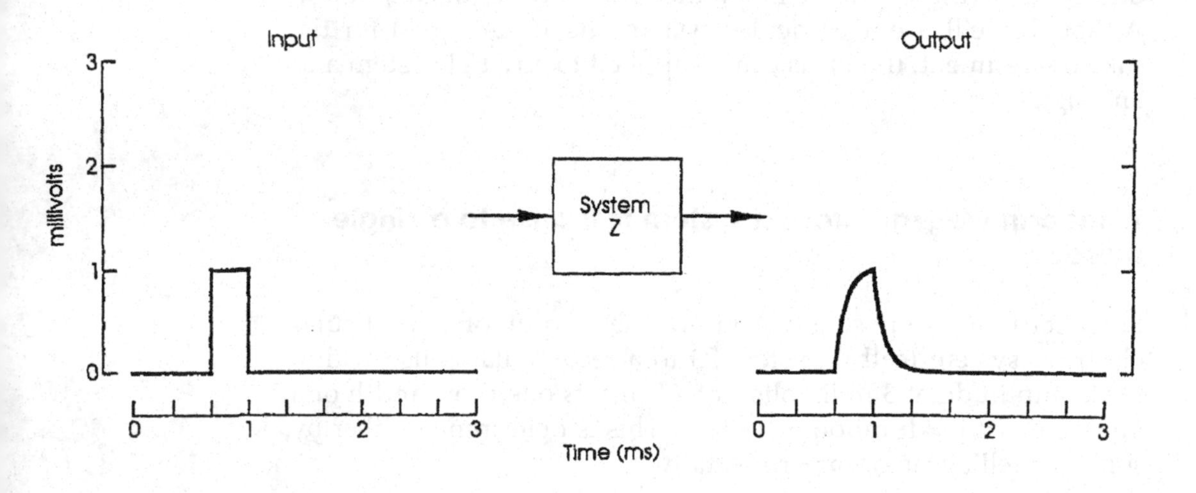

In fact, by appropriate scaling, we know the system output to a pulse of this duration and *any* voltage.

We also know the response of System Z to a 1/3 ms pulse of 3 mV delayed by, say, 1 ms. Since the system is time invariant (again see Chapter 4), the output signal is simply delayed by 1 ms, as shown at the top of the next page.

This result can be generalized, of course, to predict the response of the system to this particular pulse at *any* time. Furthermore, by using homogeneity and time invariance together, we can predict System Z's response to a 1/3-ms pulse of *any* voltage at *any* time.

LTI systems have one more property—additivity—that we have yet to exploit. Recall from Chapter 4 what additivity means:

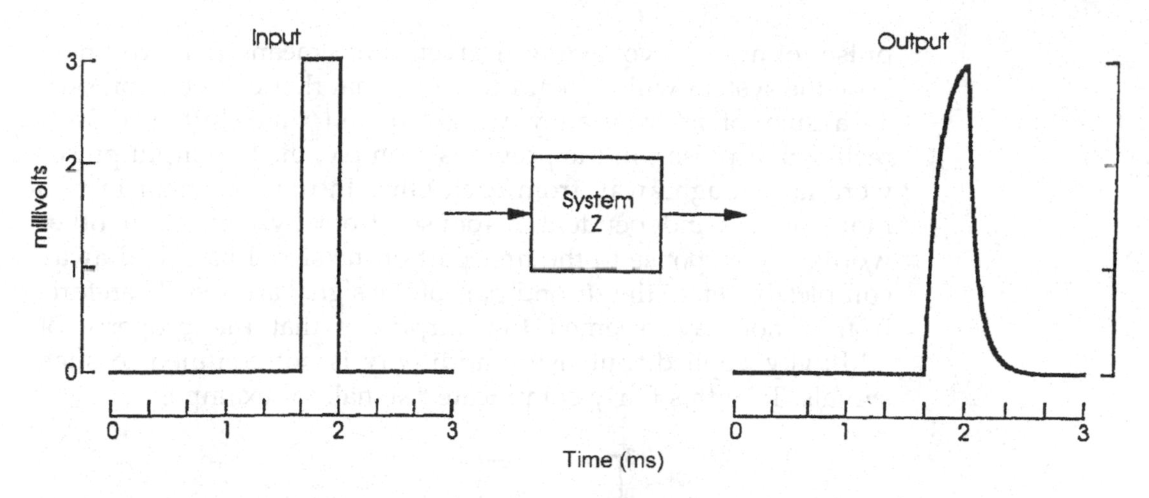

if we know the system response to two signals on their own, we know that the response to their sum is simply the sum of the outputs to each individual input. So we can predict System Z's output to the sum of the two previous inputs:

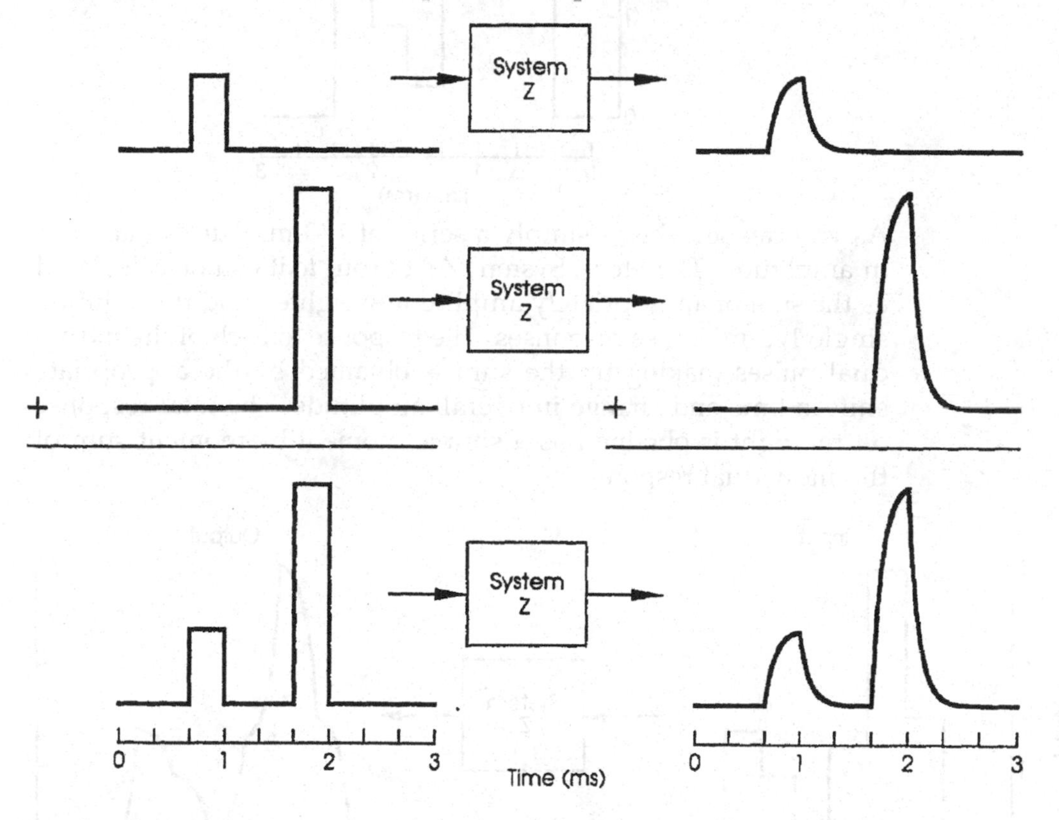

Again, this property can be generalized to the sum of any number of input signals. Therefore, we can predict the system response to a sum of 1/3-ms pulses, each of which can be of any voltage and presented at any time. In short, because the system is LTI, knowing the response of the system to one particular 1/3-ms pulse (of a given voltage at a given time) means that we know how the system will respond to any signal that can be expressed as a sum of appropriately weighted and time-shifted 1/3-ms rectangular pulses. In the previous example, the two input pulses were far enough away from each other in time to result in two more or less independent responses from System Z. In other words, the response to the first part of the signal had died away completely before the second part of the signal arrived. Therefore, it may not have seemed too surprising that the property of additivity applied. But using additivity is not confined to such signals. Take this fairly complicated signal, for example:

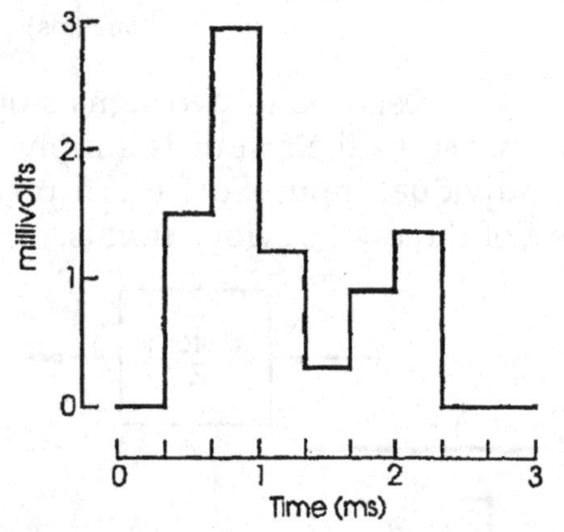

As you can see, this is simply a series of 1/3-ms pulses that vary in amplitude. Therefore, System Z's output *to* it can be calculated as the sum of appropriately amplitude-weighted and time-shifted single 1/3-ms-pulse responses. The response to each of the individual pulses making up the sum is obtained by the appropriate shift in time and change in overall amplitude. The total response on the right is obtained as a simple moment-by-moment sum of the individual responses:

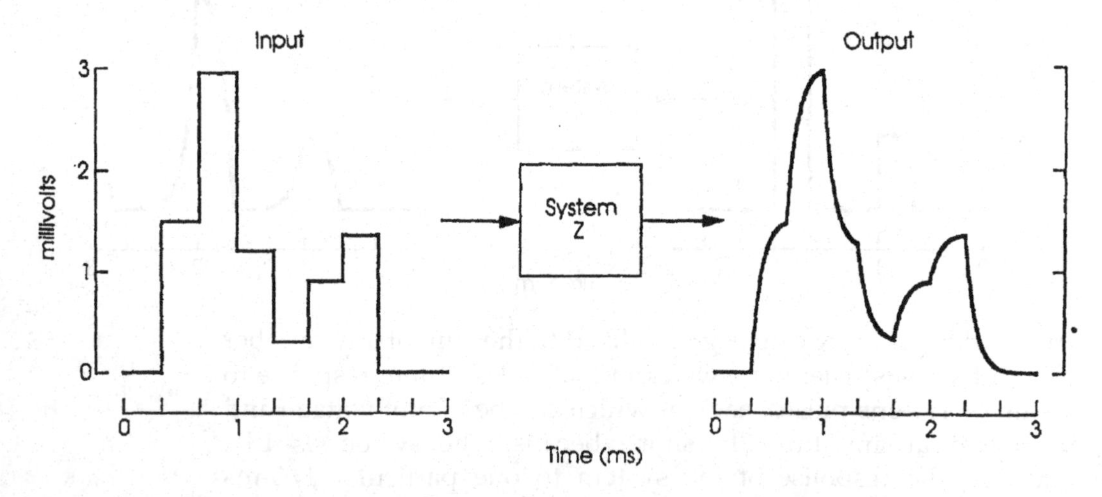

Approximating signals with rectangular pulses

Unfortunately, not every signal (nor even most of them) can be perfectly expressed by a sum of 1/3-ms pulses. Take this 1-ms triangular waveform, for instance:

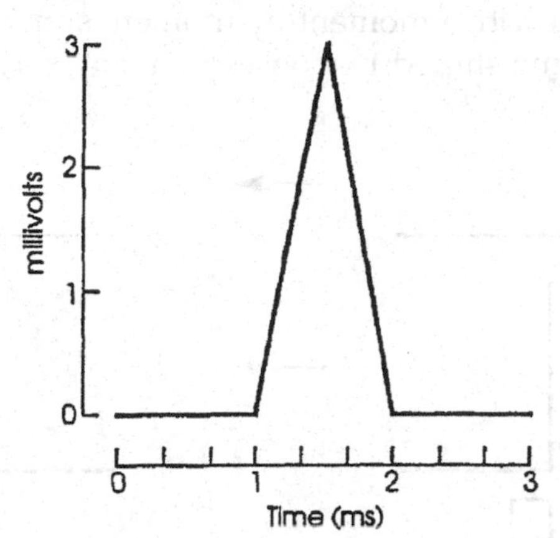

The triangle cannot be created by adding up non-overlapping 1/3-ms pulses in the way demonstrated above. This is primarily due to the fact that the triangle slopes upward and downward, and we can't add up flat-topped pulses in such a way as to mimic this.

How then can we use what we know about System Z's responses to 1/3-ms pulses to tell us what its output will be to this signal? One possibility is to *approximate* the triangular signal with a sum of three time-shifted 1/3-ms rectangular pulses of the appropriate amplitude:

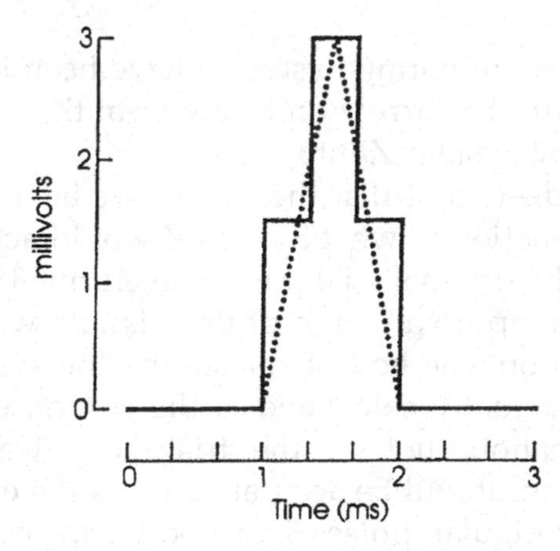

The sum of rectangular pulses (solid line) isn't a very good representation of the triangle (dotted line), but it's certainly closer to it than a single 1/3-ms pulse on its own.

The advantage of approximating the triangle in this way, rough as that may be, is that the approximation need not be put through System Z to determine what output will result. Since the approximation (unlike the original triangle) is a weighted sum of time-shifted 1/3-ms rectangular pulses, the output of the system can be calculated with a moment-by-moment sum of appropriately weighted and time-shifted responses to a single 1/3-ms pulse:

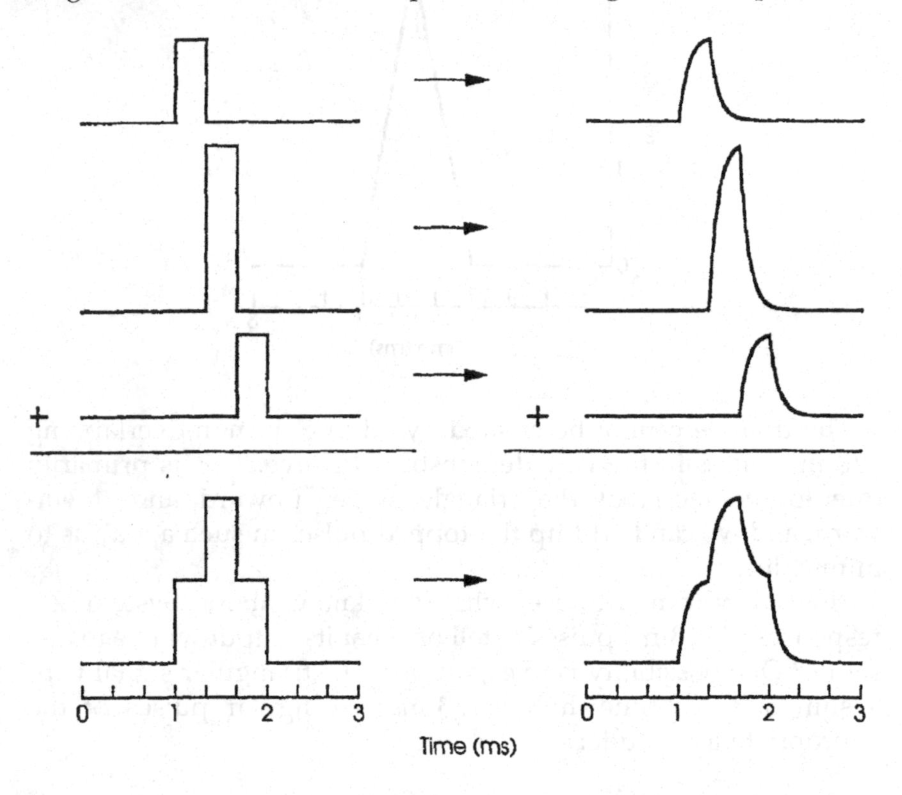

Here, the boxes representing System Z have been left out to save space in the figure. The arrow on its own can thus be read as '... is transformed by System Z into ...'.

So far, so good—except that the output we have determined is only an approximation to what System Z would actually give out in response to the original triangular waveform. The error arises from the original approximation of the triangle with rectangular pulses, not from our method of calculating the system response. What we have actually calculated is the response to the three-pulse approximation, not to the triangular waveform itself. Therefore, the output will be accurate only to the extent to which our sum of rectangular pulses is a good approximation to the original triangular waveform.

This immediately suggests a way to improve our method. If narrower pulses were used in the sum, we could better approximate the triangle, and this would lead to a more accurately predicted system output.

This is illustrated in the next figure. Along the top line is a summary of the steps we have already described in calculating the system response using 1/3-ms pulses (in the first column) as our basic waveform. The response of the system to this pulse is depicted in the second column, followed by the approximation to the triangular waveform obtained from a sum of appropriately amplitude-weighted and time-shifted pulses. The fourth column shows the sum of the responses to each of these pulses separately, which is, of course, the predicted output of the system to the approximated triangle. The three dots arranged in a triangle in the middle of the figure are a special symbol for the word 'so'. Therefore, the line can be read 'a 1/3-ms pulse is found to give a particular output, so this sum of such pulses can be predicted to give the following output'. The same series of operations is depicted for 1/7-ms pulses at the bottom:

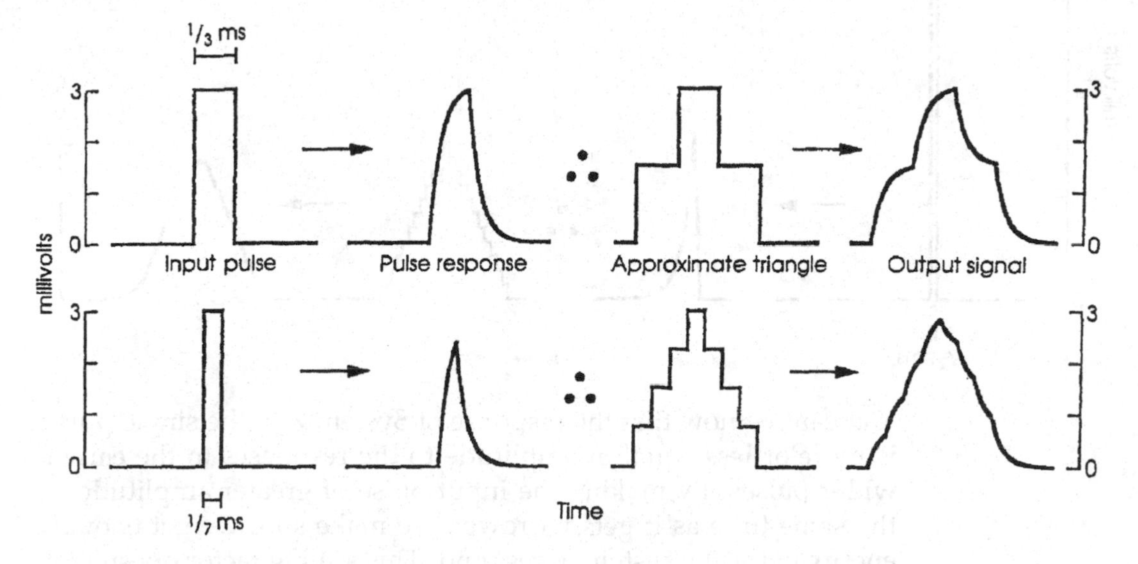

Not only does our sum of 1/7-ms pulses look more like a triangle, but the calculated output also looks a bit smoother and more like the kind of output that we'd expect from a simple triangular input.

Why stop there? We can get even better approximations with even narrower pulses, say 1/21 ms:

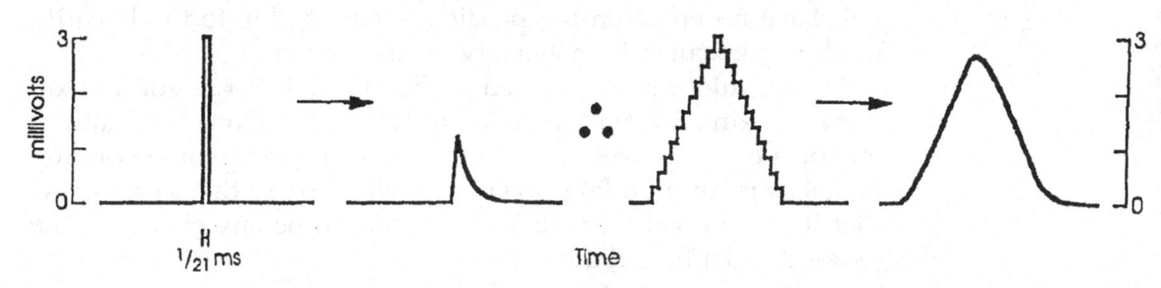

Again, the triangular waveform is better approximated and the output correspondingly smoother.

You may have noticed that the response to a single pulse on its own has been getting smaller as the pulses have become narrower. It's easy to see why this is so. If a pulse is made narrower while its amplitude is kept the same (as was done here), its energy will decrease. Therefore, the response of the system will decrease. At some point, if the input gets too narrow, the system will barely respond to the input signal. Therefore, what we normally do is make the pulse greater in amplitude as we narrow it. Here, for example, is the 1/21-ms pulse at more than twice its previous amplitude, and its corresponding output:

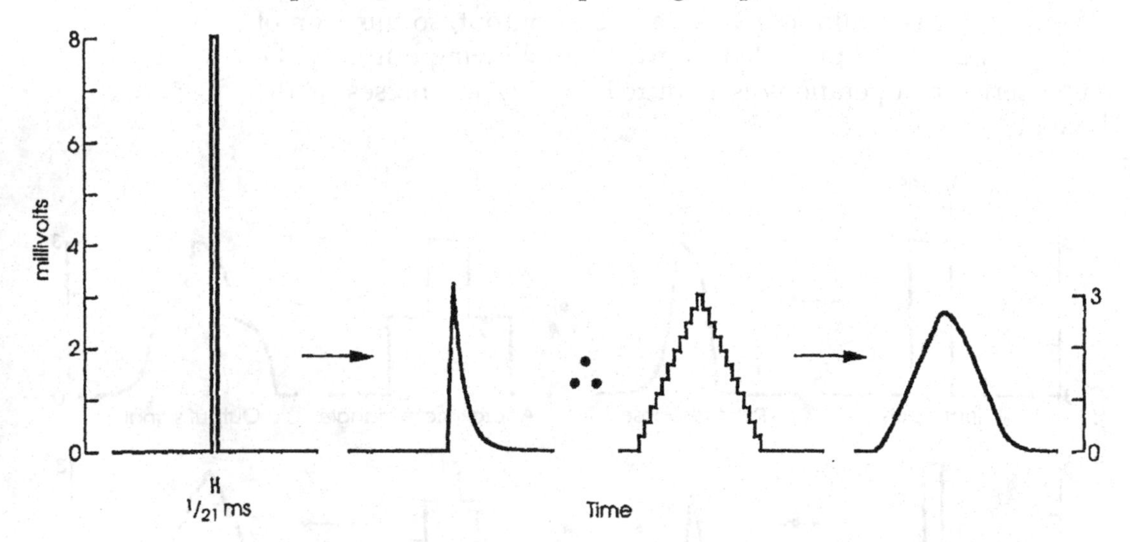

You can see now that the response of System Z to the single pulse is more or less equal in amplitude to the responses to the earlier wider pulses. By making the input pulse of greater amplitude at the same time as it gets narrower, we make sure we get enough energy in for the system to respond. This scaling factor doesn't, of course, change the shape of the output of a linear system, only its overall amplitude. After all, if we wanted to know the response of the system to a 1/21-ms pulse of 1 mV, we could use pulses of *any* voltage and divide the output by the input voltage. In short, although the pulse used to determine the system pulse response gets taller and taller as the pulse gets narrower and narrower, this will have no effect on the predicted output, due to the linearity (and, in particular, homogeneity) of the system.

As you might have guessed earlier, there is no reason to stop our approximation with pulses that are 1/21-ms long. The natural outcome of the process of further narrowing (and corresponding increases in amplitude) is to end up with a pulse that's so narrow that it has no width at all. Yet, for there to be any energy in the pulse, it must be infinitely tall!

We can't really vary the amplitude of an infinitely high pulse, so instead we talk about varying its area, or energy—two measures that are related to one another, but not directly. Common-sense notions about area are hard to reconcile with the idea that an infinitely high, infinitesimally narrow pulse can vary in area, but it is possible mathematically. As we'll soon see, we never have to worry about this eventuality in real life, where common-sense notions *can* be accurate.

With such an infinitesimally narrow pulse, we can represent the triangle perfectly and thus predict the system output exactly. To show that the input pulse is considered to have infinite amplitude, we symbolize it as an arrow with its top pointing skyward:

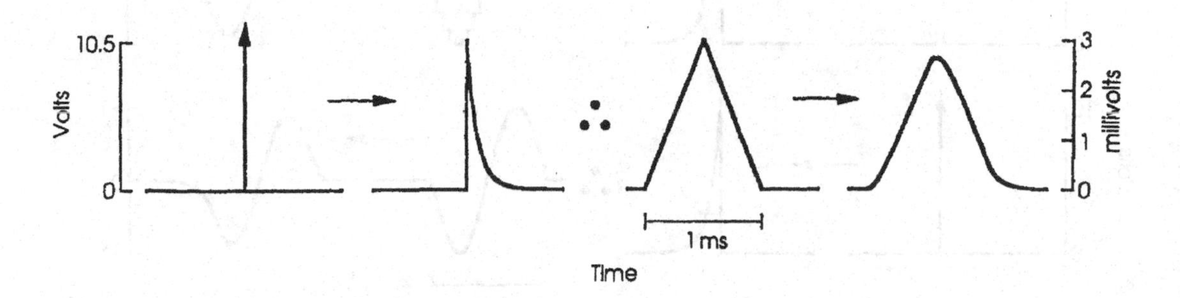

Now you know where we were going to end up, you can see why we had to make the pulse taller as it got narrower. Difficult as it is to imagine an infinitely tall and infinitesimally narrow pulse that has a finite area (and thus energy), it's easy to see that an infinitesimally narrow pulse that has any finite (that is, measurable) amplitude, no matter how big, could not contain any area (or energy) at all.

This infinitesimally narrow, infinitely tall pulse of finite energy is known as an *impulse.* True impulses are, of course, a mathematical idealization and so don't occur in the real world. Usually, though, for a given system, we can make real pulses narrow enough so that for all intents and purposes they serve as impulses. So in a real-life problem, we can always vary the amplitude of the pulse. (More detail on this issue can be found in Chapter 10).

Although we have only explored this process for one particular input signal (a triangle), it should be readily apparent that *any* signal can be perfectly expressed as a sum of appropriately weighted and time-shifted impulses. On the following page, for instance, is another input to System Z (one cycle of a 1-kHz sinusoid) being successively better approximated by narrower and narrower rectangular pulses (of greater and greater amplitude) in the same way as for the triangle.

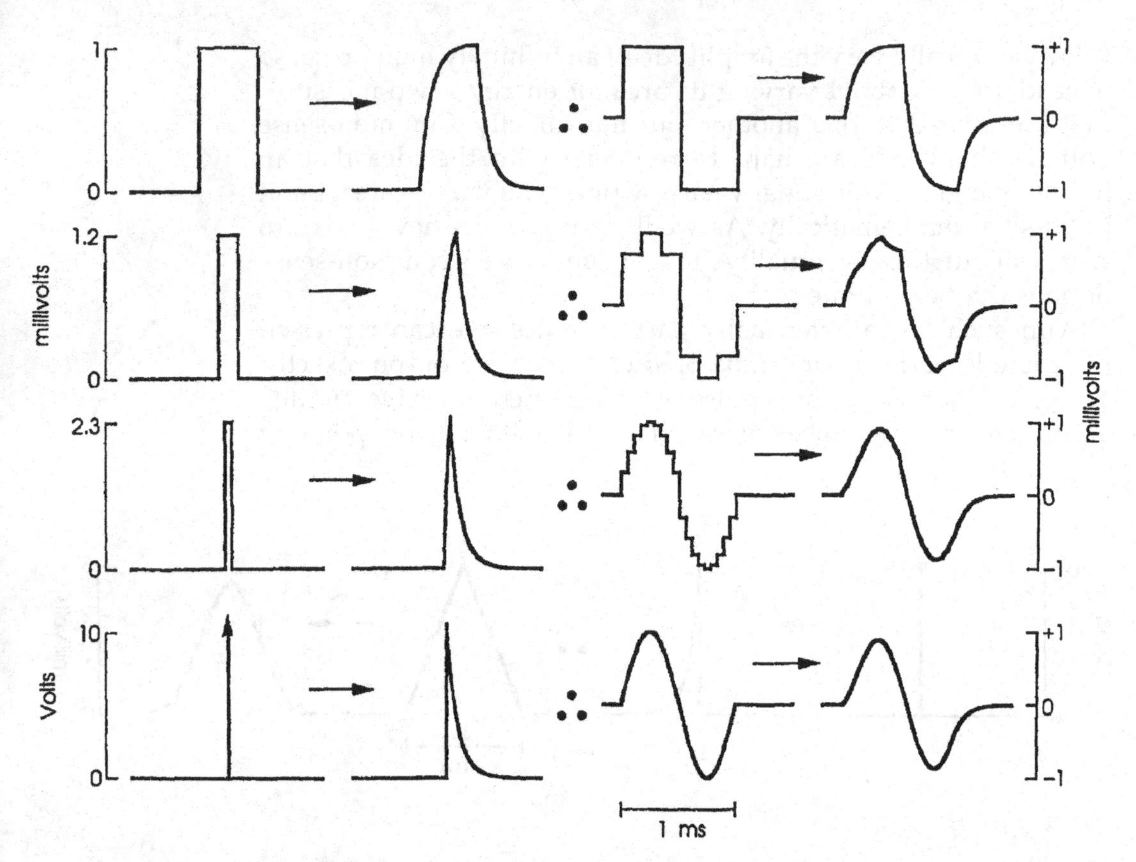

Note that, in this figure, the y-axis scales are different for each of the single-pulse inputs (first column), but are the same for each of the corresponding single-pulse outputs (second column).

Since any signal can be expressed as a sum of the appropriate impulses, knowing the response of an LTI system to a single impulse means that we can predict its output for *any* input. Therefore, LTI systems are completely characterized by their *impulse response.* Of course, we can't use exactly the technique we have illustrated above to calculate the output of a system to an arbitrary input. Since impulses are infinitesimally narrow, we'd need an infinite number of them to express any real-life input signal. There would therefore be an infinite number of separate responses to add together. Even with modern-day computers, an infinite number of additions would take a little bit too long to work out! The way around this is through a mathematical technique known as *convolution*. Convolution requires the use of calculus, however, so we'll not go into the mechanics of it. For our purposes, it is enough to know that, given the impulse response of a system, and an arbitrary input signal, one 'convolves' the input and the impulse response to determine the system output. In essence, it is just like the amplitude weighting, time shifting and addition of separate pulse responses illustrated above, but done an infinite number of times with 'genuine' impulses.

The relationship between the impulse response and the frequency response

We now have two ways of characterizing LTI systems. We can work in the frequency domain if given the frequency response or transfer function of the system (multiplying amplitude spectra and adding phase spectra), or we can work in the time domain if given the impulse response of the system (convolving it with the input signal). Therefore, the frequency response and impulse response must be giving the same information about the system, and can only be two different ways of looking at the same thing. This relationship is made clearer if we now take a frequency domain approach to the determination of the impulse response of a system.

Suppose we are given the frequency response of a system and want to know what its output will be to an impulse. Recall from Chapter 8 that to obtain the amplitude spectrum of the output, $\text{Output}_{\text{amp}}(f)$, we multiply the amplitude spectrum of the input signal, $\text{Input}_{\text{amp}}(f)$ by the amplitude response of the system, $A(f)$. In a simple formula:

$$\text{Output}_{\text{amp}}(f) = \text{Input}_{\text{amp}}(f) \times A(f) \tag{1}$$

To obtain the phase spectrum of the output, $\text{Output}_{\text{phase}}(f)$, we add the phase spectrum of the input, $\text{Input}_{\text{phase}}(f)$, to the phase response of the system, $P(f)$. That is:

$$\text{Output}_{\text{phase}}(f) = \text{Input}_{\text{phase}}(f) + P(f) \tag{2}$$

In order to calculate the output spectrum, then, we need to know the amplitude and phase spectra of an impulse. These turn out to be extremely simple. The phase spectrum is always zero while the amplitude spectrum has a constant value for all frequencies (depending on the energy of the particular impulse). In short, since the input signal we are considering is an impulse:

$$\text{Input}_{\text{amp}}(f) = k \tag{3}$$

where k is some constant which depends upon the energy of the impulse, and:

$$\text{Input}_{\text{phase}}(f) = 0 \tag{4}$$

Note that an impulse has the same amount of energy at all frequencies, just as white noise does. The only difference between white noise and an impulse is in their phase spectra. An impulse has all its component sinusoids in cosine phase (thus adding up to an infinitely high pulse where the peaks of all the cosine waves coincide) while in white noise they are added together in random phase.

To determine the spectrum of the system output to an impulse we simply substitute from equations (3) and (4) into (1) and (2):

$$\text{Output}_{\text{amp}}(f) = \text{Input}_{\text{amp}}(f) \times k \tag{5}$$

$$\text{Output}_{\text{phase}}(f) = \text{Input}_{\text{phase}}(f) \tag{6}$$

Let's put these equations into words. When we multiply the amplitude response of the system by the amplitude spectrum of an impulse (which is flat), we get back the amplitude response of the system, multiplied by a constant factor k (which won't alter the properties of the spectrum). Therefore, the amplitude spectrum of the system impulse response is simply the amplitude response of the system.

If we add the phase response of the system to the phase spectrum of an impulse (always zero), we get back the phase response of the system. Therefore, the phase spectrum of the impulse response is simply the phase response of the system.

Here, for example, are the amplitude and phase responses of System Z; they are simply the amplitude and phase spectra of its impulse response. How would you describe System Z?

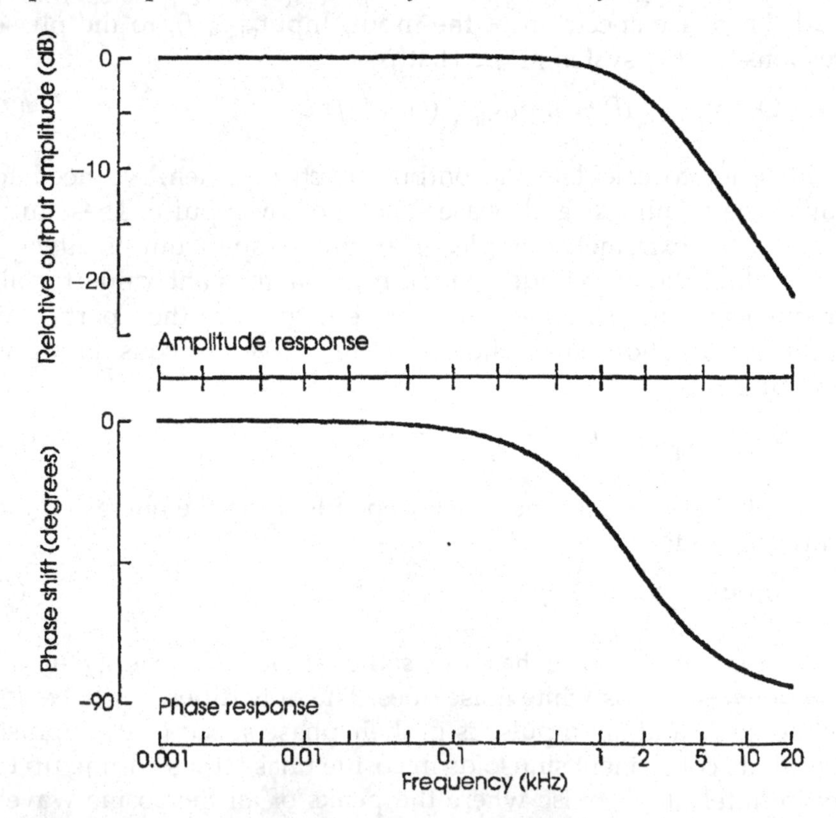

To summarize, the spectrum of an LTI system's output to an impulse is the system transfer function, or frequency response. In diagrammatic form:

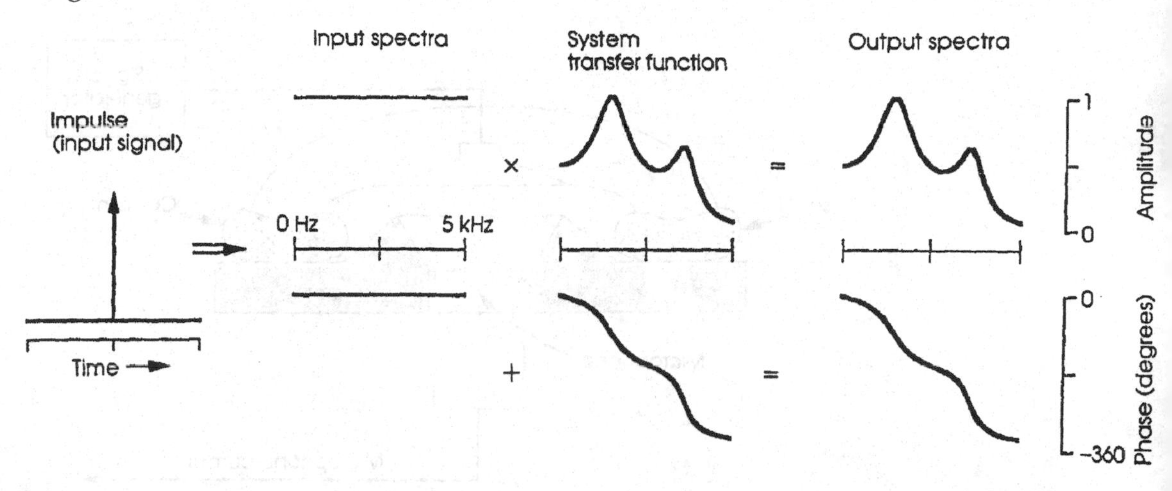

Determining the frequency response of a system: a practical example

You now know three distinct ways to measure the frequency response of a system. Firstly, you can present sinusoids at a number of discrete frequencies to the system, and calculate the ratio of the output amplitude to the input amplitude at each frequency (or the differences in phase). Equivalently, a sinusoid can be swept in frequency over the desired range. Secondly and thirdly, white noise or an impulse can be put through the system and its spectrum calculated. An important limitation of using white noise with this method (and as described towards the end of Chapter 8) is its inability to extract information about the phase response. Because the phase spectrum of the input white noise is random, so will be the phase spectrum of the output signal. The easiest way around this is simply to calculate the phase spectrum of the *input* white noise as well, and thus to calculate, for each frequency, the phase shift undergone.

It may be surprising that these quite different techniques will lead to the same results, so the three were put to a practical test. The system whose frequency response was investigated was a headphone. That is, we determined the relationship between the voltage delivered to the headphone and the sound emitted by it, as a function of frequency. Although there are many complicated issues involved in the choice of exact measuring apparatus, we used a standardized coupler that is meant to mimic some properties (but far from all) of the human ear. At least this coupler is more like a real ear than the typical 6cc coupler that is standard for calibrating audiometric headphones. Here's a sketch of our set-up, with the headphone and coupler in cross-section:

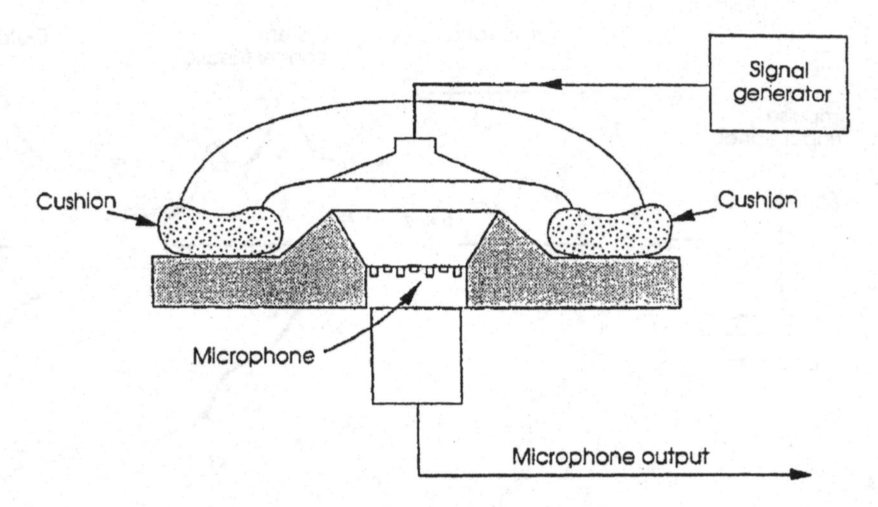

One earphone of the pair of headphones sat on the coupler, which contained a high-quality microphone at the bottom of a small cavity. The amplitude and phase response of this microphone (in transforming acoustic energy into electrical voltages) was flat enough over the frequency range investigated (20 Hz to 2 kHz) for no correction to be needed for it. There were also means to present electrical signals to the headphone, and for measuring both input and output signals.

All three possible input signals were presented to the headphones: (1) a swept sinusoid of constant amplitude; (2) a noise that was sufficiently flat over the frequency range of interest to be considered white and (3) a pulse that was narrow enough in this situation (at about 60 μs) to be considered an impulse. At the top, opposite, are the three results, after appropriate processing. Remember that the frequency response determined with an impulse was found by calculating the spectrum of the impulse response at the bottom right of the figure.

Note how closely the three sets of frequency responses agree, except at low frequencies. These small discrepancies arose as a result of low-frequency noise present in the room where the tests were done. The swept sinusoid, because it put all its energy into a single frequency at any one time (about 95 dB SPL), was able to dominate this noise. The white noise input had its energy spread over a wide range, so that the low-frequency room noise corrupted the measurement in places. Matters in this regard were worse still for the narrow pulse which had its energy spread over a wide frequency range, and over a very short stretch of time. This meant that its amplitude had to be severely limited, so as not to

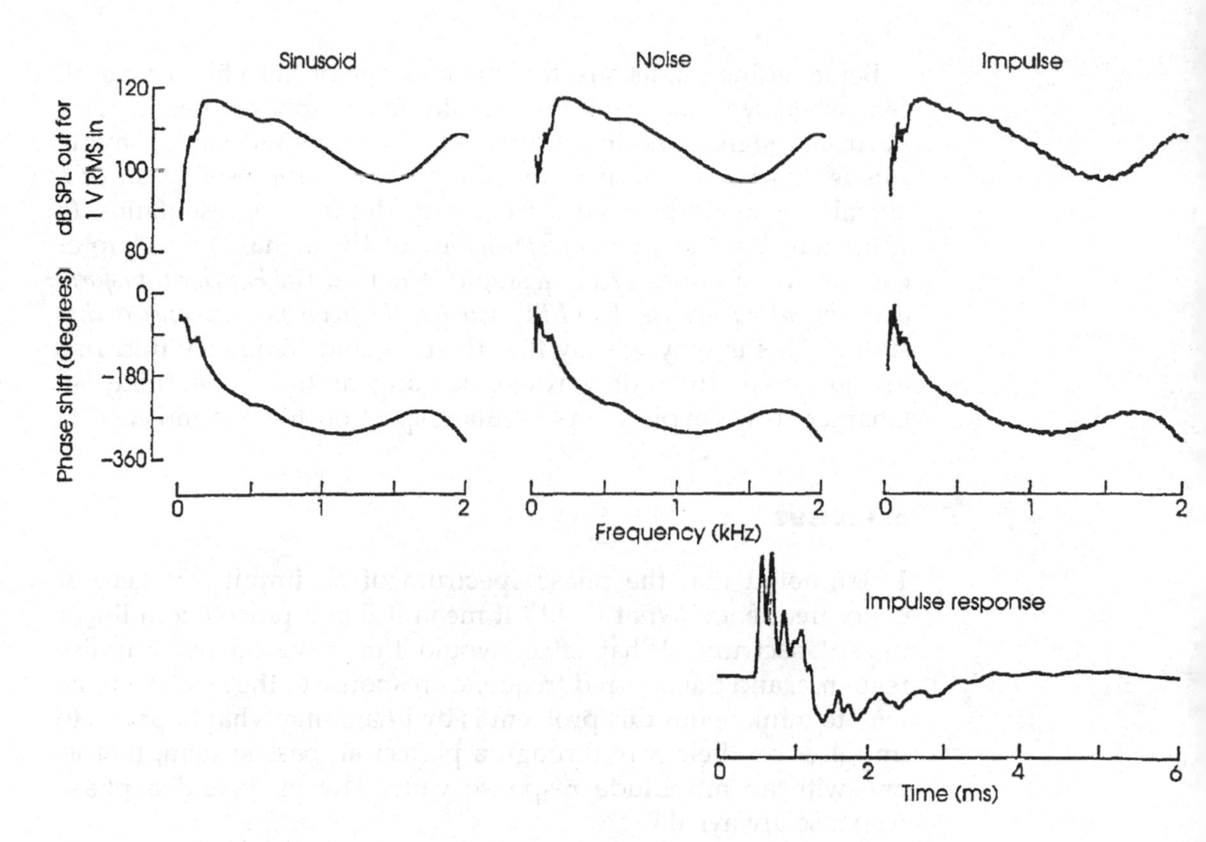

damage the headphones, leading to relatively little energy in each frequency region. Therefore, even low-level room noise corrupted the measurement somewhat, as can be seen by the general slight irregularity of the trace.

But you don't get something for nothing. In these measurements, there was a large trade-off in time versus accuracy. The measurement with a swept sinusoid took over two minutes, and that with white noise about 22 s, while the collection of the impulse response took only 80 ms. There are relatively straightforward averaging techniques, however, that could increase the accuracy of the impulse response technique, at the cost of longer measuring times.

One final aspect of the duality of the time and frequency domain representations is suggested here. We could take the amplitude/phase responses of the system (either from the swept sinusoid or white noise), pretend they were spectra and calculate the time waveform they correspond to. What would that be? We already know that the spectrum of the impulse response gives the frequency response—therefore, the time waveform corresponding to the frequency response must be the impulse response. You can thus see that all three methods give precisely the same information—it's just that the information is displayed in different ways.

Before going on, let's restate the message of this chapter. Recall that when we determine the amplitude and phase spectra for a particular signal, we do a *Fourier analysis*. Another way of saying this is that we can transform a time domain representation of a signal (a waveform) into a frequency domain representation (a spectrum) by a *Fourier transformation* of the signal. This chapter can now be summarized in a single sentence: *The Fourier transform of the impulse response of an LT1 system is the frequency response of that system.* This is why we say that the time and frequency domains are just two different ways of looking at the same thing. In Chapter 10 we explore a particular aspect of this relationship.

Exercises

1. We noted that the phase spectrum of an impulse is zero at every frequency. What would it mean if the impulse had a linear phase spectrum? What effect would this have on the impulse response, and the inferred frequency response of the system? (One way to think about this problem is by imagining what happens to impulses on their way through a perfect all-pass system, that is, one with an amplitude response value always 1, and a phase response always 0.)

2. Suppose that a system has an impulse response that is another impulse. What is its transfer function (amplitude and phase)?

3. Make up a good question for this chapter and send it to the authors. We can't think of any more! A free copy of the book will go to entries we deem good enough for inclusion in the next revision of this book.

CHAPTER 10

The Relationship Between the Time and Frequency Domains

We can now describe both systems and signals in either the time or the frequency domains. You may be relieved to know that these are the only types of representation used—there are no further 'domains' for you to conquer! Therefore, the four main characterizations that we have discussed can be conveniently summarized in a 2 by 2 table (along with the number of the chapter in which they first appeared):

	Type of representation	
	Time	Frequency
Signals	Waveform (Chapter 3)	Spectrum (Chapter 7)
LTI systems	Impulse response (Chapter 9)	Transfer function (Chapter 6)

This table is also meant to emphasize two important facts, both of which you have met before. First, not only can signals and systems be represented in time *and* frequency, but they are also described equally well in *either* domain. Whether to use one or the other depends on the particular task in hand.

Second, the relationship between the time and frequency domains is really the same for both signals and systems in that the process used to transform a time domain representation into the frequency domain is identical in the two situations. Just as we convert a waveform (time domain) into a spectrum (frequency domain) using the Fourier transform, in the same way we convert an impulse response (time domain) into a transfer function (frequency domain).

It should not be too surprising, then, given the intimate connection between the time and frequency representations, to discover that there is an interesting symmetry between the two domains. This relationship takes the form of a correspondence between the width of a frequency domain representation and the width of the associated time domain representation. Here we'll explore some of the implications of this fact. Since the relationship between time and frequency is essentially the same for signals and systems, the symmetry applies to both. Let's talk about signals first.

The spectra of rectangular pulses of varying duration

Briefly put, there is generally an inverse relationship between the widths of the time and frequency representations of a signal. Signals that are compact in time are spread out in frequency and vice versa. This relationship only holds between the amplitude spectrum of a signal and its time function, so we won't consider phase spectra any longer.

We'll illustrate this property with a signal that you have seen the spectrum of before—a single rectangular pulse. Let's begin with a rectangular voltage pulse that is 1 s long and 1 V high, and take a look at its spectrum (similar to the one that you saw in Chapter 7):

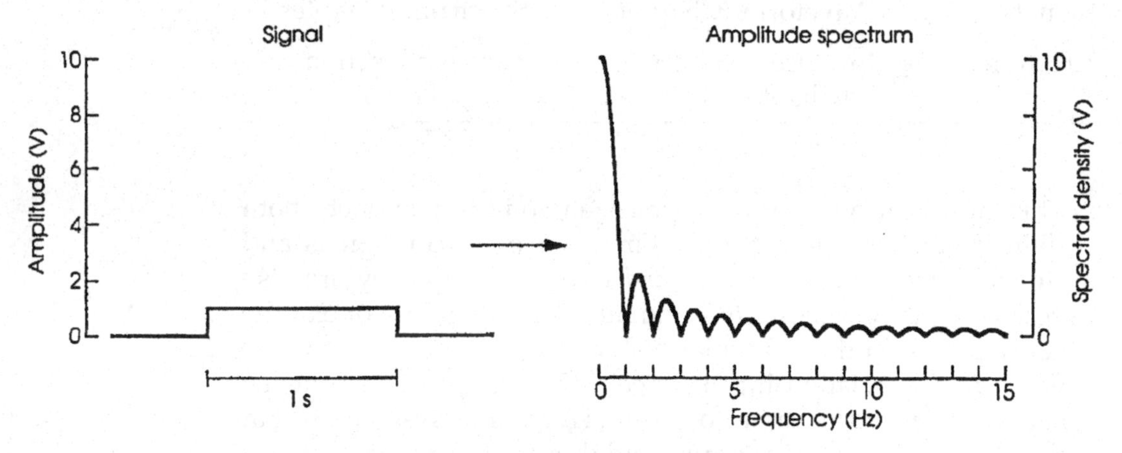

We first need to define a measure of width in each of the two domains in order to compare them. This is easily and unambiguously done for the time signal, as it only comes into existence for a short time. Nothing happens before or after the appearance of the pulse, so its width is just its duration, here 1 s.

Defining the width of the spectrum, on the other hand, takes a little more thought. Note its series of peaks which die away gradually as frequency increases. We cannot talk about the width of this spectrum in a conventional way because it never dies away completely, however high in frequency we go. Practically speaking, though, at some point the amount of energy gets so small that it won't matter if we ignore it. In other words, we'll use a measure of spectral width which only includes that frequency region where there is a lot of energy. For our current purposes, the most convenient way of doing this is to determine the frequency at which the amplitude spectrum first becomes zero. You can see that the frequency region between this point (here 1 Hz) and 0 Hz is the most important, as it has the most energy (in fact 90.2% of it). So we'll say that this spectrum has a width of 1 Hz.

What happens if we make the pulse shorter, say 1/2 s, and keep its area constant by giving it a peak voltage of 2 V?

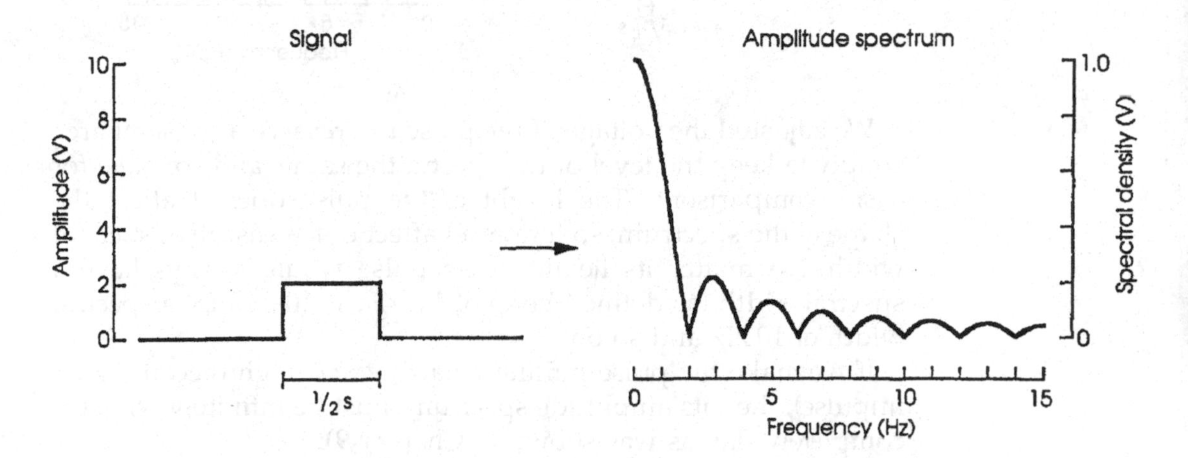

The pattern is roughly similar. There are still spectral peaks which die away gradually as frequency increases, but the peaks are wider. What's the width of this spectrum? Looking along the frequency axis to the point where the amplitude of the spectrum first goes to zero, we find a spectral width of 2 Hz. So far, then, halving the duration of the rectangular pulse has doubled the width of the associated spectrum.

We can continue this process for even narrower pulses, say 1/4 and 1/10 of a second, again keeping the area in the pulses constant by increasing the voltage appropriately, as seen overleaf. The width of the spectrum is now 4 Hz for the 1/4-s pulse and 10 Hz for the 1/10-s pulse. You can see that the spectrum for each pulse has a width that is just the reciprocal of its duration (remember, just another name for width). This is, in fact, a relationship that holds for *all* rectangular pulses. Therefore, as the pulse gets shorter and shorter, its amplitude spectrum gets wider and wider.

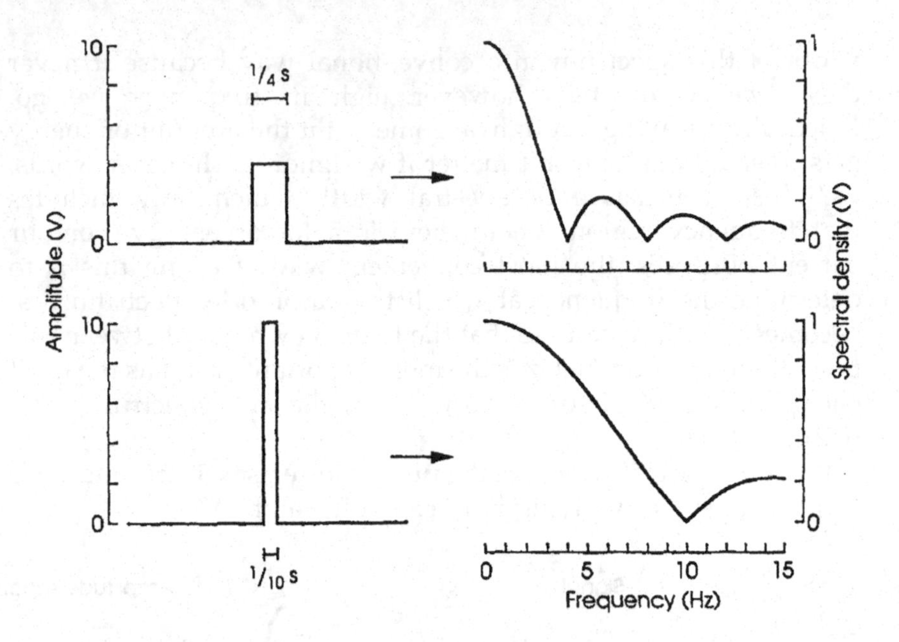

We adjusted the voltage of the pulse to preserve a constant area simply to keep the level of the spectra the same, and so make for easier comparisons. The height of the pulse doesn't affect the shape of the spectrum, so it doesn't affect our measure of spectral width. No matter its height, a 1-s pulse would always have a spectral width (as defined here) of 1 Hz, a 1/10-s pulse a spectral width of 10 Hz and so on.

If we make the pulse infinitely narrow and high (becoming an impulse), then its amplitude spectrum must be infinitely wide (or completely flat, as was shown in Chapter 9):

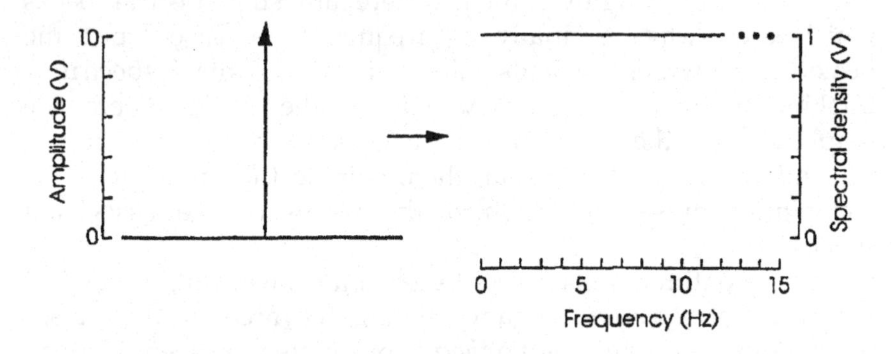

This brings us back to a practical matter. In Chapter 9, we showed that a system transfer function (both amplitude and phase responses) could be derived from the impulse response of the system. But we also pointed out that impulses are mathematical idealizations that don't exist in the real world and so aren't of much use in testing real-life systems.

In fact, we don't actually need genuine (infinitely narrow) impulses to test our systems. Approximations, based on rectangular pulses of a finite (although very small) width, will do just as well. This results from the fact that, in any real system, we would only be interested in its behaviour over a certain band of frequencies. There is always an upper frequency beyond which we can, for practical purposes, disregard the behaviour of the system, since the signals we put into it will not contain any significant amount of energy at those upper frequencies. This upper limit will, of course, be different for different applications. An audio recorder designed for recording bats, who emit high-frequency sounds, would have to handle much higher frequencies than recorders designed for recording orchestras.

All we need to do, then, when we measure the impulse response of a system, is to ensure that the pulses used are narrow enough to give a sufficiently flat spectrum out to the highest frequency necessary. Since we don't care what the system does above this limit, it doesn't matter what the spectrum of the pulse is up there. For many applications in speech and hearing, this will mean a pulse that is only a few microseconds wide. For instance, a 10-μs rectangular pulse has an amplitude spectrum that drops by only 0.6 dB from 0 Hz to 20 kHz, and so would probably be sufficiently narrow for testing any equipment designed for human listeners. Testing equipment designed to record the cries of bats would require much narrower pulses. Horseshoe bats, for example, emit cries with frequencies as high as 83 kHz and the spectrum of a 10-μs pulse changes by 14 dB over this frequency range.

The spectra of sinusoids of varying duration

We've just shown the inverse relationship between the width of spectra and the duration of signals for a particular waveform. As a rectangular pulse becomes narrower and narrower, its amplitude spectrum becomes wider and wider. In that example, the energy spreads to higher and higher frequencies, but this isn't always the case. We could have energy spreading to lower as well as upper frequencies at the same time—this is most easily demonstrated by examining what happens to the spectrum of a sinusoidal waveform when it is, say 100 ms or 4 ms long, instead of infinitely long in duration.

Let's start with a true (infinitely long) sinusoid at 1 kHz. This has a very simple spectrum—a single line at 1 kHz. There is really no width to be measured here at all, so we'll say that this spectrum is infinitely narrow. This fits in with our previous result. A signal that is infinitesimally narrow in time (an impulse) has a spectrum that is infinitely wide, while here, an infinitely long signal (a sinusoid) has an infinitesimally narrow spectrum.

When the duration of the sinusoid is shortened to 100 ms, the resulting spectrum is still fairly narrow. Nearly all its energy is near 1 kHz, as seen in the figure below. Note the stretched out frequency axis.

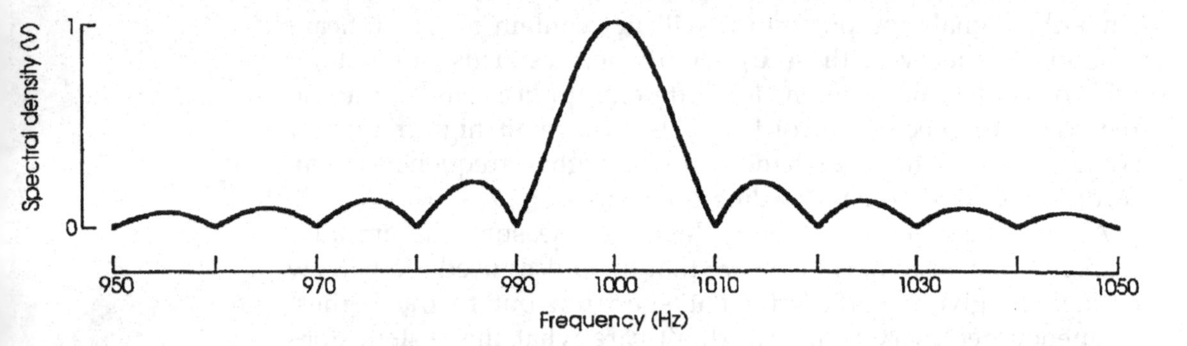

As before, the width of the time waveform is given unambiguously by its duration, but we still need to define the width of the spectrum, which extends downwards to 0 Hz and infinitely upwards. As before, a convenient measure is one based on where the spectrum becomes zero, but here we must consider two places where this happens—one above 1 kHz and one below. We will consider the width of this spectrum to be the distance (measured in Hz, of course) between these two points, which again includes most of the energy. For our 100-ms burst of 1-kHz sinusoid above, the two zero points are at 990 and 1010 Hz and so the width of the spectrum is 20 Hz.

We can now progressively shorten the duration of the 1-kHz sinusoid and see what happens to the spectrum. Illustrated on the next page are the spectra for 1-kHz sinusoids of duration 100, 40, 10 and 4 ms. The frequency axis here shows a much wider range than the previous one. That is why the spectrum for a 100-ms tone is shown again.

Note that all the spectra have their peak energy at 1 kHz, as you might expect. However, the spectra get wider and wider (in terms of our definition of spectral width) as the signal is shortened. A 1-kHz sinusoid of 40 ms duration has a spectrum 50 Hz wide, and one of 10 ms duration a spectrum 200 Hz wide. In the rather extreme case of a 4-ms duration (only four cycles of the 1-kHz sinusoid), the width of the spectrum is 500 Hz. A simple rule allows us to predict the width of the spectrum, which is given by *twice* the reciprocal of the signal duration. In a simple equation, if d is the duration of the signal expressed in seconds, and w its width, then $w = 2/d$. So, for a 10 ms window, $w = 2/0.01 = 200$ Hz.

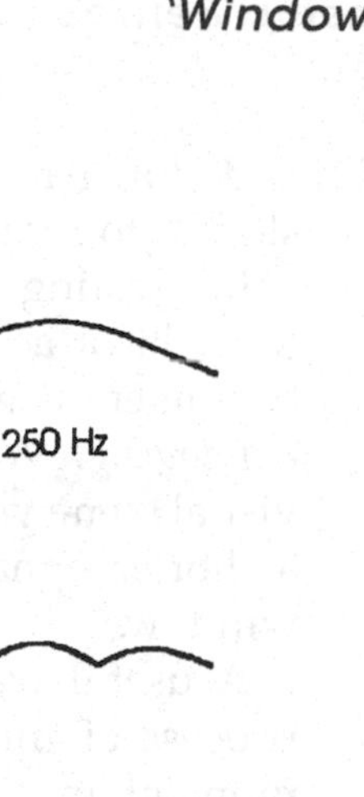
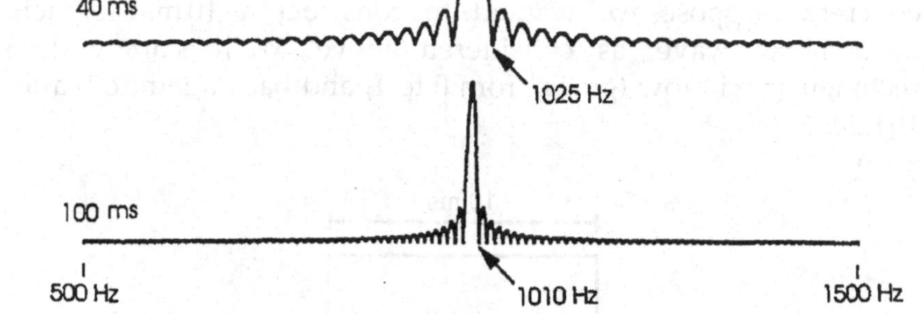

Again we see that signals that are narrow in one domain are broad in the other. Unlike the example of a rectangular pulse, here the spectral energy spreads both up *and* down the spectrum. This spread of spectral energy (often referred to as *splatter*) has important consequences in many psychoacoustic experiments which use tones to 'probe' some aspect of auditory sensation. If the probe tone is made long in order to be narrowly confined in frequency (in order to 'probe' a narrow spectral region), it will necessarily 'probe' an extended period of time. If the probe tone is made short in order to finely probe the time course of some auditory event, it will necessarily 'probe' an extended frequency region. Some compromise between these two conflicting demands must be made.

'Windowing' signals

In fact, sinusoids used in psychoacoustic experiments are almost never turned on and off as abruptly as those in the previous examples, because listeners perceive the resulting spectral splatter as a 'click' accompanying the tone. They are typically ramped up and down in amplitude a little bit (and sometimes a lot) more slowly, to reduce that spectral splatter.

In creating a finite-duration tone, what we are doing can be thought of as the selection of one short segment of sound from a longer waveform. This procedure is typically known as *windowing*. When you look out of a window, the part of the visual scene you can see is restricted by its frame; in the same way, a shorter signal can be selected out of a longer one by applying a window.

A useful way to think about windowing is to imagine it as a process of multiplying two waves against one another, at each moment in time, with one wave forming a 'window' on the other. As usual, let's take a particular example to make matters concrete. Suppose we wanted to construct a 10 ms segment of a 1 kHz wave, as considered above. We'll start with a rectangular window (going from 0 to 1, and back again to 0 after 10 ms):

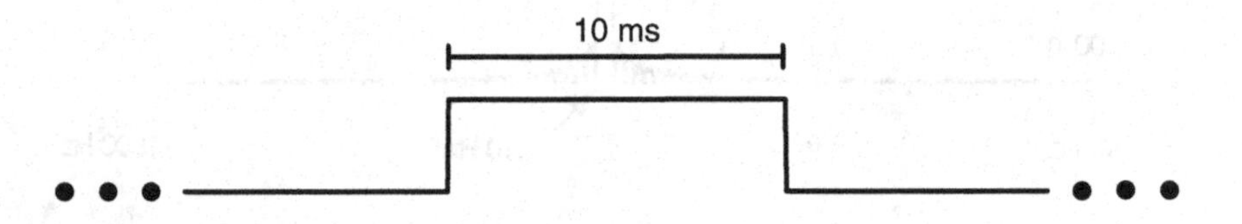

If we multiply this window by an infinitely long sinusoid (note the multiplication symbol above the horizontal line), we end up with a sinusoid that is 10 ms long:

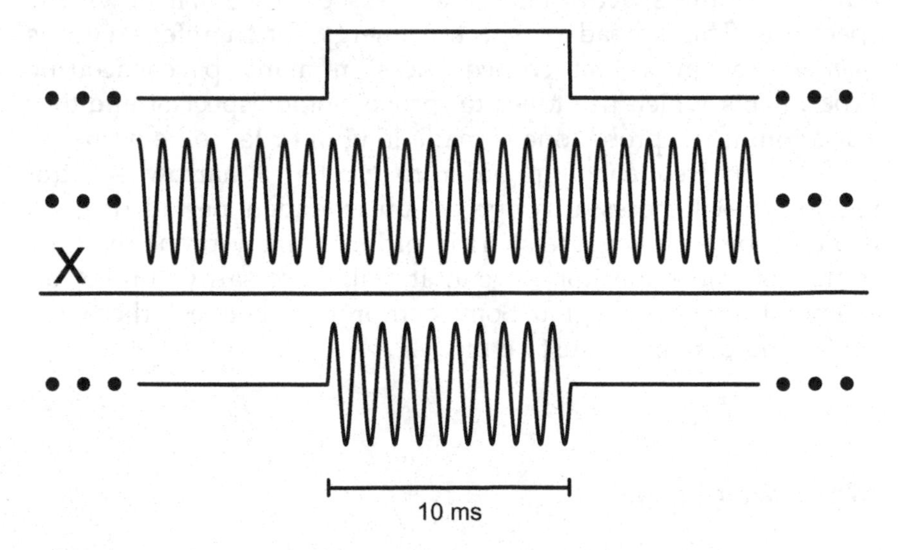

In order to change the way the sound is turned on and off, we simply vary the form of the window. Many, many different windows are used, but a common one found in psychoacoustics involves using a piece of a sine wave. Details can be found in the appropriate appendix, but all you need to know here is that this window is based on turning the wave on-and-off with 1/2 of a cycle of a sine wave. We'll start with a sine wave scaled and shifted upward so that its minimum value is 0 and its maximum value is 1. As you can see below, we have superimposed on a continuous sinusoid on- and off-ramps of 4 ms each:

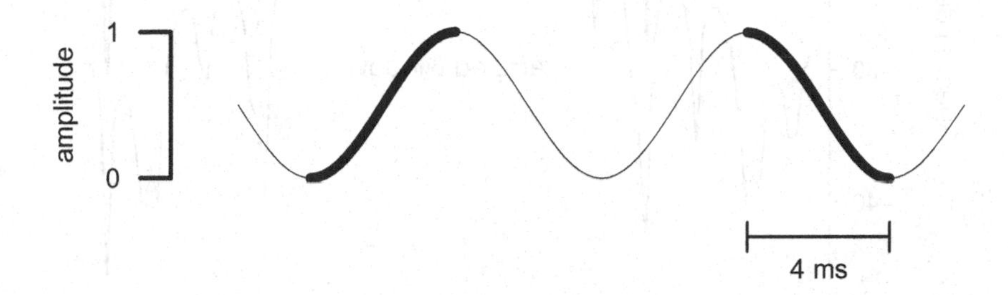

The 'on' ramp starts from its minimum value, and goes up to its peak (shown as the dark section towards the left). The 'off' ramp starts from the maximum and goes to its minimum (the dark section towards the right). By substituting these ramps for the essentially instantaneous ramps of the rectangular envelope used above, we obtain:

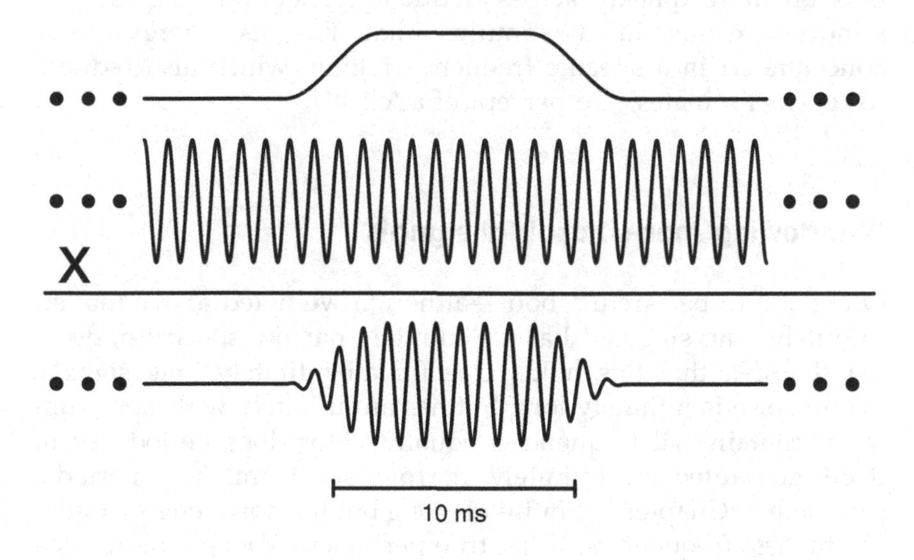

Note that in this particular instance, we have made the tone pulse a little longer by setting the midpoints of the on- and off-ramps to be 10 ms apart. The difference in the spectra from using rectangular and ramped windows can be seen here (now plotted on dB scales):

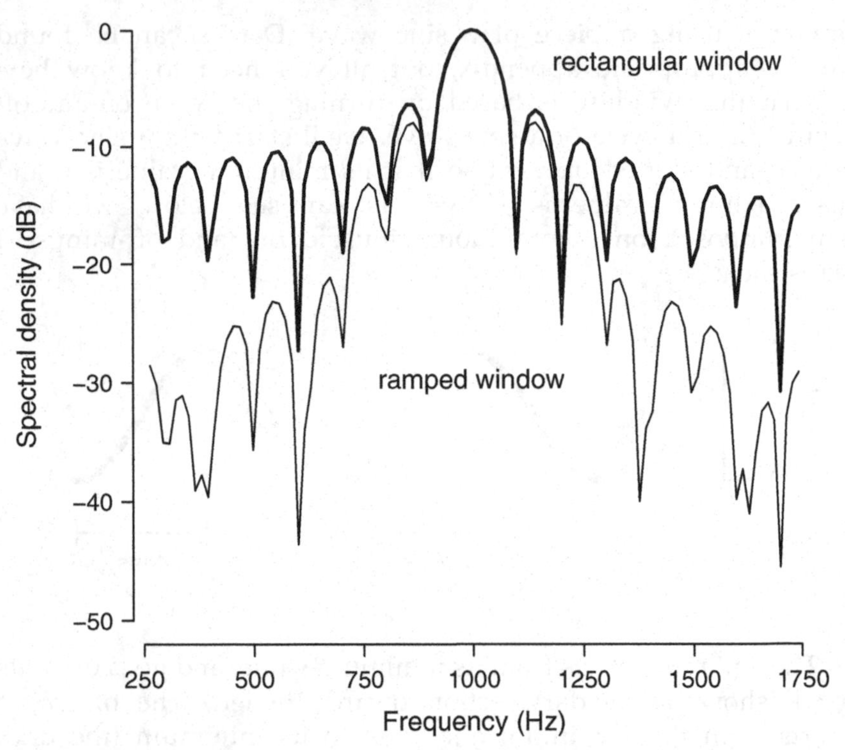

Note that the concentration of energy near the frequency of the sinusoid (the so-called *central lobe*) is the same for the two windows, but that the ramped window results in a spectrum that tails off more quickly across frequency. Therefore, the ramped window results in a stimulus that has its energy more concentrated in a specific frequency region, which also reduces (or even eliminates) the percept of a 'click'.

'Windowing' non-sinusoidal signals

One point to be careful about—although we noted above that an infinitely long sinusoid has an infinitely narrow spectrum, don't get the idea that this holds true for all infinitely long signals. White noise is infinitely long but has an infinitely wide spectrum (as it contains all frequencies equally). Nor does periodicity in itself guarantee an infinitely narrow spectrum. The periodic sawtooth of Chapter 7 is infinitely long but has some energy out to the highest frequencies. What true periodicity *does* guarantee is a line spectrum. Each of the components is, of course, infinitely narrow, but the collection of components can have any width.

This makes it obvious that only sinusoids can have infinitely narrow spectra. Any other periodic signal (even if truly periodic and hence infinitely long) would have more than one spectral component and so some measurable spectral width. Any transient or random signal would have a continuous spectrum, which couldn't be infinitely narrow either.

On the other hand, a rule that does apply to all types of waveform is that no finite-duration wave can have an infinitely narrow spectrum. As well as affecting the spectrum of any specific sinewave of limited duration (as observed earlier for the 1 kHz tone), this time-frequency relationship has similar consequences for each of the sinusoidal components that make up a complex periodic wave.

One way to get a better insight into this admittedly confusing issue is to consider the type of spectra one would see for various durations of a sawtooth waveform. Only for an infinitely long sawtooth (which is to say, one which was genuinely periodic) would a true harmonic spectrum (consisting only of spectral lines) be obtained. As shorter and shorter segments of the sawtooth are considered, each spectral line would get broader and broader, although, in some sense, the width of the entire spectrum (spreading up to infinity always) wouldn't change much. Here, for instance, are the spectra for a waveform that contains the first five harmonics of a 200-Hz sawtooth at three different durations—infinite, 100 ms and 13 ms, all ramped on and off with a rectangular window:

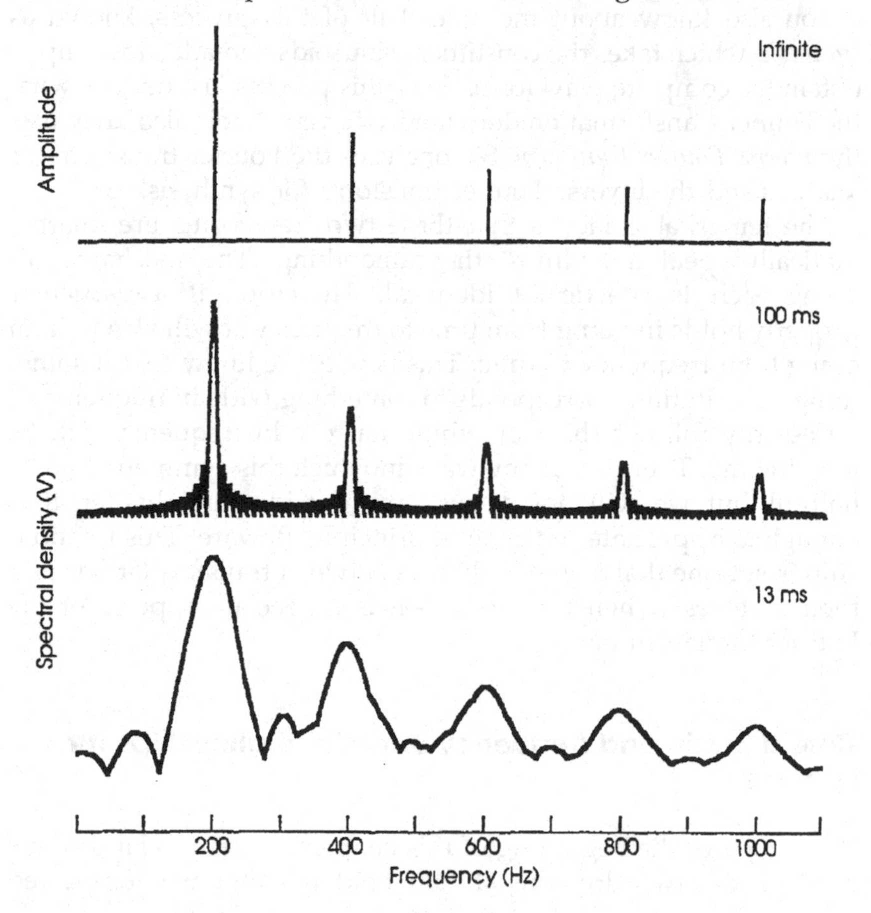

Only the spectrum at the top is a true line spectrum, as it corresponds to a truly periodic waveform. All the other spectra must be considered to arise from transient signals and hence have continuous spectra. Even so, the quasi-periodicity of the signals (the fact that they are periodic at least during the time they exist) means that most of their spectral energy is concentrated around the harmonics of the truly periodic waveform.

The relationship between the ordinary and inverse Fourier transform

Before we go on to discuss the application of these ideas to systems, it seems a good moment to at least hint at the reason for these symmetrical relationships between time and frequency. First, we need to introduce a little new terminology.

You already know what a Fourier transform is—a mathematical tool that changes (or transforms) a time representation of a signal (a waveform) into a frequency representation (a spectrum). In other words, the Fourier transform *analyzes* a waveform into its constituent sinusoids.

You also know about the other half of this process, known as *synthesis*, which takes the constituent sinusoids and adds them up to obtain the complete waveform. Since this process just undoes what the Fourier transformation does (and vice versa), it is also known as the *inverse Fourier transform*. So, one uses the Fourier transform for analysis and the inverse Fourier transform for synthesis.

The remarkable fact is that these two transforms are, mathematically speaking, almost the same thing. The mechanics of doing each is practically identical. Therefore, if a particular property holds in going from time to frequency, it will also hold in going from frequency to time. This is why we know that if something short in time corresponds to something wide in frequency, it necessarily follows that something narrow in frequency will be long in time. There are many ways in which this symmetry can be helpful, but we will not discuss any outside this chapter. It is enough to appreciate the general principle. Beware! This relationship is not one that is generally true between transformations and their inverses (when they exist)—it is a special property of the Fourier transform pair.

Time domain and frequency domain relationships for systems

We argued at the beginning of this chapter that any relationships between the two domains should hold whether we considered signals or systems. So far, we have seen that, generally speaking, signals long in time are narrow in amplitude spectrum and vice versa. This inverse correspondence has a direct analogue in systems. Systems that have a long impulse response (time domain) have a narrow amplitude response (frequency domain) and vice versa, a property which is normally expressed in a different way: systems that are good for resolving details in frequency are bad for resolving details in time, and vice versa.

The impulse and amplitude response of simple band-pass filters

To keep things relatively simple, we'll restrict our discussion to band-pass filters of the same centre frequency, 500 Hz. Suppose we measure the impulse response to two such filters by presenting a 10-μs pulse to each of them:

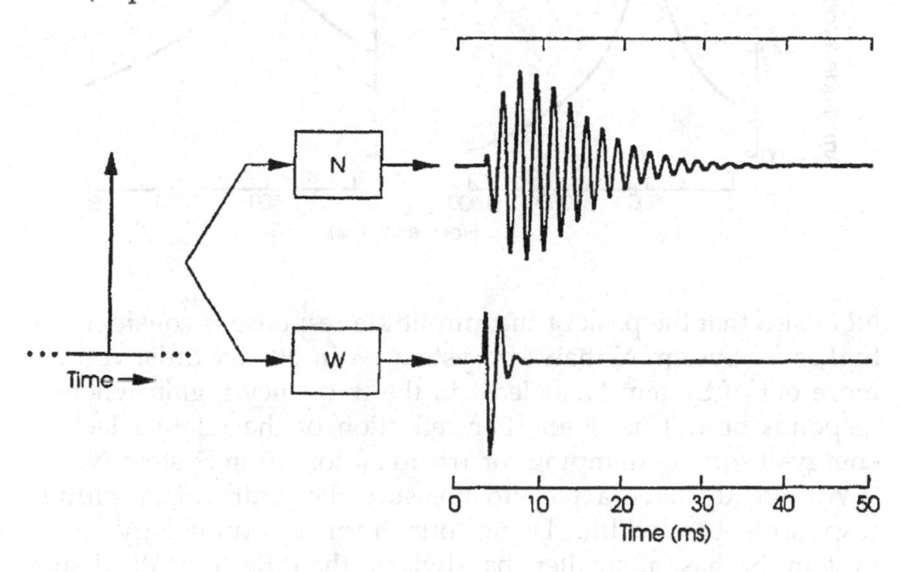

The output of each system is longer in time than the input signal. When the output oscillates quasi-periodically while dying away, as it does here, we say that the system *rings*. Clearly, the system at the top (System N) rings for considerably longer than the system at the bottom (System W). This would normally be the result of System N having smaller frictional forces (which cause the signal to die away) than System W. These frictional forces are also known as *damping*, and so System W would be said to be more highly damped than System N. How do these differences in damping (and hence duration of the impulse response) relate to differences in the amplitude responses of the two systems?

We could, of course, go back to the two systems and measure their amplitude responses in the ordinary way using sinusoids.

Or we could be clever. Remember from Chapter 9 that the amplitude spectrum of the impulse response of an LTI system is nothing other than the amplitude response of that system. Therefore, all we need to do, to obtain the two amplitude responses, is calculate the spectra of the impulse responses. Knowing that both systems are band-pass filters of centre frequency 500 Hz, and keeping in mind the relation between duration of signals and the width of spectra, what is the difference between the two amplitude responses likely to be?

Right—a difference in their width. System N, with its longer impulse response, should have the narrower amplitude response, as seen here:

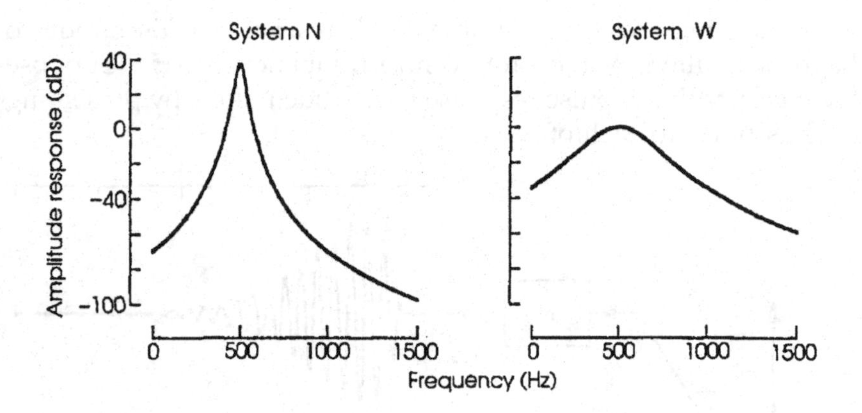

Note also that the peak of the amplitude response is considerably higher for System N than for System W. In other words, you get more out of System N, at least in the frequency region where it responds best. This is another reflection of the relative lack of energy-absorbing damping (or frictional forces) in System N.

We already have a way to measure the width of amplitude responses—bandwidth. Using our normal terminology, then, System N has a smaller bandwidth than System W (hence Narrow and Wide).

To summarize, systems with a narrow bandwidth will ring longer (that is, have a longer impulse response) than systems with a wide bandwidth. This can also be expressed in terms of damping: relatively highly damped systems have relatively wide bandwidths, while relatively lightly damped systems have relatively narrow bandwidths.

This result is of little interest to us on its own. It is only when we come to use band-pass filters to analyze signals (the subject of Chapter 11) and want to understand the limitations of such techniques that these properties come to the fore.

Resolution in frequency for different bandwidths

Probably the most common use we make of band-pass filters is to separate out the spectral components in a signal, in order to enhance or reduce them, or just to see what they are. For instance, we may have a recording of some speech that is very difficult to understand because there is a lot of low-frequency hum and high-frequency noise mixed in with the signal. One way we might try to improve the intelligibility of the speech would be to band-pass filter the recording so as to exclude the extraneous noises but retain the speech. As long as the speech and extraneous noises were mostly in different frequency regions, this could be done with LTI filters. For instance, we might band-pass filter the speech with a filter that had its cutoffs at 200 Hz and 4 kHz, since this region is most important for speech intelligibility. Hum from the electricity supply (at 50 or 60 Hz) and high-frequency noise above 4 kHz could be attenuated in this way.

Alternatively, we might just want to separate out closely spaced spectral components, to see how large each one of them was. It should be apparent that a narrow-bandwidth filter will do a better job of separating spectral components than a wide-bandwidth filter. One way to appreciate this intuitively is to think of a filter as a sort of microscope that can be used to examine the spectrum of a signal. Changing the centre frequency of the filter is like scanning across the frequency axis. The narrower its bandwidth, the smaller the frequency region that will fill the eyepiece; hence, the greater detail that will be seen. A wide-bandwidth filter will fill the eyepiece with a more extensive frequency region so small details will be obscured.

This can be shown explicitly if we consider analyzing a simple signal that consists of only two, relatively closely spaced, spectral components:

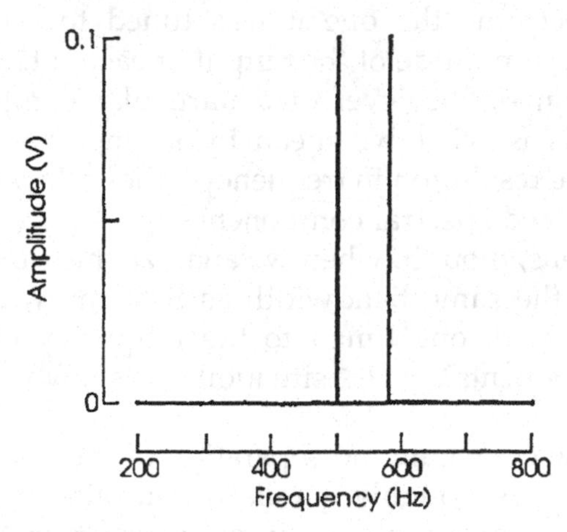

We'll pass this signal through two band-pass filters with the same bandwidth as System N, the narrow-band system. The only difference between the two filters will be their centre frequencies; one will be tuned to the frequency of the lower spectral component (at 500 Hz) and one will be tuned to the upper (at 580 Hz). To obtain the spectra of the two outputs, all we need to do is multiply the input spectra by the amplitude responses (as long as all are expressed in linear terms—that is, *not* in dB):

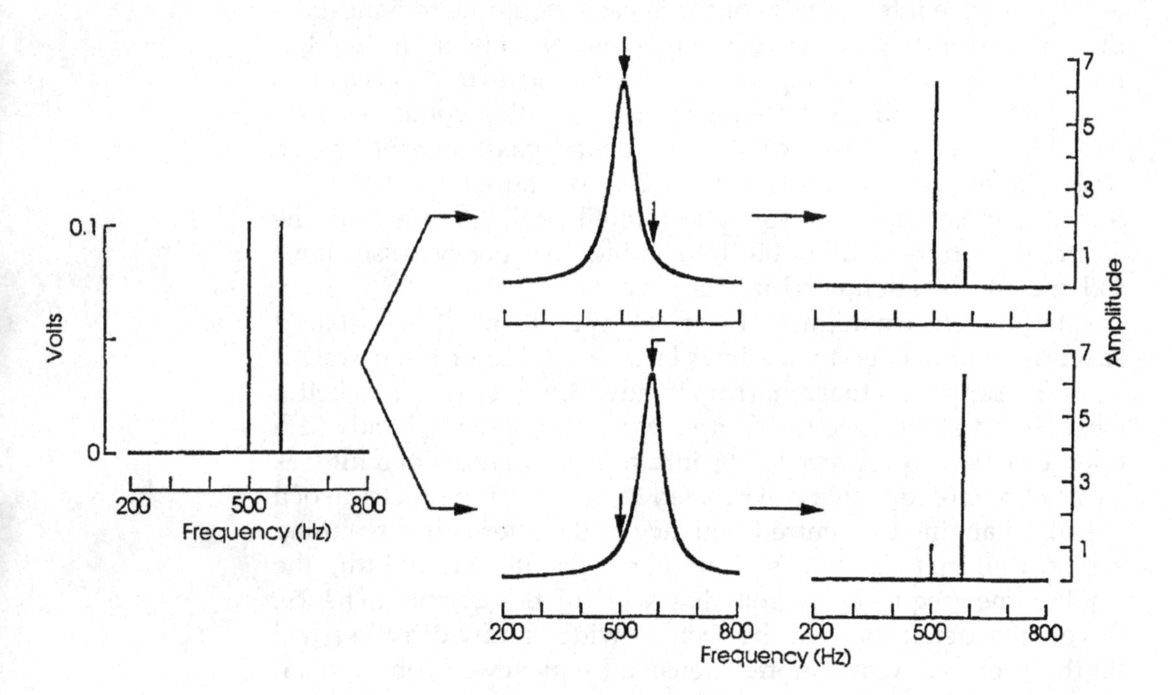

Here the arrows show where each of the two spectral components fall on the filter amplitude response curve. Note how the output spectrum from each filter contains essentially one component of the input spectrum—the one it was tuned to. Thus, a simple measure of the amplitude of the output of each filter would give a good indication of the level of a particular component in the spectrum. This is what we mean by saying that narrow-band filters have fine resolution in frequency—they allow us to separate out closely spaced spectral components.

What happens, though, when we analyze this same signal with two filters of the same bandwidth as System W, the wideband system, again with one tuned to the frequency of each of the spectral components? This situation is shown schematically overleaf.

Here there is essentially no separation of the two components. Each filter shows roughly similarly outputs—a sum of the two sinusoidal components. A simple measure of the amplitude of the output of each filter would only indicate the overall level of

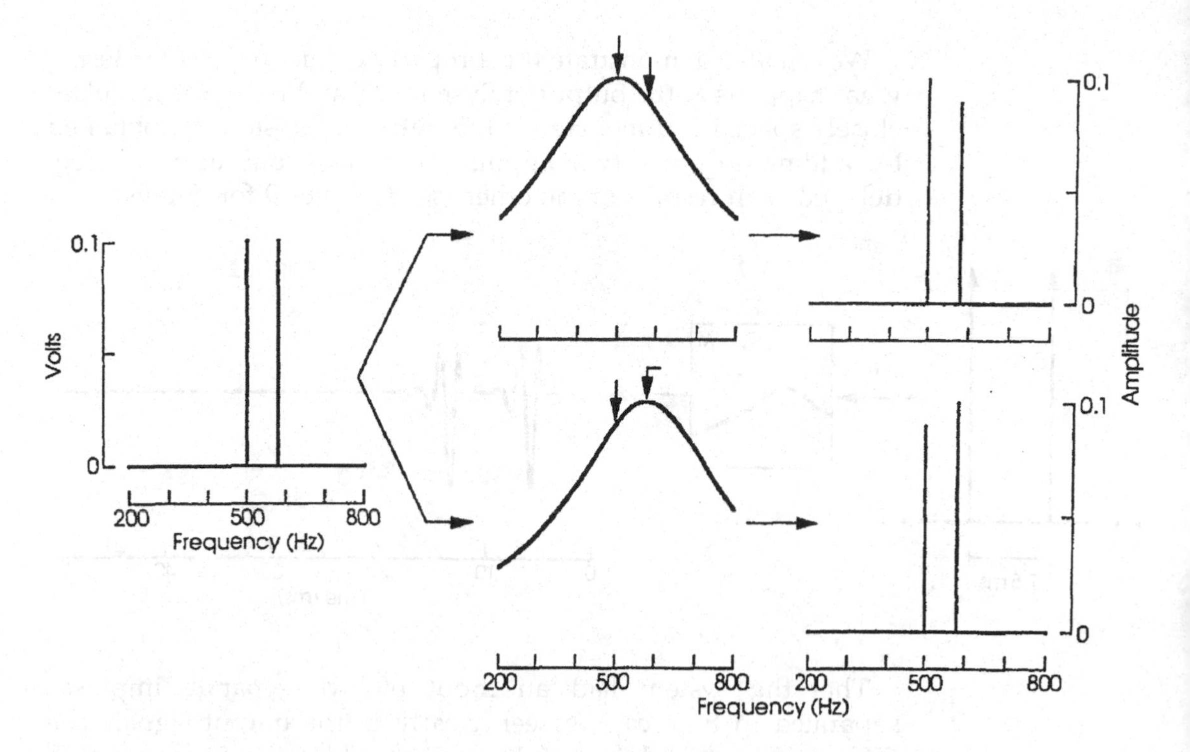

the spectrum without indicating much about the level of the particular component it was tuned to. Thus, wide-band filters have relatively poor resolution in frequency.

We can extract a general rule from this example. *In order to separate out two spectral components using band-pass filters, the bandwidths of the filters must be small relative to the spacing in frequency between them.* In the example here, where the spectral spacing was 100 Hz, the narrow filters (of bandwidth 45 Hz) separated the two components, whereas the wide filters (of bandwidth 300 Hz) did not.

Resolution in time for different bandwidths

Given the superiority of narrow-band filters in resolving spectral detail, you may wonder why they aren't used *all* the time, no matter what the task. The reason, as we mentioned above, is that a gain in *frequency* resolution is matched by a loss in *time* resolution.

Again, you may be able to appreciate this intuitively from what you already know. Narrow-band systems, with good frequency resolution, ring for a longer time than wide-band systems. This ringing makes determining the exact time of the input impulse uncertain. Therefore, if we're interested in the timing of events in a particular filter, we'd be better off with a system that didn't ring very much: a wide-band system.

We can also demonstrate this property explicitly by considering what happens at the output of Systems W and N to two impulses closely spaced in time. Here's the output of System W, obtained by adding together two impulse responses, one appropriately delayed with respect to the other (see Chapter 9 for details):

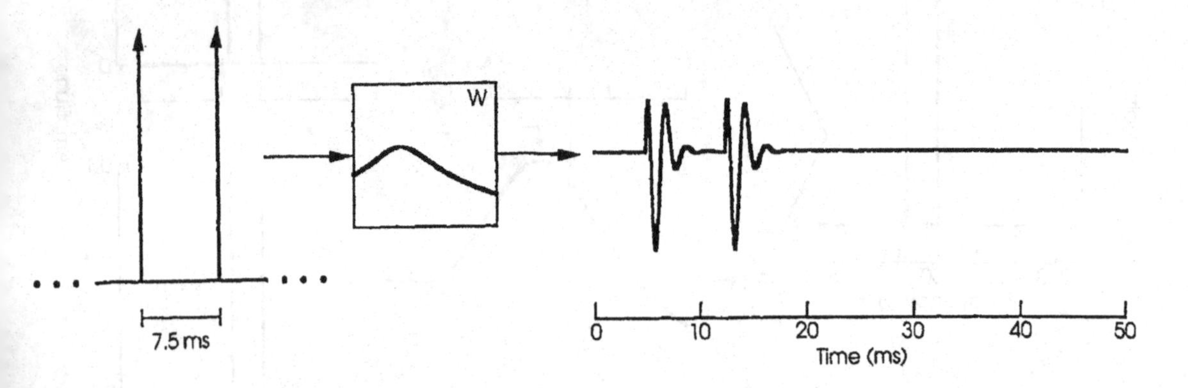

That the system had an input of two separate impulses separated in time can be seen clearly in the output signal. The difference in arrival time of the two would be easy to measure. If, however, we put this same signal through System N, we get quite a different result:

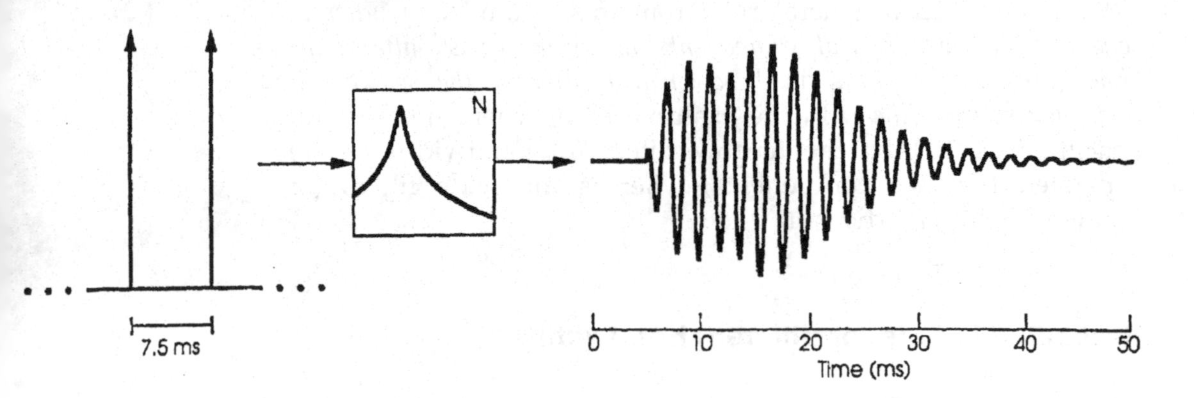

Now it's not so easy to see that there were two inputs separated in time because the impulse responses run into one another. It would be fairly difficult to determine, from the output of this system, the difference in the time of arrival of the two input impulses.

This result leads to another general rule, analogous to the one we stated above for the separation of spectral components. *In order to separate out in time two impulses passing through a band-pass filter, the duration of the impulse response of the filter must be short relative to the duration between them.*

Bandwidth versus resolution: a summary

The relationship between resolution in time and frequency, and between the duration of the impulse response and bandwidth for band-pass filters, is summarized here:

Bandwidth	Time resolution	Frequency resolution	Impulse response
Wide	Good	Bad	Short
Narrow	Bad	Good	Long

These relationships will be crucial in the next chapter, where we discuss one of the most commonly used tools in speech research, the sound spectrograph.

Exercises

1. Suppose that the impulse response of an LTI system is a rectangular pulse of duration 500 ms. Sketch its amplitude response. How would you classify this system? (Hint: See the figure on page 185 and Chapter 9).

2. Sketch the spectra of a 4-kHz sinusoid of durations 5, 20, 100 ms and infinitely long.

3. Describe what analyses you would perform using band-pass filters to reveal the properties of a signal consisting of two closely spaced sinusoids in noise (a) for the structure of the noise and (b) for the periodic component.

4. A sound consists of two frequencies of 1000 and 1050 Hz. This is picked up by a microphone in an environment that has lots of low-frequency hum (for practical purposes, with all its energy below 500 Hz). Describe an equipment layout that would allow a psychoacoustician to select the low- or high-frequency sinusoid alone and lead the low-frequency one to the left ear and the high-frequency one to the right ear. What analyses would you perform to validate that the equipment was operating properly?

5. In Chapter 4, in the figure on page 58, the output of a tape-recorder was shown. Describe specifically how you would (a) extract the sinusoidal component from the noise and (b) the noise from the sinusoidal component.

6. The following diagram shows the impulse response for two resonators. Which would be more appropriate for the analysis of fine detail in frequency?

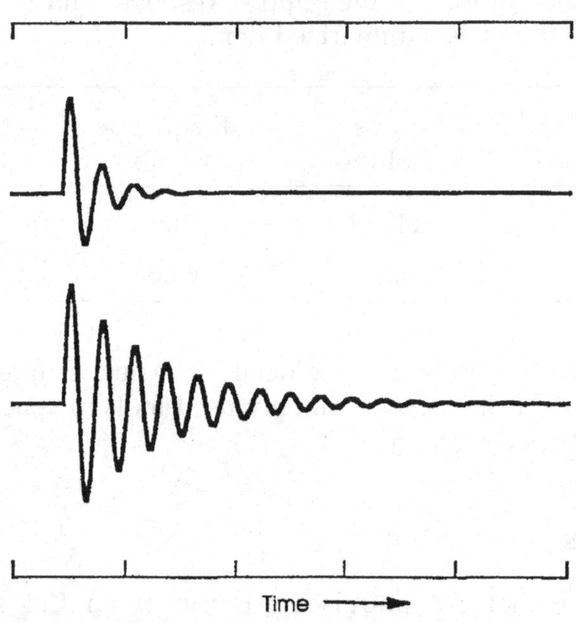

7. What would the amplitude spectra of a 1-kHz square wave look like for sections of (a) infinite, (b) 100 ms and (c) 13 ms duration?

8. The following diagram shows the impulse response for two low-pass filters. Which of these has the lower cutoff frequency? Present the reasons for your choice.

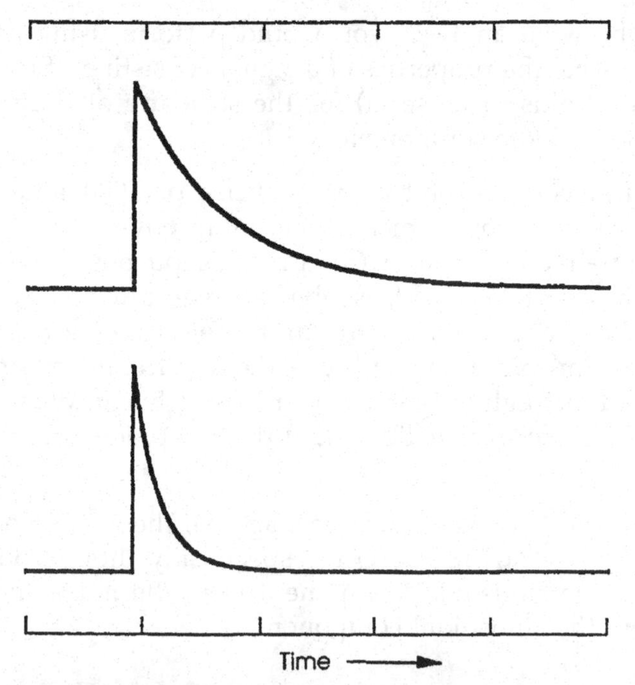

CHAPTER 11

The Spectrogram

We have now come to the end of the theoretical aspect of our work. From now on, we'll deal with rather more practical problems. Don't be tempted into thinking, though, that you can forget what's been done up until now! We'll be applying *all* the concepts previously developed.

The basic problem: determining the spectra of real-life signals

The particular practical problem that we want to solve is one of central importance—how to derive the amplitude spectrum of a real-life signal. Because the phase properties of acoustic signals influence auditory percepts relatively little, most effort has gone into determining amplitude spectra. So, in this chapter, when we talk about determining the spectrum of a signal, we will implicitly mean its *amplitude* spectrum.

In previous chapters, many of the signals whose spectra we've shown have been ones that have straightforward mathematical formulae. Therefore, it was possible (if not necessarily easy) to *calculate* their spectra using a mathematical tool, the Fourier transform. But what happens when we want to find the spectrum of a signal in the real world, one that isn't defined by a simple formula? The Fourier transform only works with abstract symbols on paper. On its own, it's of no use in determining, for example, the spectrum of a spoken vowel recorded on to a computer.

Until the advent of digital computers, there had been, in essence, only one practical way to determine the spectra of real-life signals. This technique (still very much in evidence) relies on the use of *band-pass filters*. For this application, the crucial property of band-pass filters is that they selectively attenuate energy at frequencies both above and below the region to which they are most sensitive. Therefore, if we want to know how much energy there is in a particular frequency region, all we need do is put the signal through a band-pass filter tuned to that frequency region and see what comes out the other end. If nothing comes out, obviously there was little or no energy in the signal in the frequency region that the band-pass filter was most sensitive to. If a lot comes out, there must have been a lot of energy in that particular frequency region.

Of course, the use of *one* filter only allows us to see what is happening in *one* frequency region. If we wanted to look at the spectrum of a signal over a wider range of frequencies, we'd need lots of filters with lots of different centre frequencies. Such a collection of band-pass filters is known as a *filter bank*. So, by using a filter bank, with the centre frequencies of the individual filters spread across the frequency range over which the analysis is desired, we could see how much energy is in the signal over the entire range of frequencies of interest.

Analyzing a sawtooth with a filter bank

We'll illustrate frequency analysis with band-pass filters on a signal you have come to know and love—a periodic sawtooth with a period of 5 ms (corresponding to a fundamental frequency of 200 Hz because $1/0.005 = 200$). You already know where this waveform has its energy, but perhaps you want to check the relative amplitude of its harmonics. Our analysis will be done with a bank of relatively narrow band-pass filters (45 Hz wide) whose centre frequencies are spaced densely over the frequency range from 0 Hz to 1100 Hz, say every 1 Hz. Any particular filter in that bank will have a frequency response like this but with different centre frequencies:

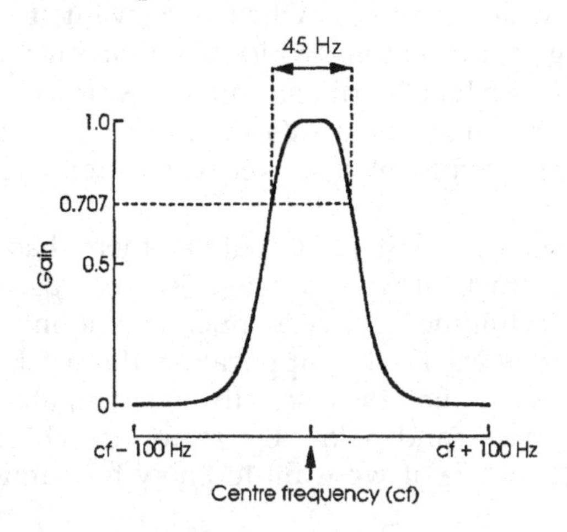

Clearly, the energy in the sawtooth is concentrated at multiples of its 200 Hz fundamental frequency, as that is where the harmonics are located. Note that there will be very little output from band-pass filters whose centre frequency is much below 200 Hz, as there are no spectral components in that lowest frequency region. So let's look first at the output of a filter whose centre frequency is 200 Hz. As you can see, quite a lot comes out of this filter:

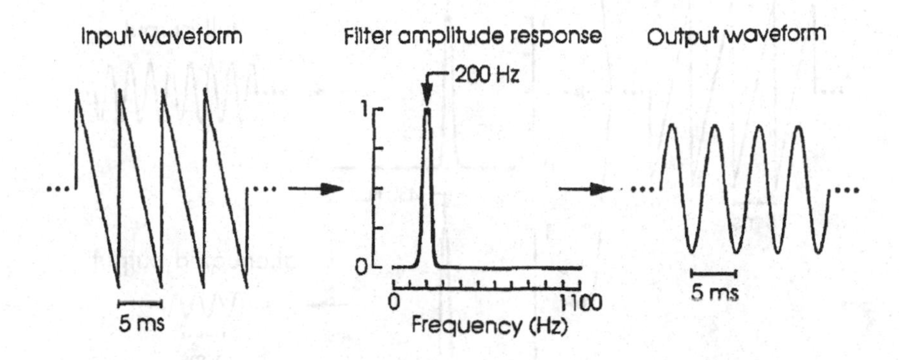

It would also be nice to see that there is no energy *between* the harmonics, so we can also look at the output of a filter tuned to a centre frequency of 300 Hz, where we see practically no output:

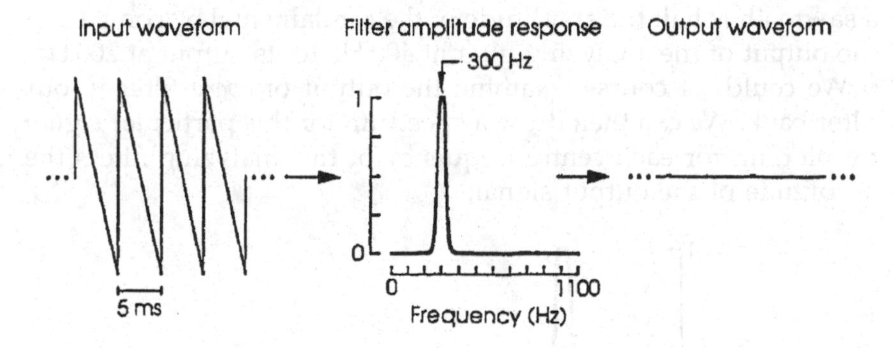

Because the bandwidth of the analyzing filter is not infinitely narrow, it isn't necessary for it to be tuned exactly to the frequency of a harmonic for there to be a significant output. Generally speaking, you can expect to see a significant output from a filter as long as its centre frequency is within about one bandwidth of the frequency of a spectral component in the signal (depending upon the filter slopes). Here, for example, are the outputs for a filter centred at 400 Hz, where there is a harmonic, and at 30 Hz on each side of this second harmonic frequency:

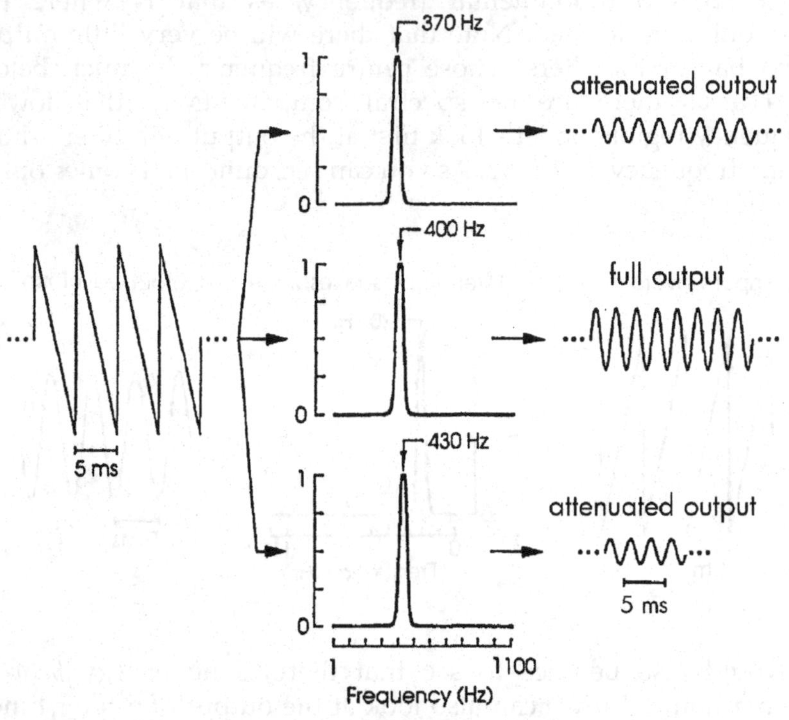

Note that all the output waveforms have a periodicity of 2.5 ms corresponding to the 400-Hz frequency of the second harmonic. You can also confirm our earlier claim that the second harmonic of a sawtooth is half the amplitude of the fundamental by comparing the output of the analyzing filter at 400 Hz to its output at 200 Hz.

We could, of course, examine the output of *every* filter in our filter bank. We can then draw a spectrum for this particular signal by plotting for each centre frequency of the analyzing filters the amplitude of the output signal:

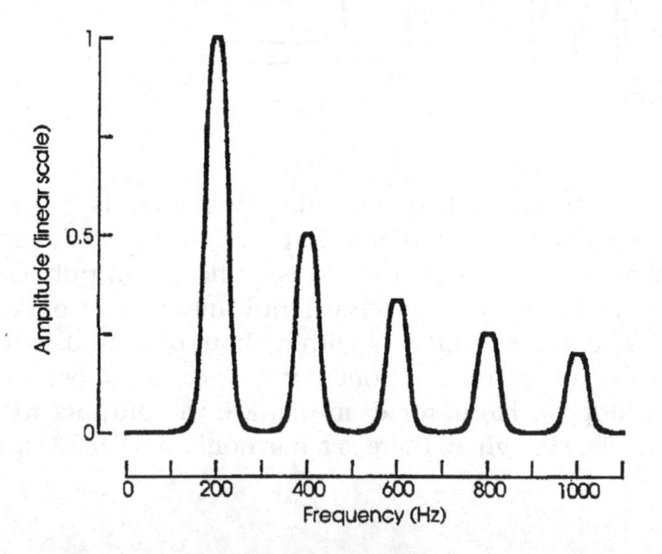

The only difference between this spectrum and the theoretical one seen in Chapter 7 is that the harmonics no longer appear as narrow spectral lines, but as broadened regions of energy. This arises from the fact that the analyzing filter must have some bandwidth—it cannot be infinitely narrow. The wider the analyzing filter, the wider a harmonic will appear in the filter-derived spectrum, an effect which we'll discuss in much more detail later in this chapter. For the moment, it is enough to appreciate in a general way that band-pass filters can be used to determine the spectrum of a waveform.

Signals that change in time

We now have some idea of how to do practical spectral analysis (another name for determining the spectrum of a signal). Our concept of a spectrum is still too restrictive, however, in one important way. Up until now, we have almost always made (at least implicitly) an important simplifying assumption about the signals that we have analyzed—namely, that they can be considered to be absolutely unchanging over an infinite period of time. Thus we've considered, for example, eternal sinusoids and the outputs of frozen vocal tracts. There are only three instances where we've considered signals that don't go on forever: the spectra of transient waveforms in Chapter 7, the impulse introduced in Chapter 9 and in the exploration of time and bandwidth trading in Chapter 10.

Even in Chapter 10, we were not so much interested in signals that changed in time as in ones that were off, were then turned on for a while and then turned off again. During the period in which those signals existed, they did not change their character in any way. We could still regard them as non-varying signals that existed for all time, although they were zero before some time and zero after some other. Therefore, we only needed one spectrum to characterize each of them. If the signals had changed in some respect over the interval in which they existed, we would not be able to know that simply by looking at their amplitude spectra.

Consider, for example, a simple whistling sound, consisting of a sinusoid that rises in frequency from 500 Hz to 800 Hz over 400 ms. Here is a single spectrum taken over the complete duration of the whistle:

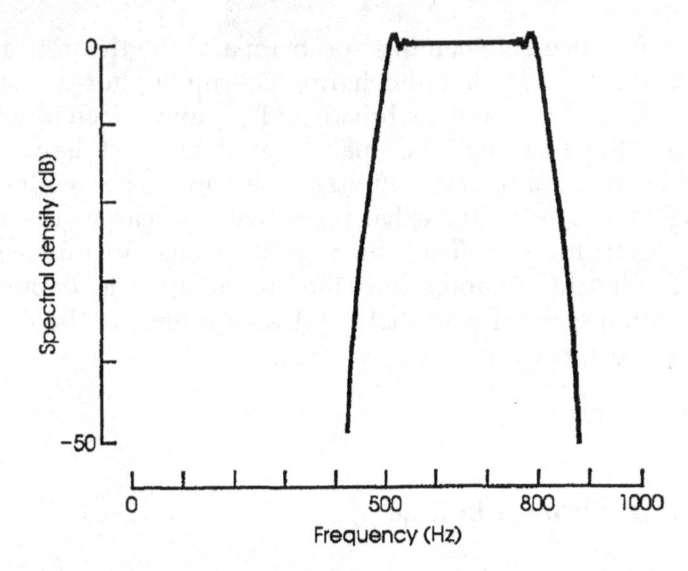

As would be expected, the energy in the whistle is concentrated in the frequency region between about 500 Hz and 800 Hz. What this spectrum doesn't capture is anything related to our experience of the whistle—that it starts low in pitch and goes up. In fact, it turns out that this whistle reversed in time (going from high to low pitch) would have exactly the same long-term amplitude spectrum.

The inability to examine how a signal's spectrum changes in time is obviously a severe restriction on the types of signals we can usefully analyze. Our present techniques may be sufficient for the steady-state tones used in some 1960s avant-garde musical compositions! They are clearly not so for most music, and even less so for the ever-changing signals we're particularly interested in—speech.

How, then, do we go about adding the dimension of time to our representation of signals in the frequency domain? There are a number of ways to approach this problem, but they all have the same aim, which is a practical system for the spectral analysis of dynamically changing acoustic signals. Here we'll describe three somewhat different approaches. Two of these are extensions of the filter-bank approach outlined above for steady-state signals. The essential idea is to analyze the signal with a filter bank, and display in some manageable form not only the amount of energy in each individual filter, but also how it varies with time. After describing in detail the mechanics of the filter-bank approaches, and their implications, we go on to detail quite a different way of constructing spectrograms, one which only became practical once computers were readily available. This 'third-way' is based upon a direct computation of the spectrum of short sections of the time waveform, similar to the 'windowing' described at the end of Chapter 10. But we'll stick to filter banks first, not least because this was the route that historically led to the first practical methods of dynamic spectral analysis.

Whistling through a single band-pass filter

As usual, for the sake of concreteness, we'll develop our ideas by trying to analyze a particular signal—the whistle mentioned above. As noted previously, this is a sinusoid that changes smoothly in frequency from 500 Hz to 800 Hz. Sinusoids that vary in frequency are, in general, known as *frequency-modulated* or FM tones. 'Modulation' is really just a synonym for a change, and so a frequency-modulated tone is just one that changes in frequency over time.

This change in frequency (from 500 Hz to 800 Hz) is equivalent, of course, to the period of the sinusoid changing from 2 ms to 1.25 ms over its 400-ms duration as seen here:

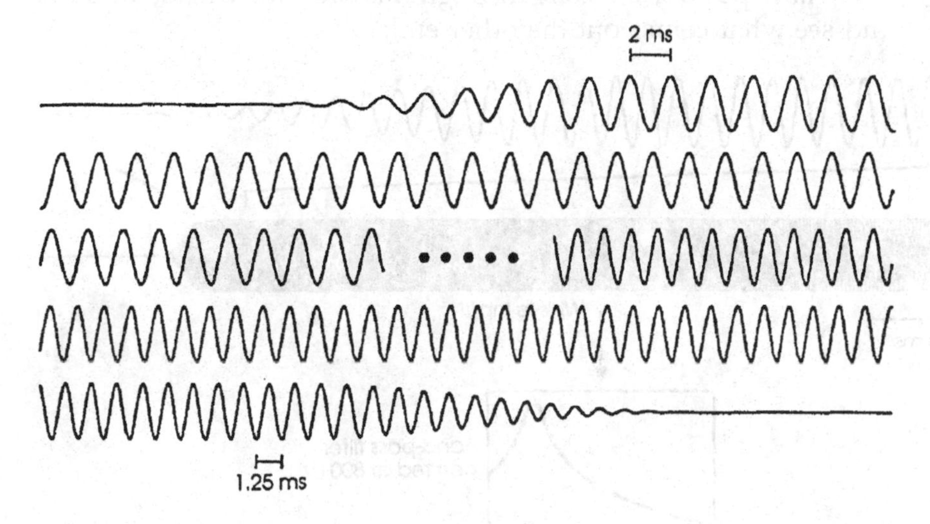

Because the signal is long enough to contain many periods, it's necessary to split it across a few lines in order to see any of the detail in the waveform. We've also left out part of its middle. In any case, you can see clearly that the period is changing smoothly from 2 ms to 1.25 ms over the course of the sound.

Let's suppose that we're interested in the amount of energy in this whistle near 800 Hz. Taking note of the suggestion above, we can put it through a band-pass filter that is most sensitive to an 800 Hz input. Remember, though, that such filters need to be characterized not only by their centre frequencies, but also by their bandwidth. We'll use a bandwidth of 45 Hz for the moment, giving a frequency response for our filter that looks like this:

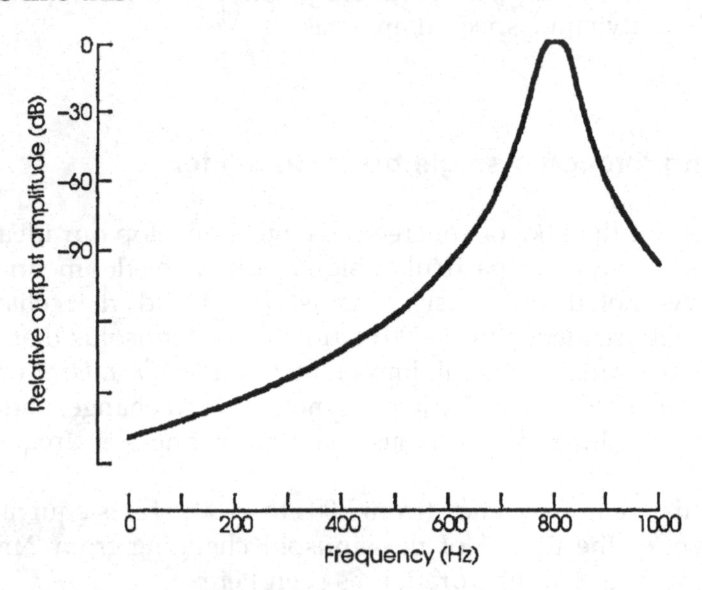

We now put our FM tone through this filter (or whistle into it!) and see what comes out the other end:

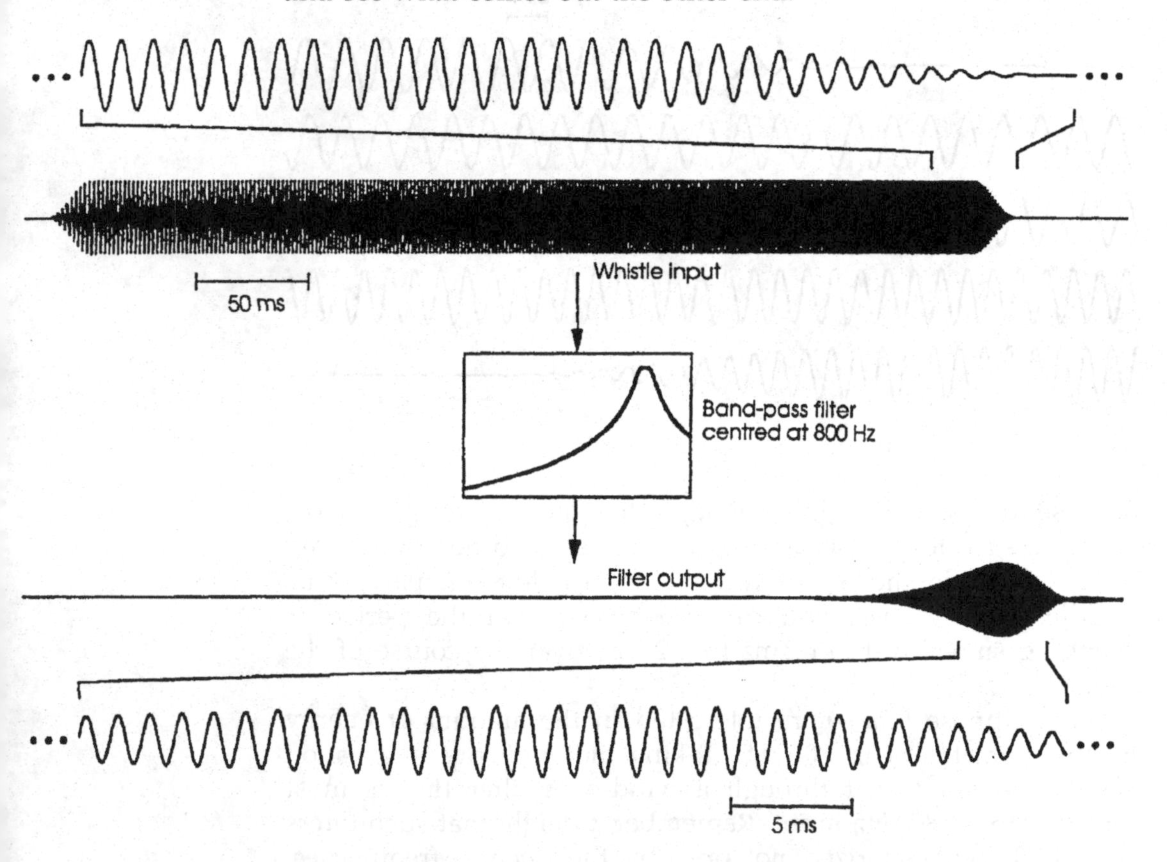

Again, showing the entire whistle means that individual cycles of the waveform cannot be seen. Therefore, small sections of the input and output waveforms near their finish have been expanded in time at top and bottom.

The output from this filter tells us two things. First, it indicates that there is energy in the whistle near 800 Hz, as something has come out of the filter. Second, it tells us *where* in the input signal this energy is. At the beginning of the whistle, where the frequency is about 500 Hz, nothing much comes out of the filter. The finish of the whistle, which is at 800 Hz, leads to a big output. So we know that the whistle has a lot of energy near 800 Hz at its end. Here, of course, we defined the signal and so already knew this to be the case. But if we hadn't known beforehand, this analysis would have told us.

Note the delay between the input and output signals—the peak amplitude of the output signal occurs slightly after the input signal has started to decrease in amplitude. Don't be overly concerned about this. All filters introduce some delay.

This single analysis only tells us about the energy in the signal near 800 Hz. If we wanted to know about other frequency regions (say near 500 Hz), we'd have to put the whistle into a filter most sensitive to 500 Hz (again, say, with a 45-Hz bandwidth). Now we see that the whistle also has energy near 500 Hz, but only at its start:

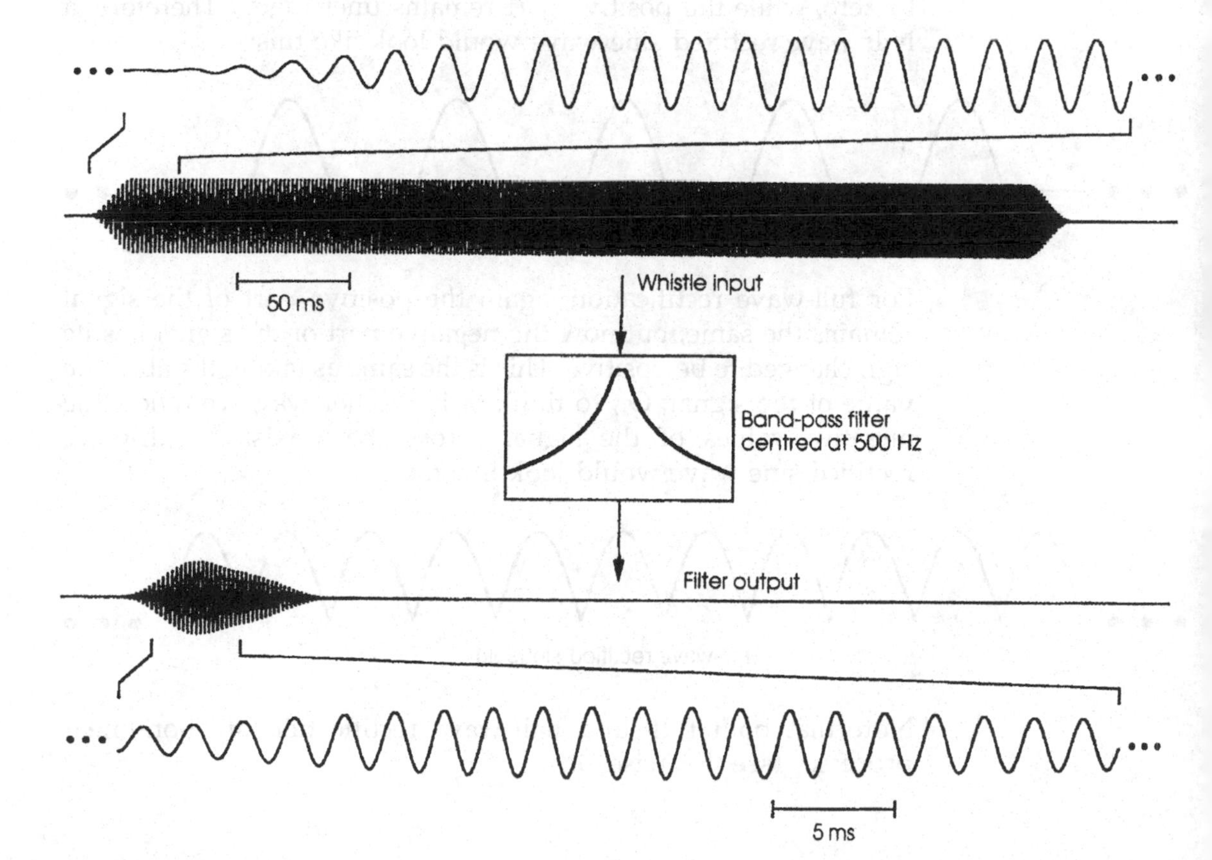

Getting rid of detail: rectification and smoothing

You may have noticed in the two previous figures that the output of the filter has the same sinusoidal shape as the input, at least when the output can be seen. These output waveforms also have the same periodicity as the input signal. Although in some situations this might be useful, for our current purposes, this intricate temporal detail is redundant. Much of it simply reflects the frequency content of the signal which, since we've passed it through a band-pass filter, we already have a good idea about. All we really want is an overall measure of the level of the signal coming out of the filter instead of its detailed shape. There are many ways of doing this. We could, for instance, take the peak-to-peak amplitude of the signal (as discussed in Chapter 3), which is probably the way we judge it by eye. A more convenient way, when it comes to actually building electronics, is to *rectify* and *smooth*. These two processes erase the very fast fluctuations of the filter output, thus giving an indication only of the relatively slow way in which the energy passing through the filter is rising and falling.

Rectification can be of two types: *half-wave* and *full-wave*. To half-wave rectify, any part of the signal which goes negative is set to zero, while the positive part remains unchanged. Therefore, a half-wave rectified sine wave would look like this:

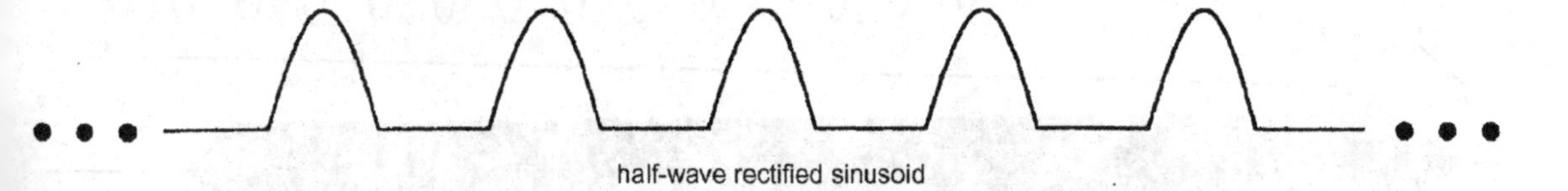

half-wave rectified sinusoid

For full-wave rectification, again the positive part of the signal remains the same, but now the negative part of the signal has its sign changed to be positive. This is the same as taking the absolute value of the signal. Or, to think of it another way, we reflect the negative values of the signal across the x-axis. A full-wave rectified sine wave would look like this:

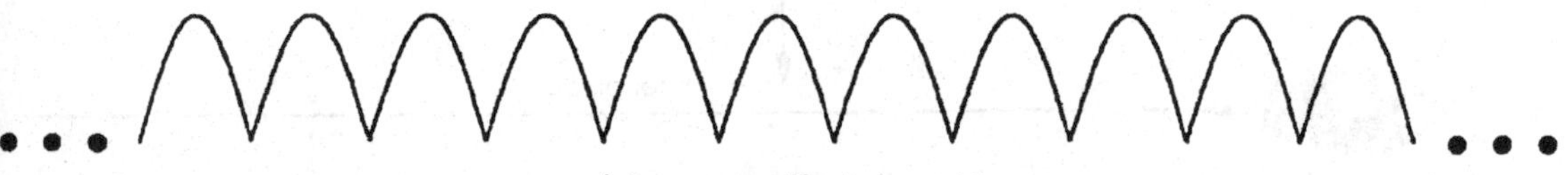

full-wave rectified sinusoid

Note that both full- and half-wave rectification are non-linear processes (see 'Exercises').

One way to think about rectification is through an extension of the notion of an input/output function that we introduced in Chapter 4. Instead of plotting the overall level of the inputs and outputs (as peak-to-peak or rms values), we can plot the *instantaneous* values of the input and output. For the simplest case of an amplifier with a gain of 1, in which the input and output have the same values, the input/output function would look like that on the left:

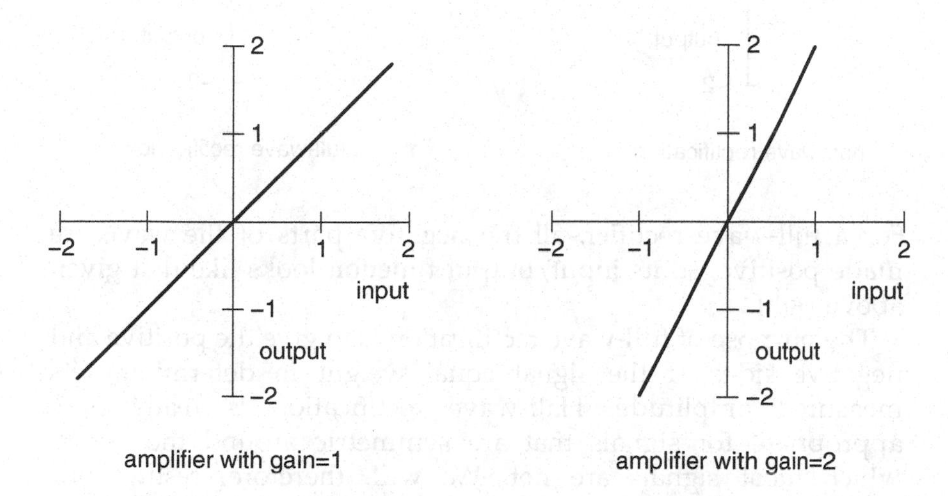

For an amplifier with a gain of 2, which doubled the size of the input, you would get the graph like that above right.

A crucial difference between these input/output functions and those you have seen before, is that they extend to negative values. Our previous input/output functions only had positive values, because they plotted a summary measure of the amplitude of a sinusoid (peak-to-peak or rms levels) which can only be zero or greater. The input/output functions introduced here describe how each *instantaneous* value of a wave (which can be positive or negative) is transformed. Even given this difference, you can clearly see that the amplifier with a gain of 2 still has a steeper line describing its functioning than an amplifier with a gain of 1.

In order to depict half-and full-wave rectification, then, we need to graph the way in which the negative part of the signal is changed (as the positive parts would undergo the same change as would happen for an amplifier of gain equal to 1, which is to say, no change). For a half-wave rectifier, every point in the original wave that is negative goes to zero, so its input/output function looks like that on the left:

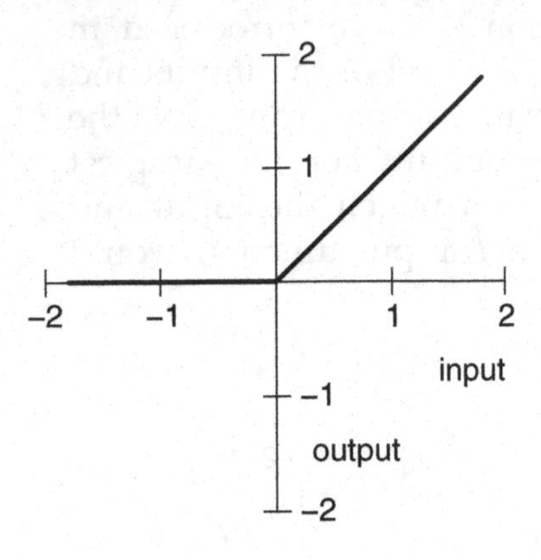

half-wave rectification

full-wave rectification

For a full-wave rectifier, all the negative parts of the wave are made positive, so its input/output function looks like that given above right.

The purpose of full-wave rectification is to give the positive and negative sides of the signal equal weight in determining its measured amplitude. Half-wave rectification is really only appropriate for signals that are symmetric around the *x*-axis, which most signals are not. We will, therefore, restrict our discussion to the more commonly used full-wave rectification. Look now at a section of the 500-Hz band-pass filtered whistle, before and after it is full-wave rectified:

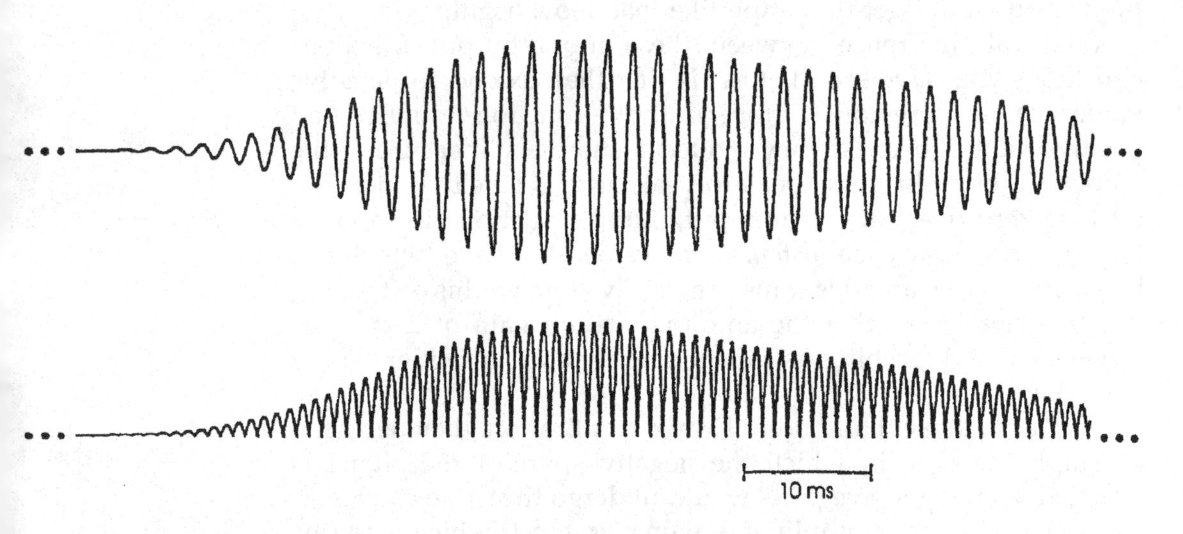

As you can see, there is still a lot of detail in this waveform. We haven't yet obtained a simple measure of the overall level of the signal coming out of the filter. What we really want is something related to the *amplitude envelope* of the signal, which is to say, what we would get by simply connecting together the peaks of the waveform. (This is exactly analogous to the spectral envelope discussed in Chapter 7, which touched the tops of the individual harmonics.)

This is where *smoothing* comes in. Smoothing gets rid of the fast fluctuations of the waveform, leaving only the more gradual ones. There are many ways to smooth a signal, but one simple way to do it is simply by low-pass filtering. After all, fast changes in a waveform are due to the high frequencies it contains. Removing those with a low-pass filter will leave only the slow changes, resulting in a good enough approximation to the amplitude envelope of the signal. Here, then, is the previous full-wave rectified filter output, before and after smoothing:

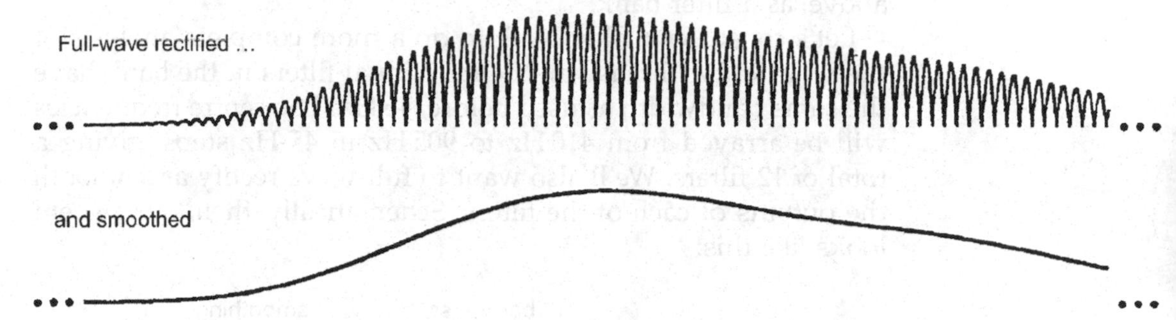

In short, rectification and smoothing give us a simple (and dynamic) representation of the size of the signal coming out of the band-pass filter.

Summary: filtering, rectification and smoothing

Let's stop for a moment to review our progress. We set out to find a way of doing a spectral analysis of a signal that could be easily implemented, and which would also give us the time course of any changes in the spectrum. Our first step was to pass the original signal through a band-pass filter which would only let through those parts of the signal in a selected frequency region. The problem with the raw output of the filter was that it was too complicated. We wanted a way to characterize this output that was more in keeping with the spectra that we've been examining all along. After all, in those spectra, if we wanted to know how much energy there was in a signal at a particular frequency region, we only needed to look at the height of a curve. Therefore, we wanted a simple measure of the size of the signal coming out of the filter. This was obtained by rectification and smoothing. In essence, this threw away the information in the filter output that told us its frequency content, information that was no longer necessary because we knew what frequency region the band-pass filter was sensitive to.

Looking across a range of frequencies with a filter bank

Of course, we've only completed this process for one narrow band of frequencies, near 500 Hz. By rectifying and smoothing the output of our 800 Hz band-pass filter, we would also know the time course of the 800-Hz energy in the signal. These particular centre frequencies happen to coincide with the initial and final frequencies of the whistle, but what we'd like to know is the time course of the energy in the signal at *all* the frequencies it contains. Clearly, the way to do this is to have a number of band-pass filters with their centre frequencies arranged over the frequency range of interest. Such an assembly of filters is known, as we mentioned above, as a filter bank.

Let's construct a filter bank to do a more complete analysis of the whistle. We'll make all the individual filters in the bank have the same bandwidth as used before, 45 Hz. The centre frequencies will be arrayed from 410 Hz to 905 Hz in 45-Hz steps, giving a total of 12 filters. We'll also want to full-wave rectify and smooth the outputs of each of the filters. Schematically, then, our system looks like this:

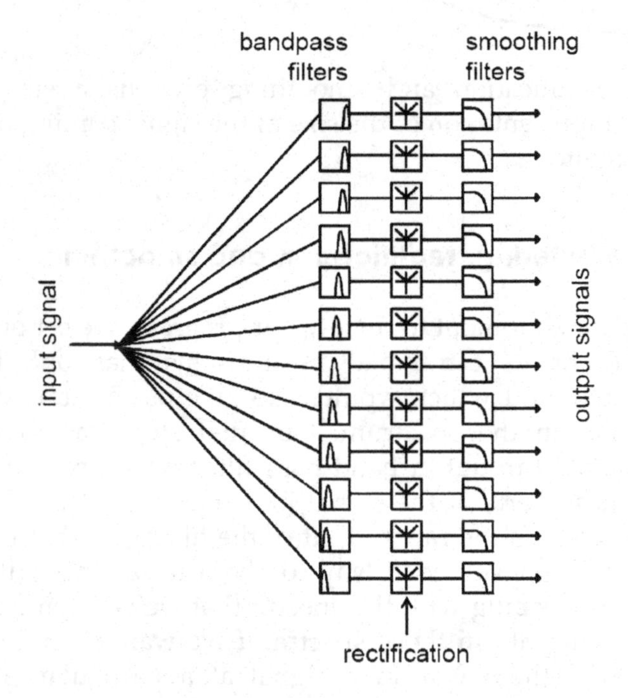

Note that this system has many outputs for a single input. The input is arranged such that all the filters are fed from the same point, so it's not necessary to route the input individually to each filter. Such an arrangement is standard practice for filter banks. Also note that rectification (symbolized by the appropriate instantaneous input/output function) and smoothing (symbolized by the frequency response of a low-pass filter) are conceptually distinct from the band-pass filters making up the filter bank. Therefore, the filter bank itself may be considered as a collection of LTI systems, even though the rectification which follows is a non-linear process.

Let's put our whistle through this system and observe the outputs (note the whistle's amplitude envelope at the top):

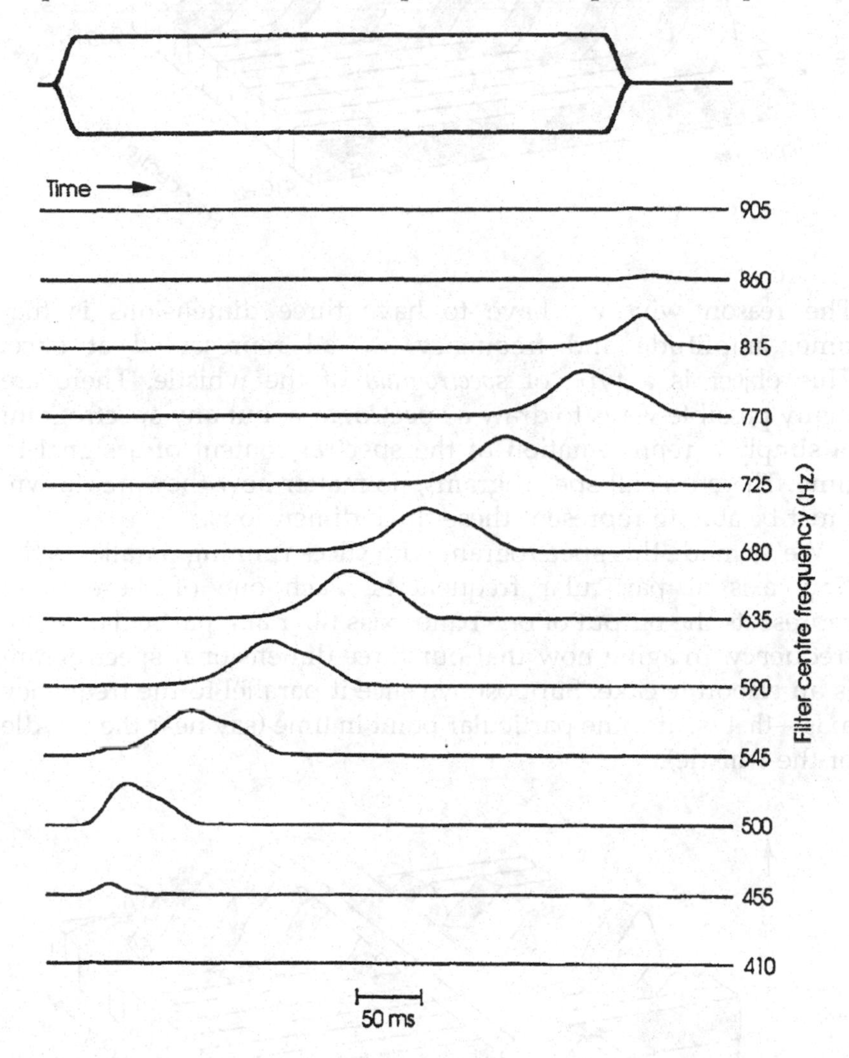

It's now easy to see how the spectrum of this signal evolves. As time advances, the peak energy in the signal moves to higher and higher frequencies.

Constructing a spectrogram

There is a more natural way, however, of combining all this information into a single plot. Imagine cutting out the shapes given by the outputs of the 12 filters and mounting them on thick cardboard. If we then stuck them together, side by side, we would end up with a three-dimensional object like this:

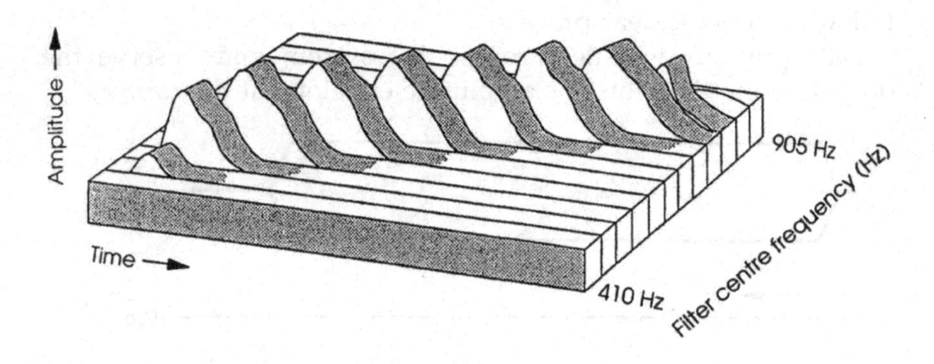

The reason why we have to have three dimensions is that time, amplitude and frequency are all represented at once. This object is a type of *spectrogram* of the whistle. There are many possible ways to draw a spectrogram, but any spectrogram is simply a representation of the spectral content of a signal in time. Therefore, all spectrograms, no matter how they are drawn, must be able to represent these three dimensions.

We've made this spectrogram with slices running parallel to the time axis at particular frequencies. Each one of these slices represents the output of one band-pass filter at a particular centre frequency. Imagine now that our three-dimensional spectrogram is an elaborate cake. Suppose we slice it parallel to the frequency axis—that is, at some particular point in time (say near the middle of the whistle):

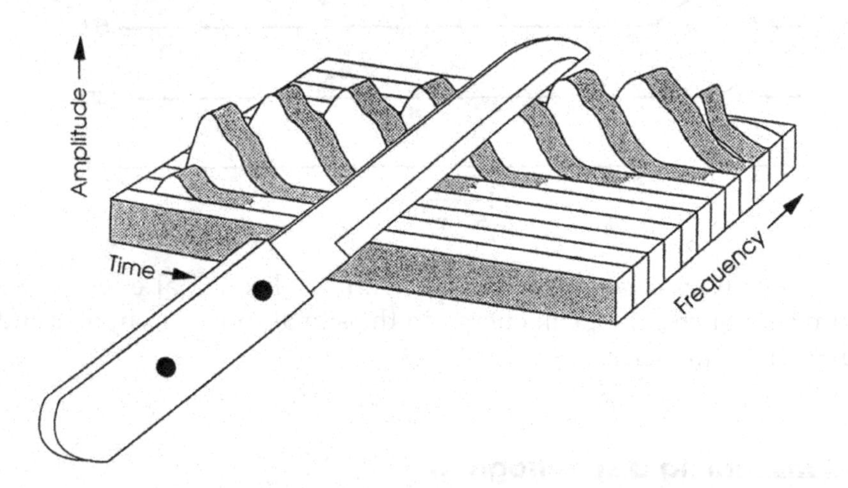

If we pull the cake apart and look at it end on, what do we see?

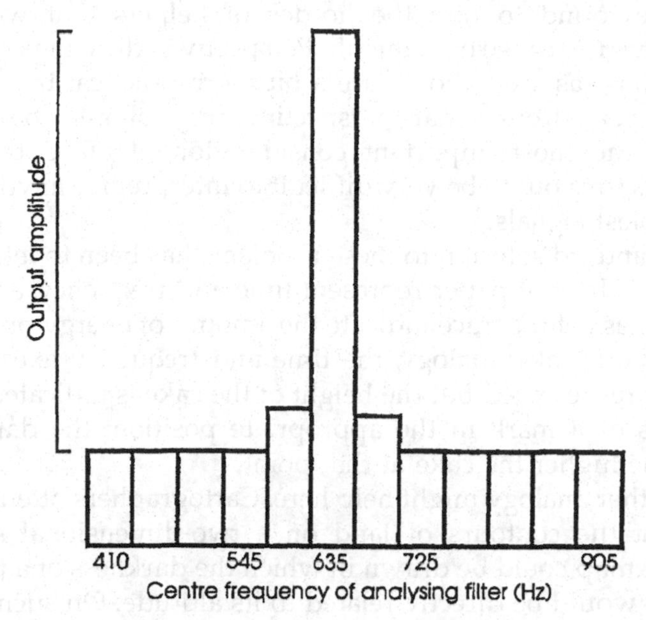

An important clue can be found by identifying the axes. Frequency runs along the bottom and amplitude up the side. Therefore, this picture represents the amount of energy at different frequencies in the signal. In other words, this is nothing but a spectrum, but a spectrum with a difference. It is the spectrum of the signal around a particular point in time. This is known as a *short-term spectrum* to distinguish it from an ordinary spectrum which is normally considered to analyze a signal over all time (or, equivalently, its total duration). Therefore, slicing the cake at different points on the time axis (parallel to the frequency axis) is equivalent to looking at the short-term spectrum of the signal at different points in time. Slicing the cake at different points on the frequency axis (parallel to the time axis) is equivalent to looking at the time course of energy in the signal near a particular frequency, as such a slice represents the output of one band-pass filter.

Displaying spectrograms in a convenient way

Three-dimensional spectrograms may be revolutionary when applied to cake design—they have serious drawbacks as a means of representing signals. First of all, three-dimensional representations are hard to slip in a folder, even if the money could be found to hire the hordes of helpers that would be necessary to construct them! Perspective drawings in two dimensions (as seen above) are a bit more practical but it is not always clear from what perspective they should be drawn. Perhaps the most important consideration of all is that such drawings turn out to be very difficult to interpret for anything but the simplest signals.

The standard solution to these problems has been to let the two sides of a sheet of paper represent time and frequency, and to let the darkness of the trace indicate the amount of energy present. In terms of our cake analogy, the time and frequency axes remain spatially represented but the height of the cake is indicated by the darkness of a mark in the appropriate position: the darker the mark, the higher the cake at that point.

One other analogy might help here. Cartographers often need to represent the contours of land on a two-dimensional sheet of paper. A map could be drawn in which the darkness of a point on the map would be directly related to its altitude. On such a map, Mount Everest would be jet black and Death Valley very white indeed. Using dark shading to represent amplitude, then, shows that our whistle has a spectrogram like this:

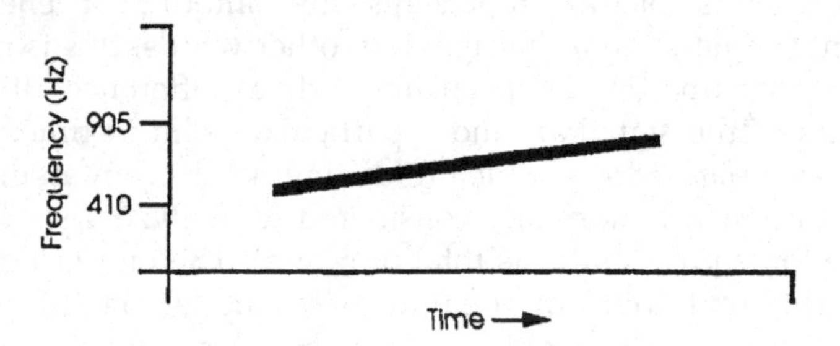

Spectrographs

The techniques that we have just described are, in fact, very similar to the principles embodied in the first practical machine for making spectrograms. Developed during the 1940s, this machine was known as a *spectrograph*. Don't confuse the two, rather similar, words. A spectrogram is the picture on the piece of paper, while the spectrograph is the equipment that makes it. It's exactly the same principle with the words telegram and telegraph. You use a telegraph to send a telegram, so you use a spectrograph to make a spectrogram. Here's what spectrographs used to look like:

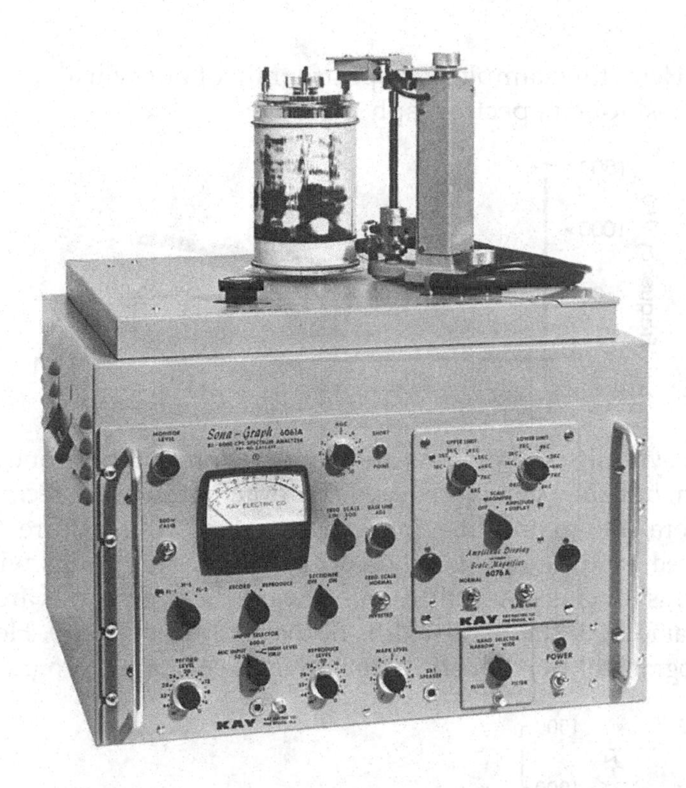

Of course, a method for making spectrograms generally would have to be a lot more versatile than the 12-channel version described above. Let's first deal with the filter bank that's needed. Obviously, we don't want to restrict ourselves to analyzing signals in the frequency region 410 Hz to 905 Hz. Spectrograms typically represent signals over a frequency range up to 8 kHz or higher. Also, 45 Hz is a little too wide a spacing for the centre frequencies of the band-pass filters—about 20 Hz would be better. This would necessitate at least 400 filters.

In fact, the original spectrograph didn't use a filter bank because it would have been too expensive to construct one with this many filters. It had a single band-pass filter whose centre frequency could be adjusted over a wide frequency range. Therefore, the analyses described above had to be done over and over again for each centre frequency of the filter. This meant that making spectrograms was very slow, taking a few minutes for each section of speech lasting only about 2.5 s. It was also a very smelly job, because a stylus passed an electrical current through a special kind of paper, burning it darker at regions in which there was a lot of energy in the signal, and less so for regions in which there was less energy. Still, these spectrographs were a great advance on previous techniques, and it would be hard to overestimate the impact they had on the development of speech science. The quality of a well-made spectrogram from these machines is typically as good as those currently made on computers, and often better. Here, for example, is a spectrogram of our whistle, made on an old-fashioned spectrograph machine:

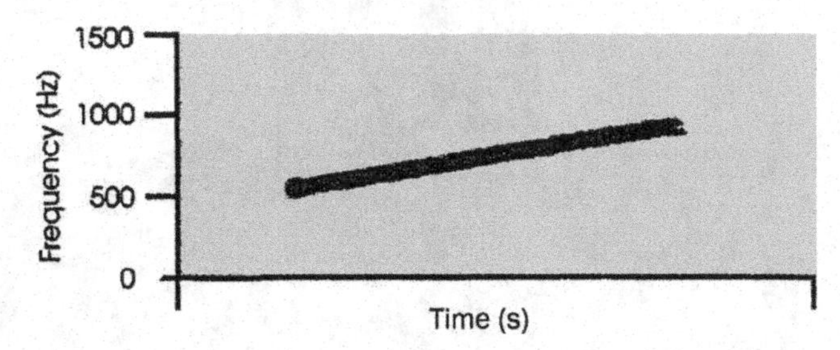

Nowadays, all spectrograms are made by computers of one sort or another, but they can be made using exactly the same techniques. Therefore, no matter the internal workings, the picture that is produced by different methods of making spectrograms will look similar, as long as the settings that govern the procedures are equal (for example, in the bandwidths of the analyzing filters). Here is a spectrogram of our whistle as made by a computer program:

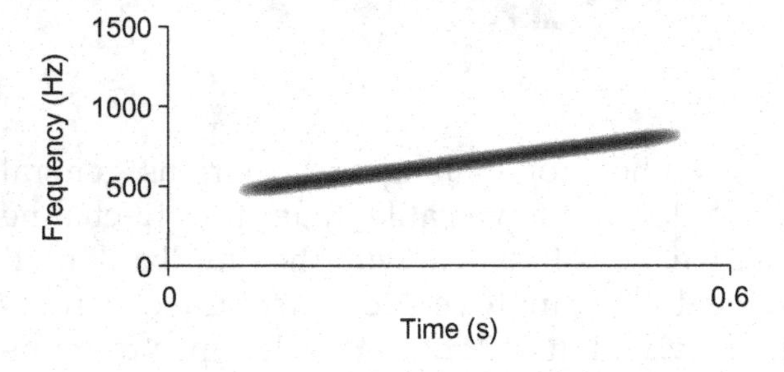

Another approach to making a spectrogram

In trying to comprehend a difficult concept like a spectrogram, it's often helpful to have a different angle on the problem. Therefore, we will go through another way of making a spectrogram, based on a fairly common piece of hi-fi equipment—a *graphic equalizer.* This approach also depends upon the use of a filter bank, so is essentially equivalent to the one just described. The main difference is in the stage of the process at which we re-code levels of the spectrum into the darkness of a trace.

The graphic equalizer

Many of you will have seen a graphic equalizer before, either as a separate piece of equipment or as incorporated into a hi-fi amplifier. Here is a high quality stereo one, with 10 adjustable controls for each of the left and right channels at either end:

In essence, a graphic equalizer is nothing more than an elaborate version of the tone controls even cheap hi-fi amplifiers have (usually labelled bass and treble). You can think of a graphic equalizer as an LTI system with an adjustable amplitude response. They are known as *graphic* because you get a direct picture of the amplitude response you program in (by the position of the sliders), and as *equalizers* because they are used to correct for (or *equalize*) deficiencies in other parts of the sound system (for example, the loudspeakers).

Graphic equalizers are based on a filter bank with an adjustable gain (or volume control) following each band-pass filter. As you can see in the diagram below of a simpler four-channel equalizer, the electrical wave is sent independently to each of the four band-pass filters, as is the case for more or less any filter bank. What makes a graphic equalizer do its job is the ability to increase or decrease the level in each frequency channel. After this adjustment, all the channels are added together again for output to whatever piece of equipment is next in the chain, typically an amplifier. By adjusting the relative gain in each channel, it is possible to vary the amplitude response of the equalizer. So, for example, it is possible to 'turn up the bass' by boosting the level of the signal in the lowest frequency channel and turning down the other three channels.

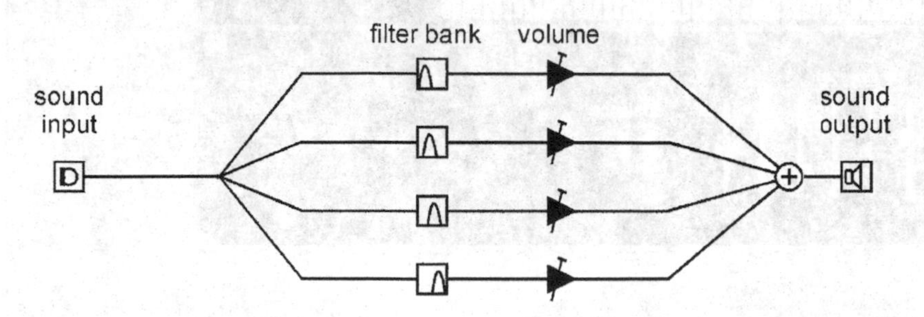

A spectrum displayed as a bar graph

Although graphic equalizers are designed to allow an adjustable amplitude response, it is another typical aspect of their function that is important for us here. Many (but not all) graphic equalizers are equipped with a real-time display of the amplitude of the signal within each channel of the filter bank, typically displayed as a bar graph. Many computer media players have such displays as they play sounds. Here is one from Windows Media Player which is based on 50 filters, one for each bar. Note too the 10-channel graphic equalizer controls in the bottom left-hand corner. This listener likes bass!

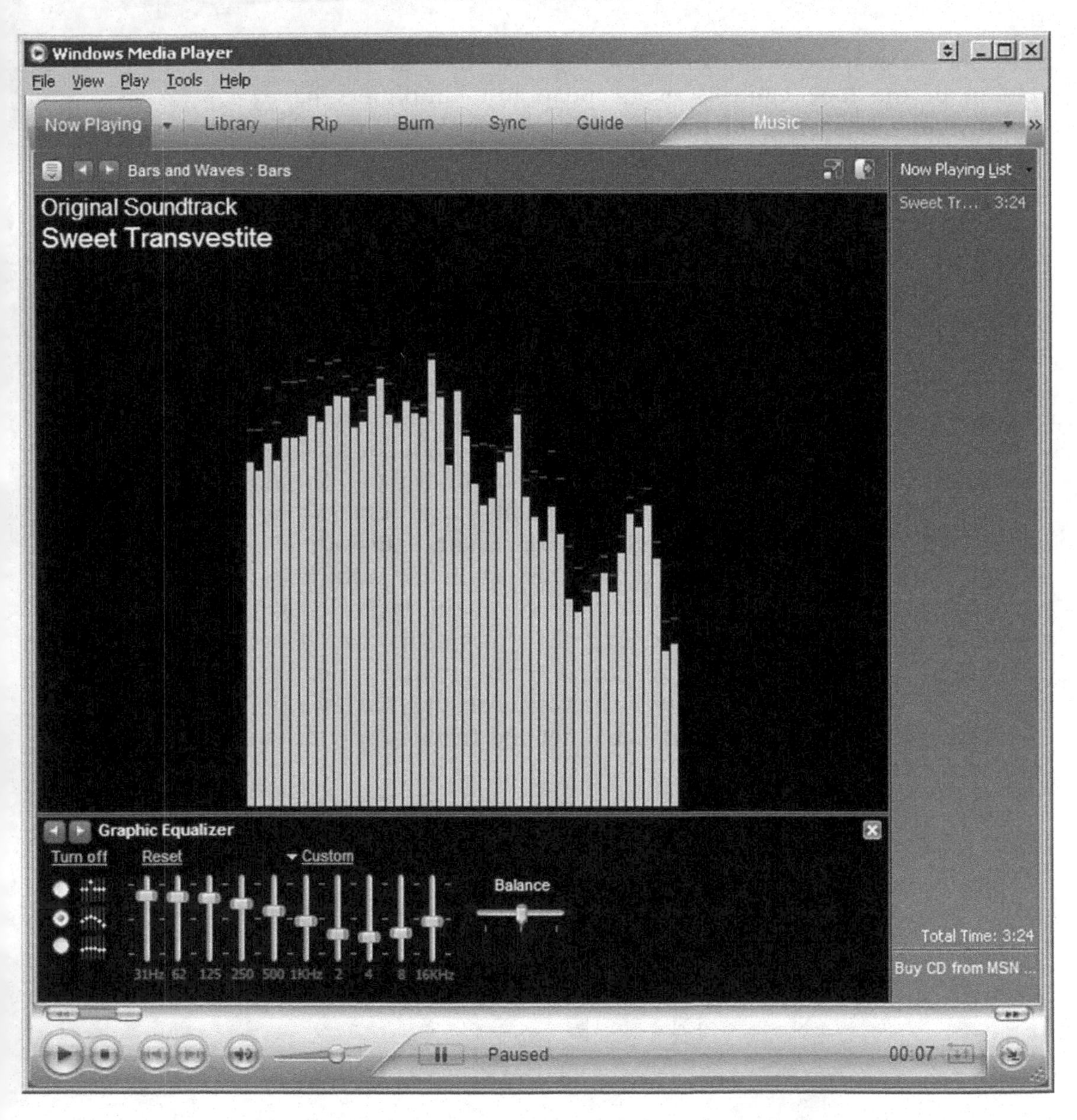

Because graphic equalizers already have a filter bank in them, it is a simple matter to take the outputs from each of the filters, rectify and smooth them, and use those levels to control the height of a number of bars on a display. Note that each bar on the graph represents the amount of energy in a particular frequency channel, determined solely by the centre frequency of the band-pass filter:

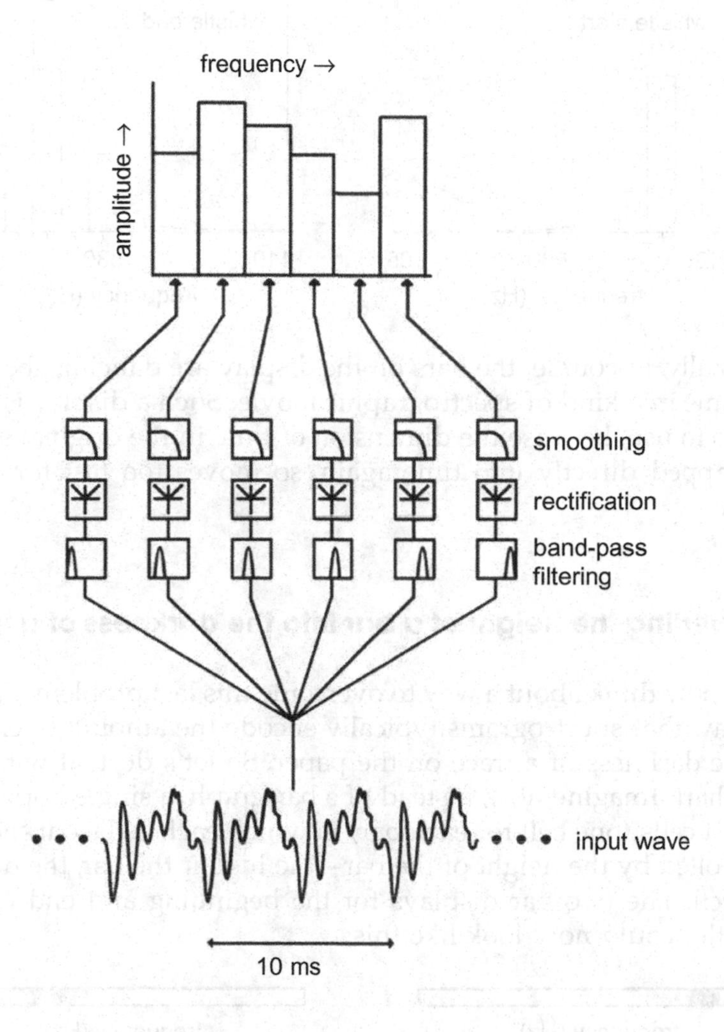

At any moment in time, this display reflects the spectrum of the sound coming into it around that time. In fact, this is nothing more than a *short-term spectrum*, just like the one you saw earlier when we sliced our spectrographic cake. Just like any spectrum, the horizontal axis represents frequency while the vertical axis represents amplitude. This spectrum is different from those in Chapter 7 in two ways. Being a short-term spectrum, it represents the spectrum of one section in time, rather than the whole, of the signal. Also, being based on a relatively small number of filter channels (each of which is relatively wide), it does not retain the spectral detail a true spectrum would have.

Let's now suppose we had a graphic equalizer based on the filter bank we used earlier for the whistle, with 12 channels. This is what we might see at the start and end of our whistle:

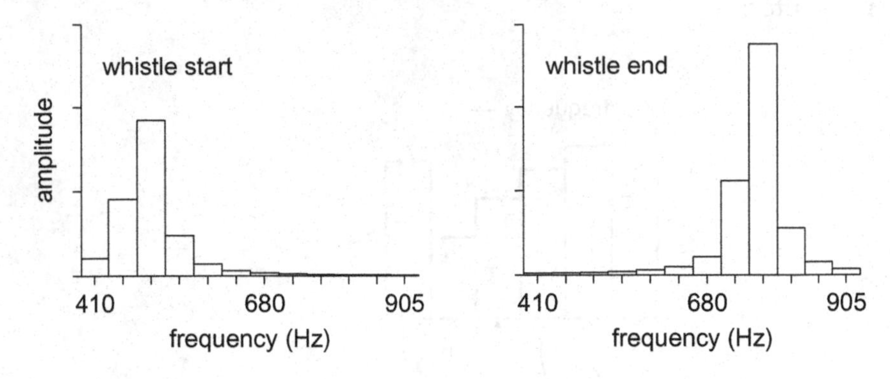

Normally, of course, the bars on the display are dancing about all the time in a kind of spectrographic movie. Such a display is hard for us to use, because the dimension of time in the original signal is mapped directly into time again, so moves too fast for us to study.

Converting the height of a bar into the darkness of a trace

Let's now think about a way to overcome this last problem. Earlier we saw that spectrograms typically encode the amount of energy as the darkness of a trace on the paper. So let's do that with our bar chart. Imagine now, instead of a bar graph, a single horizontal row of cells (one cell for each bar) in which each cell's darkness is controlled by the height of the bar—the higher the bar, the darker the cell. The two bar displays for the beginning and end of our whistle would now look like this:

Converting the time dimension into the *x*-axis on a piece of paper

This display is, of course, still a movie. Instead of the heights of the bars dancing up and down, the brightness of each cell would be changing constantly. We still want to see the whole pattern at once on a sheet of paper. One possibility is simply to line up a whole set of these cells, say one every 10 ms, going down the page. The spectrum of the start of the signal will be at the top, and that of the end of the signal at the bottom:

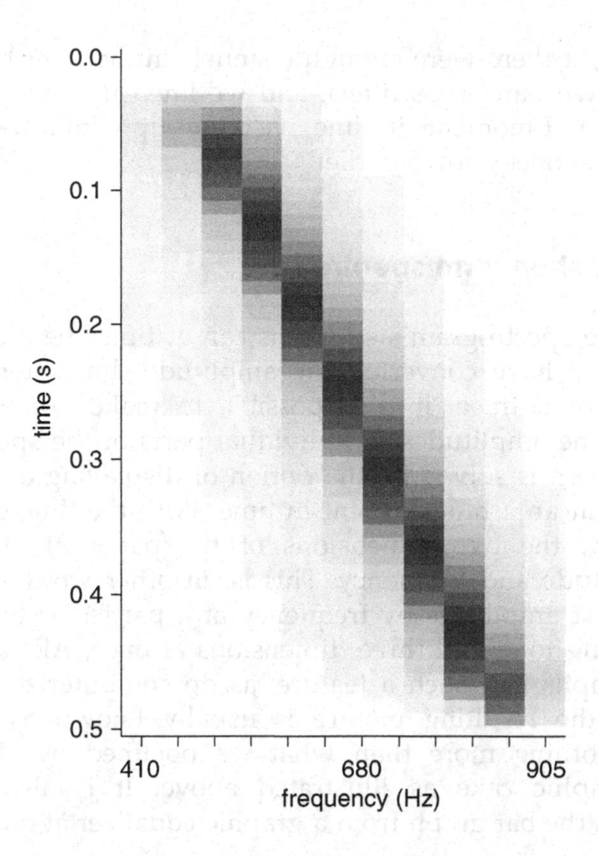

You should be able to see that this is nothing more or less than a spectrogram just like the one we constructed earlier. Here time is running down the page, whereas frequency increases from left to right. To make this spectrogram look like our others, we need to rotate it by 90° counter-clockwise.

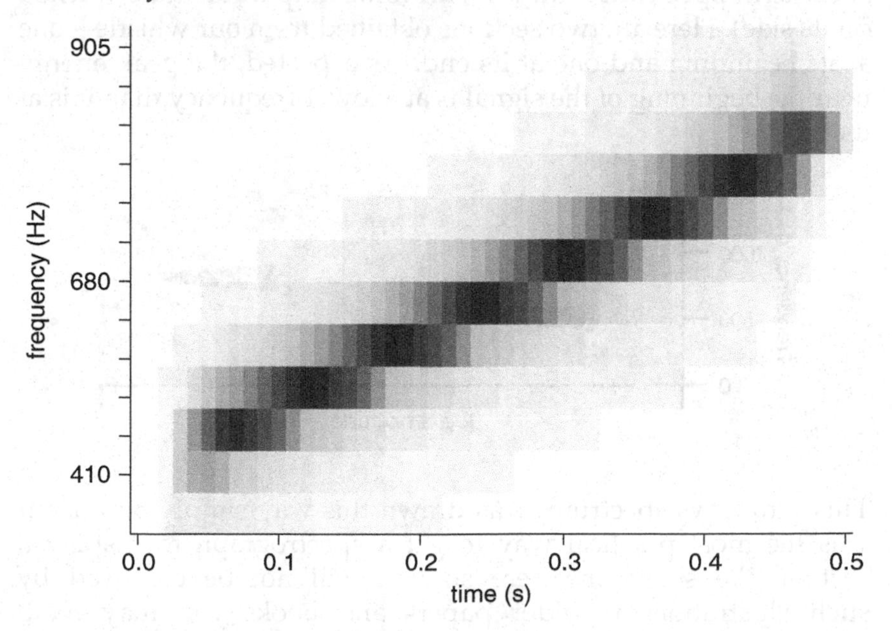

Of course, if there were enough channels in the filter bank (each with its own band-pass filter), and we lined up enough spectra from different moments in time, we'd end up with a spectrogram just like the ones shown earlier.

Sections: short-term spectra

One of the spectrogram's strengths can at times be a drawback. Because we have converted the amplitude dimension into the darkness of a trace, it is impossible to make exact measurements of the amplitudes of individual parts of the spectrogram. This problem is solved by the option of displaying a short-term spectrum at an arbitrary point in time. With the time dimension eliminated, the two dimensions of the paper are then used for amplitude and frequency. This is, in other words, a way to display just amplitude by frequency at a particular time, rather than having to see all three dimensions at once. All commercial spectrographs had such a feature, as do computerized versions as well—the resulting picture is usually known as a *section*. This is nothing more than what we obtained by slicing our spectrographic cake as illustrated above. It is also just like looking at the bar graph from a graphic equalizer at one moment in time.

Sections used to be drawn so that the frequency scale was oriented identically to that of the ordinary spectrogram, and amplitude (linear in decibels) displayed along what is normally the time axis. In other words, a section was nothing more than a short-term spectrum with a logarithmic amplitude scale (turned on its side). Here are two sections obtained from our whistle—one at its beginning and one at its end. As expected, the peak energy near the beginning of the signal is at a lower frequency than it is at its finish:

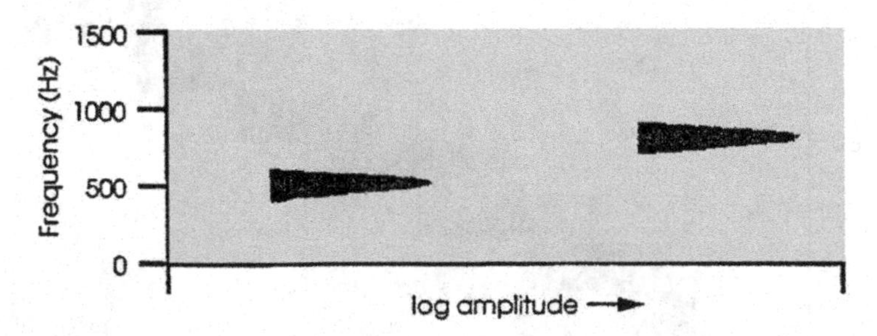

This 'sideways spectrum' was drawn this way simply because it was the most practical way to get a spectrograph to display a section. We show one here so you will not be confused by such illustrations in older papers and books you may read.

Nowadays, with computer spectrograms, sections are typically drawn 'right-side up':

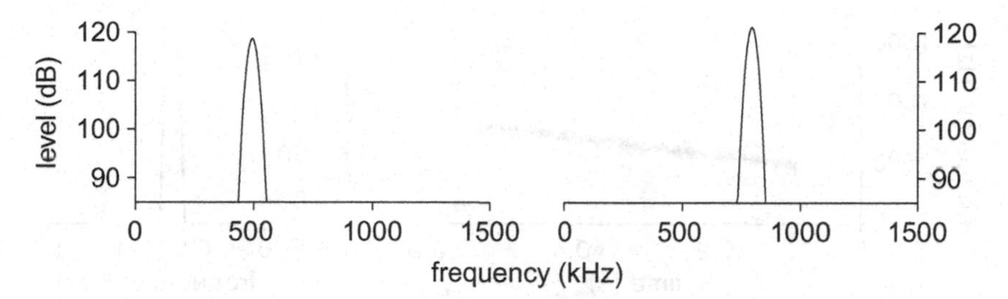

Note the amplitude scales. Because these signals are artificially generated, the dB scale does not have a meaningful reference. The absolute values on the scales are, therefore, arbitrary, although each tick mark does represent a change of 10 dB.

The choice of filter bandwidth

In Chapter 10 we explored in some detail the complementary aspects of wide- and narrow-band analysis. We showed that narrow-band filters, which gave a fine measurement in frequency, were not very good for resolving details in time, and, conversely, that wide-band filters, with their fine resolution of temporal detail, gave relatively rough measurements in the frequency domain. These relationships also apply, of course, to our filter-bank analysis.

It may therefore have surprised you that we've only discussed making a spectrogram with a filter of a single bandwidth. For the analysis of speech signals, 45 Hz is about the narrowest bandwidth that is useful. Therefore, although our determination of the frequency content of a signal will be good, the determination of its time structure will be poor. In order to make fine measurements in time, we need to be able to make wide-band spectrograms (which will, of course, lead to poorer resolution in frequency). Spectrographs thus typically used to have the option of a narrow-band analysis (45 Hz), as we have been using, or a wide-band one (300 Hz). Neither of these specific bandwidths is necessarily appropriate for any particular analysis. Different tasks call for different bandwidths. Computer-generated spectrograms can, in theory, use *any* bandwidth but experience shows that the two traditionally used bandwidths are sufficient in most cases. Note that our discussion here applies to any way of making spectrograms based on a filter bank, so we will not talk separately about the two ways discussed so far.

Let's compare a narrow- and wide-band analysis of our original whistle. Here you see spectrograms (at left) and mid-signal sections (at right) using both analyzing filter bandwidths:

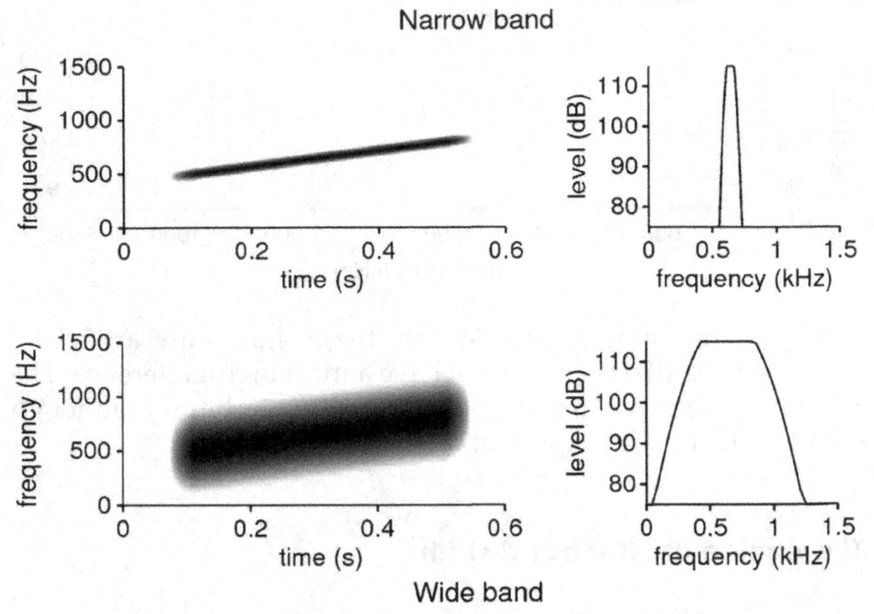

We've already seen the narrow-band analysis. The spectrographic wide-band analysis at left and bottom shows a rather fatter trace. Since frequency is represented vertically on a spectrogram, the poor frequency resolution of wide-band analysis is reflected in a trace that's thicker, which leads to more difficulty in determining its exact frequency. Neither the darkness of the trace, nor its time course, are much affected by the differences between narrow- and wide-band analysis.

Similarly, the wide-band section shows a wider distribution of energy across frequency. Wide filters respond to signals that are relatively remote from their centre frequencies; therefore, even a filter at, say, 700 Hz will respond pretty well to energy at 600 Hz. On the other hand, the narrow-band filter at 700 Hz will not respond much to energy at 600 Hz. Therefore, the spectrum derived from wide-band filtering will be broader than that derived from narrow-band filtering. We often talk of a wide-band analysis as *smearing* information across frequency—another way of saying that the determination of details in the frequency domain is not very good in this case.

Determining a spectrum with filter banks of wide- and narrow-band filters

Smearing in frequency is easier to appreciate by a consideration of the spectrum of a pure tone that is derived from wide- and narrow-band filter banks. We'll ignore the time dimension for the moment, and just assume an infinitely long sinusoid of frequency 1 kHz.

These calculations are easiest done in the frequency domain, since the spectrum of the output of a system can be obtained by multiplying the spectrum of its input by the amplitude response of the system. Let's start with the broader filter which, for computational simplicity, is assumed to have the following simple amplitude response:

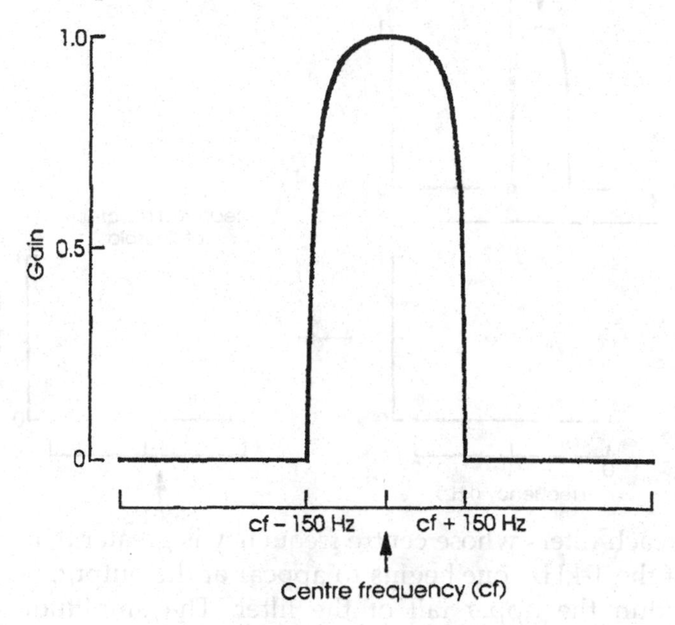

We'll say that this filter has a bandwidth of 300 Hz, although measured in the standard way at 3 dB down from its peak response, it's a little narrower. The spectrum of the input sinusoid is, of course, a single line at 1 kHz.

We begin our analysis with the filter in the filter bank whose centre frequency is lowest, here assumed to be 50 Hz (a rounded up version of the 45 Hz we typically use). We'll assume a very dense spacing of filter centre frequencies, so as not to concern ourselves with jumps in the centre frequency of each filter as we move along the filter bank. For each centre frequency, we plot the amplitude of the signal appearing at the output of the filter. This amplitude is obtained by multiplying the input spectrum by the filter amplitude response. Assuming an amplitude of 1 V for the input, the amplitude of the output is simply the value of the amplitude response at 1 kHz.

Clearly, no energy from the sinusoid will appear from any filter whose centre frequency is less than 850 Hz, since the edge of the filter only extends 150 Hz from its centre frequency. In this region, therefore, the amplitude response is always zero at 1 kHz—resulting in no output signal. Here is a pictorial representation of the situation at this point, when the filter centre frequency is 850 Hz:

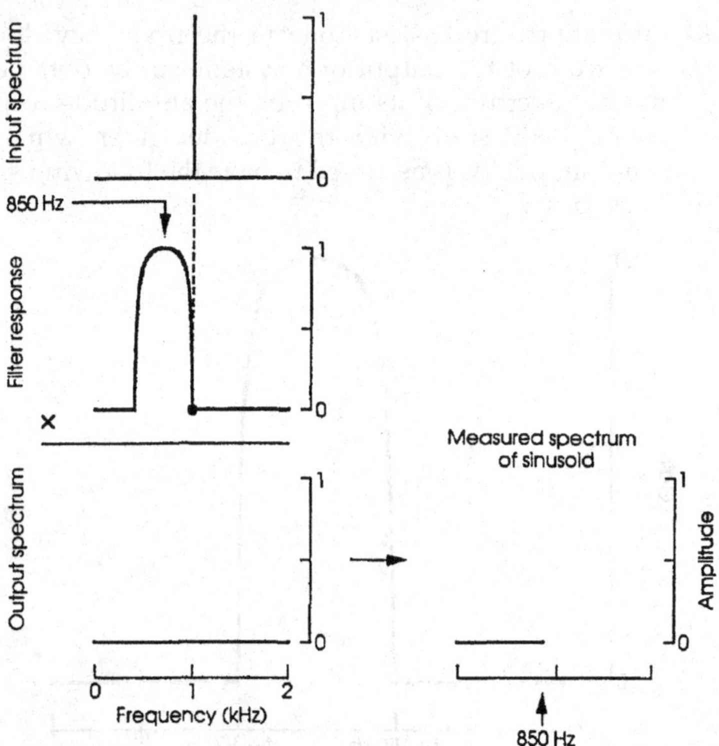

As we reach filters whose centre frequency is greater than 850 Hz, some of the 1-kHz tone begins to appear at the output, because it falls within the upper tail of the filter. The amplitude of this output increases as the filter centre frequency increases, until we reach a centre frequency of 1 kHz:

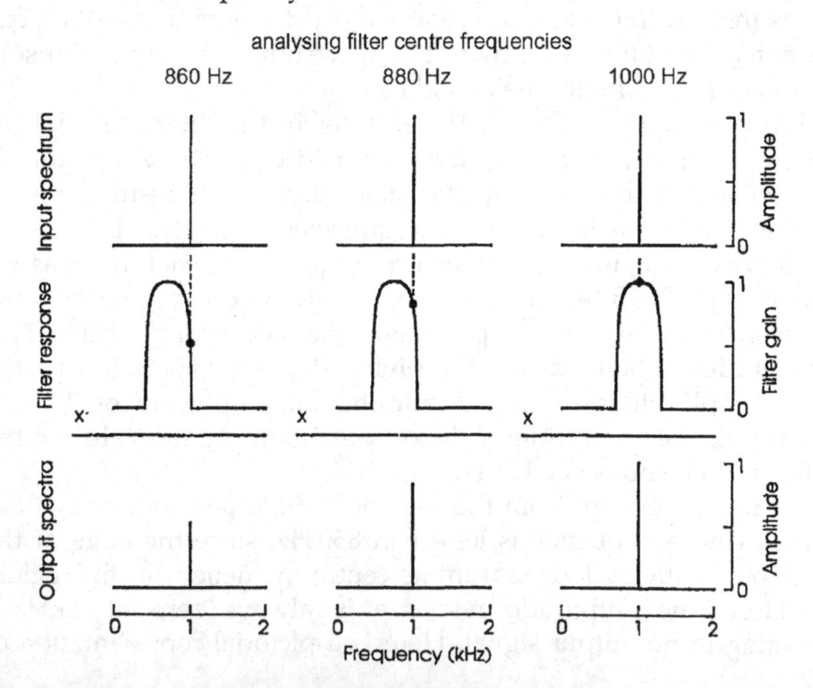

Here are the results of the filter-bank analysis so far, with the points we explicitly calculated marked with filled circles:

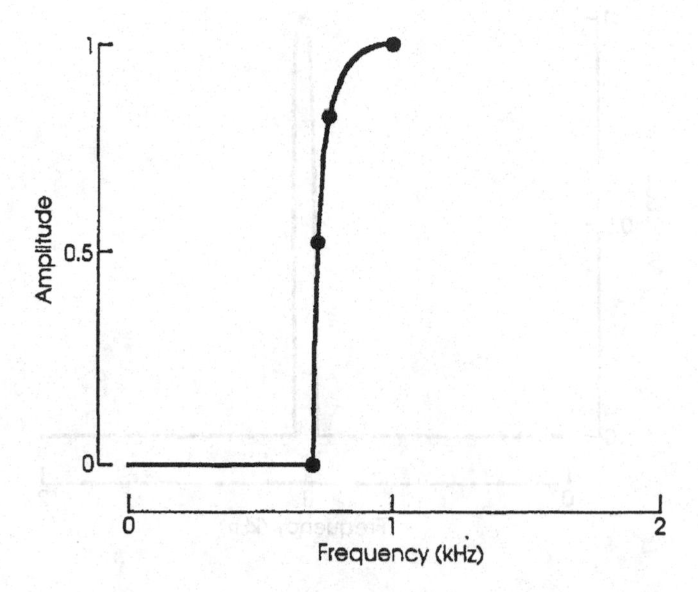

As filter-centre frequencies increase above 1 kHz, the amplitude of the output will decrease, as the input signal falls onto the lower tail of the amplitude response of the analyzing filter. This decrease will continue until a centre frequency of 1150 Hz is reached, at which point the signal will just pass out of range of the filter. For centre frequencies from here upwards, the output of the filter will be zero. Therefore, the final outcome of our spectral analysis becomes:

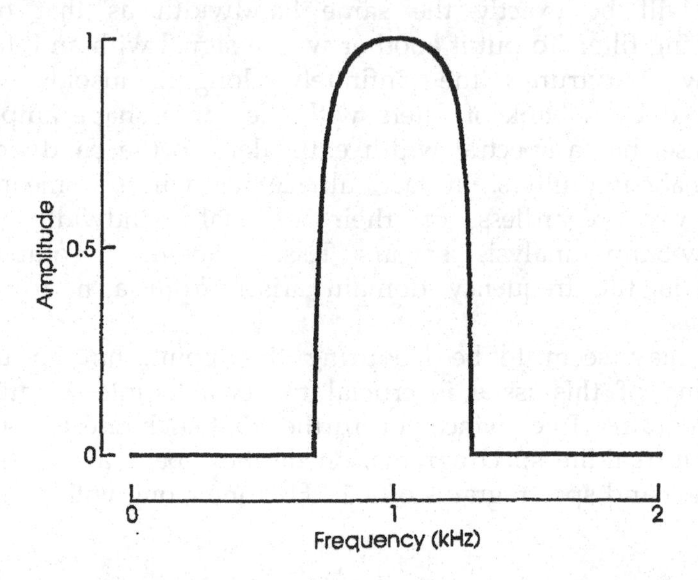

If we now perform this same analysis with a filter of the same shape but with a bandwidth of 45 Hz, we obtain:

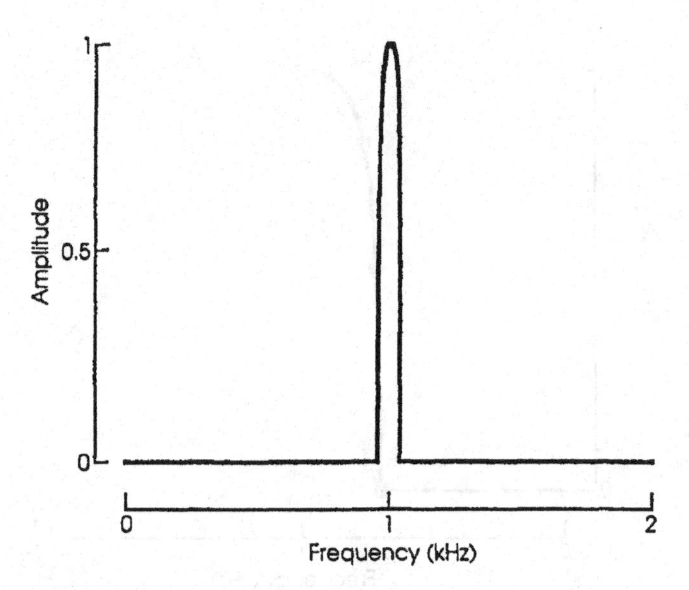

Clearly, the narrow-band analysis leads to a more tightly packed spectrum. We might also say that the narrow-band analysis *smears* the frequency spectrum less. A careful look will also show that the shapes of the spectra obtained by the filter-bank analysis are exactly the same as those of the filter amplitude responses. This is not generally true and here is a result of the symmetry of the filter shapes (see 'Exercise 5' at the end of the chapter). It is always true, however, that the bandwidth of the spectrum of a sinusoid derived using a filter bank will be exactly the same bandwidth as that of the analyzing filter. To put it another way, a signal with an infinitely narrow spectrum (the infinitely long sinusoid), when analysed by a bank of filters with the same shape amplitude response, has a spectral width equivalent to the bandwidth of the analyzing filters. In fact, all signals will be smeared in this way, regardless of their original bandwidth. Since narrow-band analysis smears less, a better resolution of detail in the frequency domain arises from a narrow-band analysis.

We may seem to be labouring this point, but an understanding of this issue is crucial to sensible interpretation of spectrograms. To convince you further that such effects really do occur in real-life spectrograms, on the next page are wide- and narrow-band spectrograms of a 1-kHz tone along with sections:

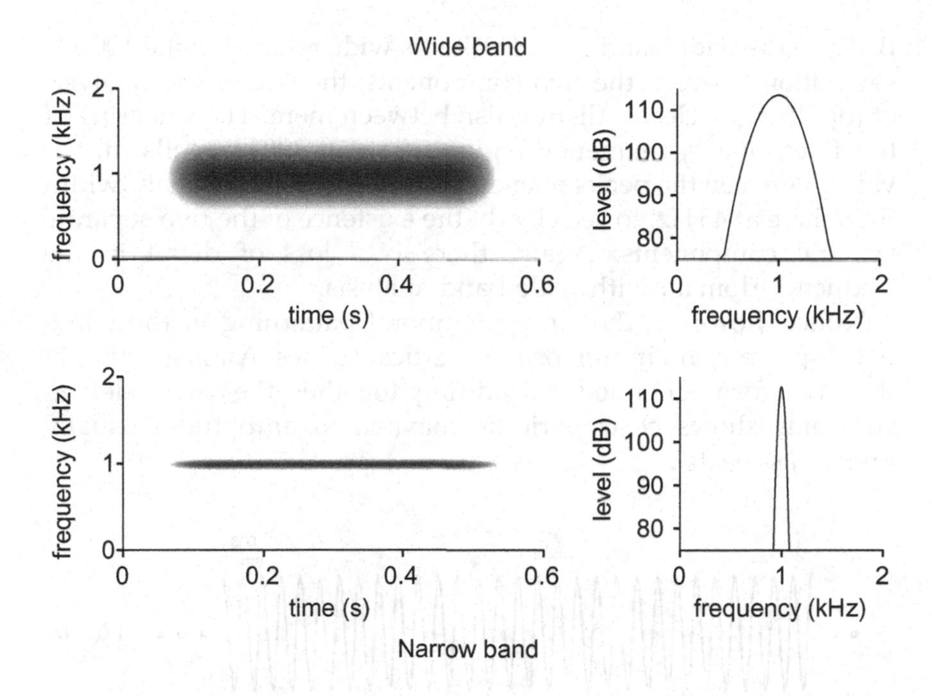

Resolving two spectral components close in frequency

The effects of spectral smearing can be more clearly seen in analyzing a signal that is the sum of two sinusoids relatively close in frequency. Here are wide- and narrow-band spectrographic analyses of a signal consisting of two equal amplitude sinusoidal components at 1.0 kHz and 1.1 kHz:

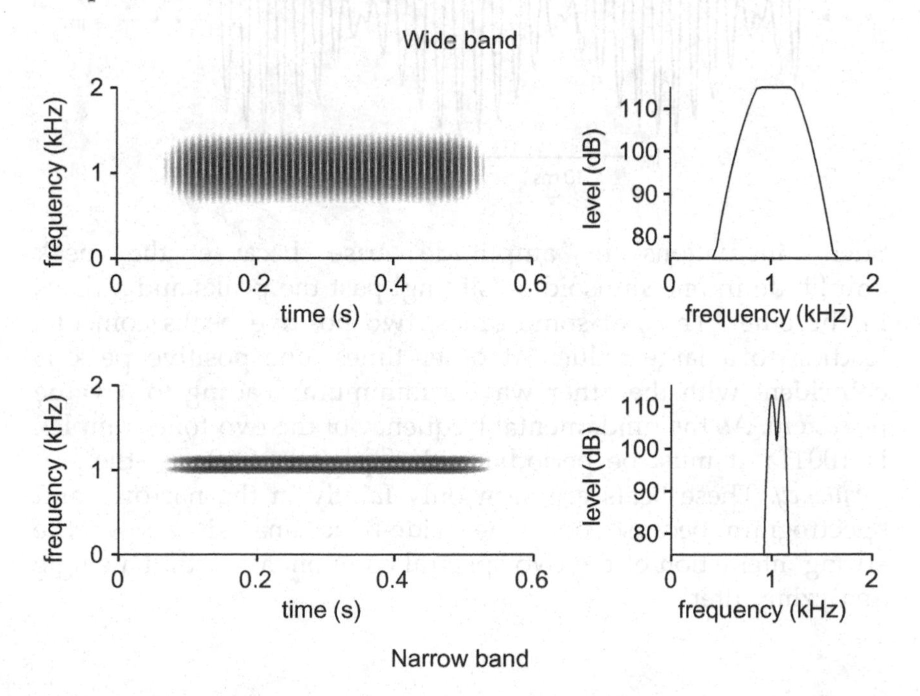

Because the wide-band 300-Hz filter is wide relative to the 100-Hz separation between the two components, the wide-band analysis at top does not clearly distinguish between them. The smearing of the filter, over a frequency region of about 300 Hz, 'fills in' the valley between the peaks of energy. It takes the narrow bandwidth filter, here at 45 Hz, to see clearly the existence of the two separate spectral components. Again, there is a loss of detail in the frequency domain with wide-band analysis.

Notice, however, the strong temporal patterning in the wide-band spectrogram in the form of vertical stripes. An inspection of the waveform obtained by adding together the two relevant sinusoids shows clear periodic maxima in amplitude, usually known as 'beats':

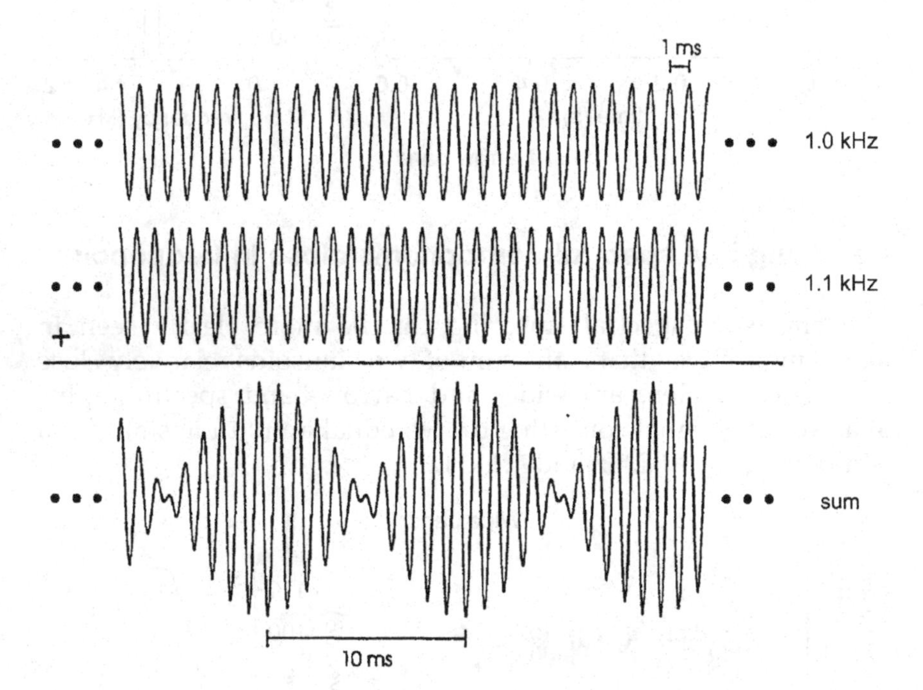

Such fluctuations in amplitude arise because the peak amplitude in one sinusoid is 'sliding' past the peaks and valleys in the other. Thus, at some times, two positive peaks coincide, leading to a large value. At other times, one positive peak is coincident with the other wave's minimum, leading to a value near zero. As the fundamental frequency of the two-tone complex is 100 Hz, it must be periodic with a period of 10 ms—the *beat frequency*. These beats are seen only faintly in the narrow-band spectrogram because only the wide-band analysis allows the strong interaction of the two spectral components within a single analyzing filter.

Wide- and narrow-band spectrograms of an impulse

So far in this chapter we've only talked about the drawbacks of wide-band analysis—smearing in frequency. We well know, however, that what we lose in the frequency domain we gain in the time domain. Since we determined the relative fineness of frequency resolution using a signal infinitely narrow in frequency (a sinusoid), it makes sense to investigate the degree of temporal resolution with a signal that is infinitely narrow in time (an impulse). True impulses are, of course, an idealization, but we can use a relatively narrow rectangular pulse that is 1/2-ms wide and see what results:

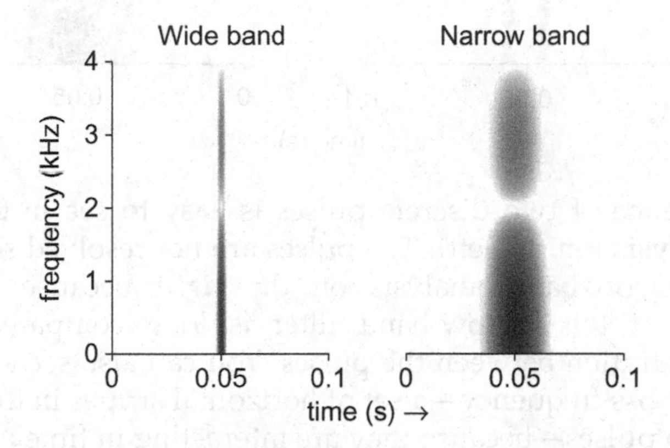

First note that both analyses show a minimum of energy, near about 2 kHz. This is a reflection of the fact that the spectrum of a rectangular pulse goes to zero near the reciprocal of its duration (see Chapter 10). More importantly, the narrow-band analysis on the right leaves a trace that is much more spread out in time than the trace obtained from wide-band analysis. Hence we say that a narrow-band analysis *smears* in *time*. This is easy to understand if we recall from Chapter 10 the relationship between a filter's impulse response and its bandwidth. Filters with a narrow bandwidth 'ring' for a relatively long time compared to filters with a wide bandwidth. Therefore, the spectrographic trace for narrow-band analysis is more spread out than that for wide-band analysis.

Resolving two pulses close in time

This smearing in time is more easily appreciated by examining the spectrographic traces obtained from the analysis of two short pulses close together in time—here separated by 10 ms:

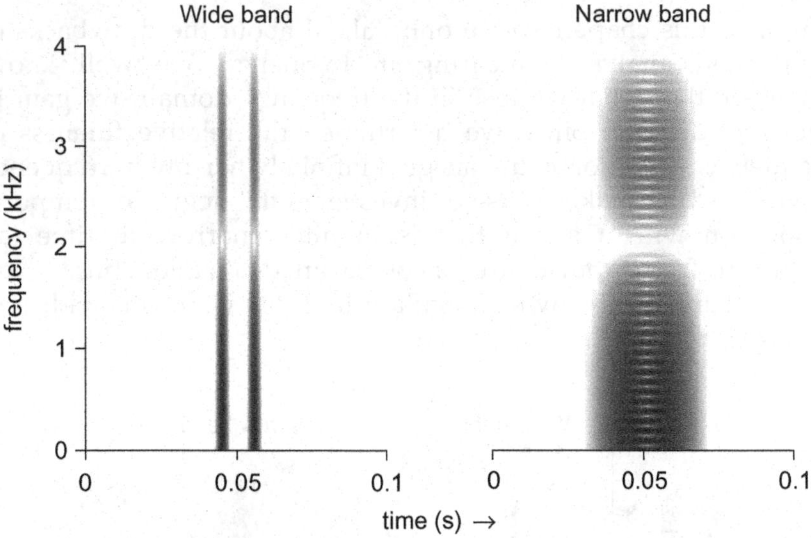

The existence of two discrete pulses is easy to see in the wide-band analysis (on the left). The pulses are not resolved separately in the narrow-band analysis on the right because the time smearing of the narrow-band filter is long compared to the 10-ms separation between the pulses. You can also see a complex pattern across frequency—a set of horizontal stripes in the middle of the two pulses—because they are interacting in time in a single filter (just as above there was an interaction pattern for closely spaced frequencies when wide-band filtered).

Wide- and narrow-band spectrograms of quasi-periodic pulse trains

Spectrographs would hardly be necessary if only signals as simple as these required analysis. Consider now a rather more complicated signal than we have looked at previously: a train of narrow pulses that changes smoothly in repetition rate from 75 Hz to 150 Hz over about 1 s. Before looking at spectrograms, let's think a little bit about how the spectrum of this pulse train changes in time. Because the pulse repetition rate is changing relatively slowly, each small portion of the signal can be treated as if it is periodic. Therefore, at the beginning of the signal, the spectrum will be similar to that of a periodic pulse train at 75 Hz with components at the fundamental (75 Hz) and at its multiples (150 Hz, 225 Hz, 300 Hz and so on). Similarly, at its finish, the spectrum of the signal will be similar to that of a pulse train at 150 Hz with components at 150 Hz, 300 Hz, 450 Hz and so on. For the entire signal, the overall spectral envelope will be fairly flat over the 1.5-kHz range being analyzed, because the pulses are fairly narrow (0.2 ms in duration). Look first at the narrow-band spectrogram along with sections from near the beginning and end of the signal:

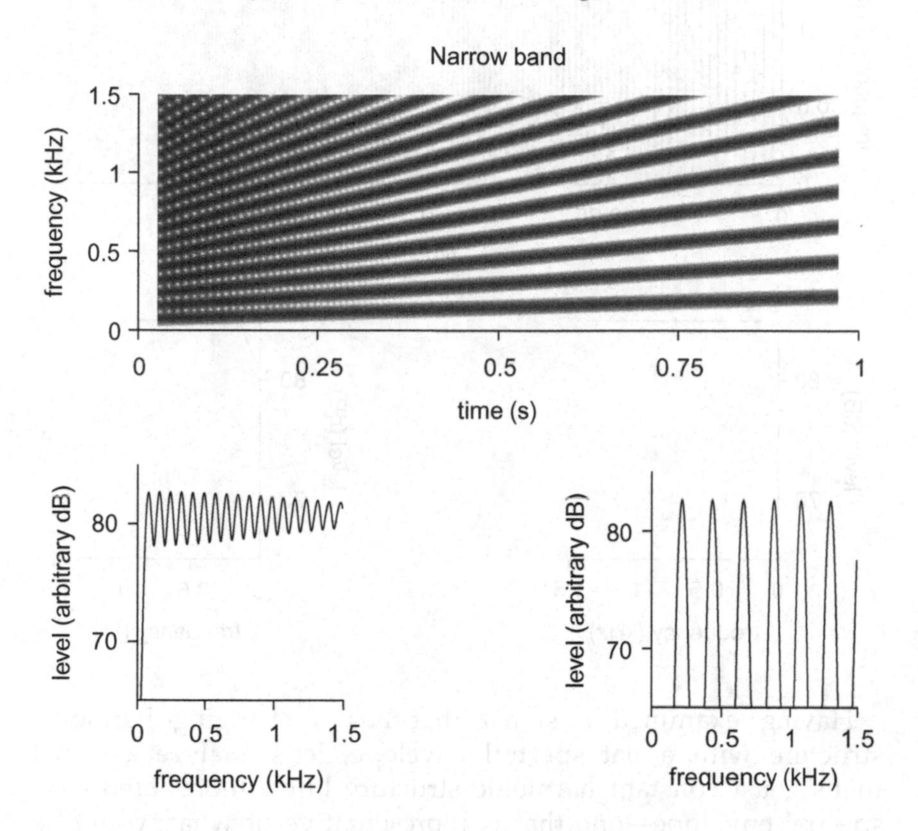

In the spectrogram, the individual harmonics are clearly seen as they double in frequency over the duration of the signal. The sections show the harmonics reflecting the approximately 75-Hz periodicity in the spectrum near its start, and the approximately 150-Hz periodicity near its finish, as well as the relative flatness of the spectral envelope. On the other hand, the smearing across time means that there is no indication in the spectrogram of the times at which the individual pulses arrived. They have all been joined together into continuous harmonic lines.

The wide-band analysis on the next page presents a completely complementary picture. Here, the good time resolution allows the spectrogram to show each pulse separately. The increase in pulse repetition rate (equivalently, the shortening of the time between pulses) is reflected in the shortening of the distance between each pulse (represented as a vertical line). Information is badly smeared in the frequency domain, however, so that the individual harmonics which make up the signal are not visible. The sections are just one great mass with little change visible over the time course of the signal.

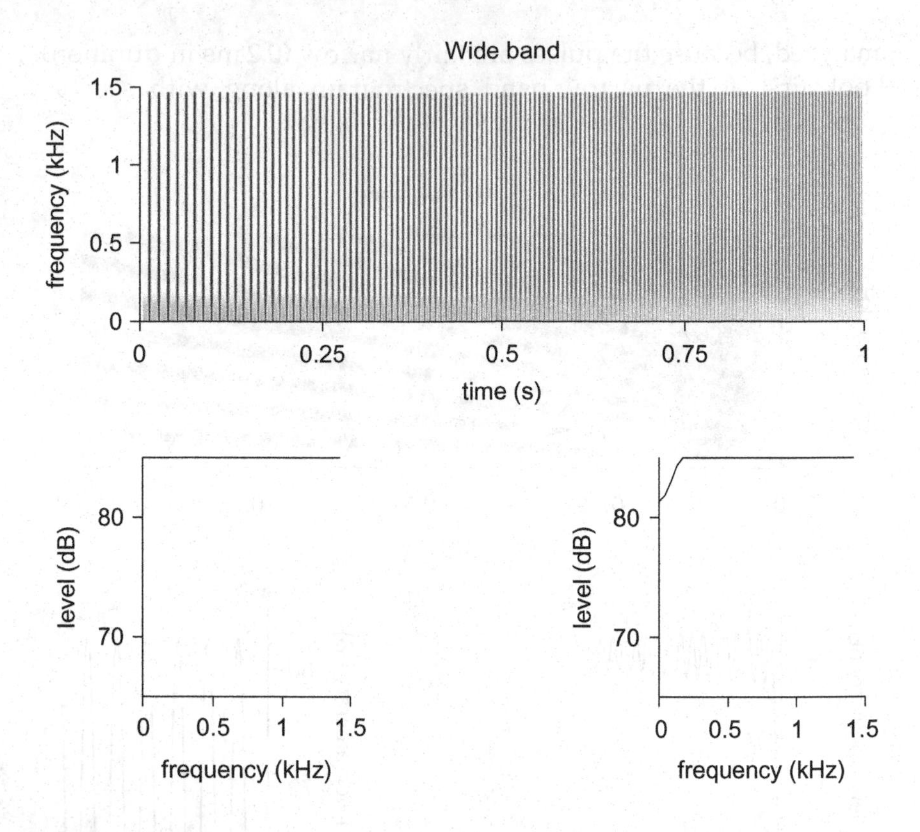

Having examined a signal that has a changing harmonic structure with a flat spectral envelope, let's analyze a signal that's got a constant harmonic structure but a more interesting spectral envelope—one that is representative of what would be found in a sustained vowel sound. Such a signal can be generated by putting a 100-Hz train of narrow pulses—considerably narrower than those used above and so narrow enough to be treated as impulses—through a filter with two resonances (at 700 Hz and 2200 Hz), creating *formants* (bands of increased energy):

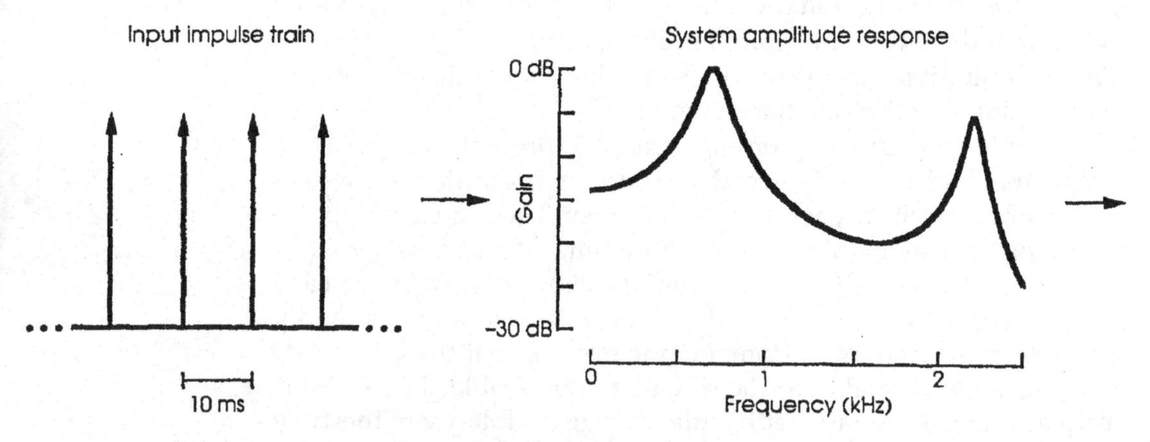

As before, the wide-band spectrogram (at top) shows each individual pulse and the narrow-band spectrogram shows the individual harmonics of 100 Hz:

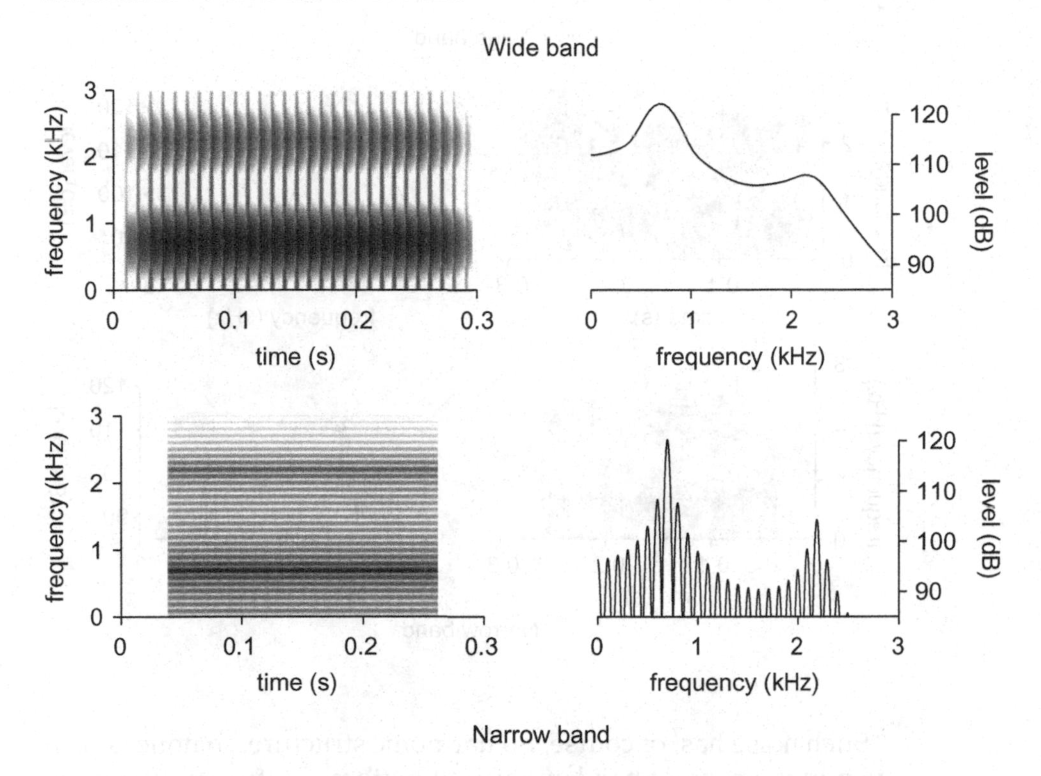

However, the overall spectral envelope of the signal is just as clear in the wide-band analysis as in the narrow-band analysis (as the sections clearly show) even though the narrow-band analysis also shows the individual harmonics. This is only true because the width of the peaks in the spectral envelope (the formants) and their separation in frequency are greater than the bandwidth of the wide-band filter. This situation typically occurs in speech sounds, so a wide-band analysis is often well-suited to the analysis of spectral envelopes there. It is, therefore, wrong to think that wide-band analyses are useful only for analyzing the temporal structure of signals—they can display spectral structure of the appropriate sort.

Wide- and narrow-band spectrograms of random signals

Up until now, we've only considered spectrographic analysis of signals that are at least quasi-periodic or transient. Many signals of interest, however, are truly random. Here are spectrograms of a random signal, created by filtering white noise with the system just used to filter 100-Hz pulse trains:

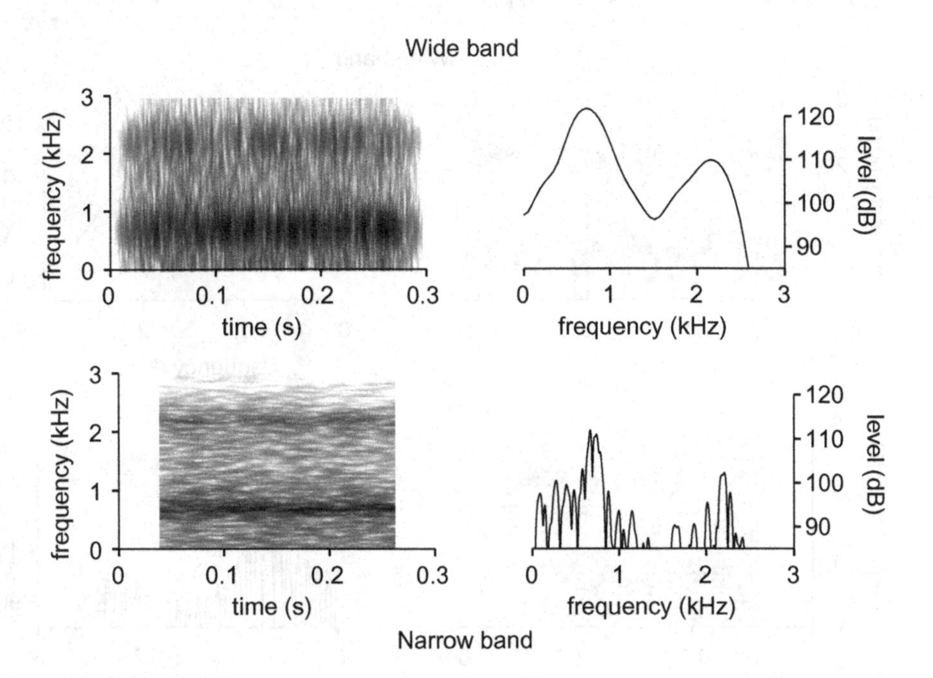

Such noise has, of course, no harmonic structure, so none is seen even in the narrow-band analysis at bottom. On the spectrograms we see instead a random pattern of markings that tend to cluster in distinct frequency regions. On average, there is greater energy in the noise at certain regions than at others, due to the action of the filter. Within this region, though, the energy varies from moment to moment. This is the basic nature of noise and is responsible for the randomness of the markings in time. Note, though, that the random marks in the wide-band spectrogram tend to be oriented in a more vertical direction than in the narrow-band spectrogram, whose marks are more horizontal. This results from the difference in the temporal resolution of the two sets of filters, the narrow-band filter 'stretching' little peaks of noise over time.

The overall spectral shape is apparent in the sections obtained with both bandwidths. However, the wide-band analysis gives a smoother curve that is much more representative of the amplitude response of the filter used to create this signal than does the narrow-band analysis. This results from the frequency smearing of the wide-band filter doing a sort of averaging across frequency. Thus, wide-band analyses can be advantageous for the determination of spectral envelopes and are not (as mentioned above) only appropriate for the analysis of temporal structure.

Making spectrograms in the time domain

You should now have a pretty good idea of how a filter bank can be used to construct a spectrogram, and how choices about the properties of that filter bank (in particular, the bandwidth of the filters used) will be reflected in the spectrograms obtained. You may now be surprised to learn that spectrograms made by computer typically do *not* use the filter-bank approach we have so thoroughly detailed here. Instead they use a technique of dividing up the time waveform into short sections, and then calculating the spectrum of each of those short sections directly. Once the amplitude variations are recoded from the height of a graph into the darkness of a trace, it is then a simple matter to construct the spectrogram in a way analogous to what was done using a graphic equalizer (see pp. 226–227).

Such techniques were only practical once computers were readily available. We will now explain this approach, not only because of its ubiquity in much software, but also because it will give you another way of thinking about how a spectrogram is made. In the end, however, the spectrograms that result are interpretable in precisely the same way, no matter what technique is used to generate them.

Dividing up a signal into sections

In order to demonstrate this approach, we'll return to the whistle we analyzed previously. Let's be clear again about exactly what it is we want to do—to display the spectral characteristics of a wave as it evolves in time. We'll start by considering just the beginning and end of the whistle, by extracting sections that are 30 ms long. As was discussed in Chapter 10, it will be much better for us to *window* the signals so that their onsets and offsets are ramped up and down somewhat gradually:

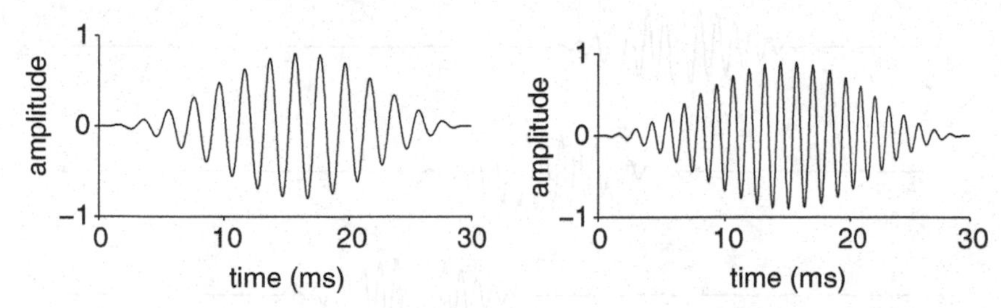

You can already see in the waveforms themselves that the frequency of the sinusoid is lower at its beginning than at its end. The most straightforward way to look at the spectral properties of these segments is, of course, simply to calculate their spectra. You may recall, from the end of Chapter 9, that we do this using a *Fourier transform,* so replacing filters with mathematical manipulations only practical on a computer:

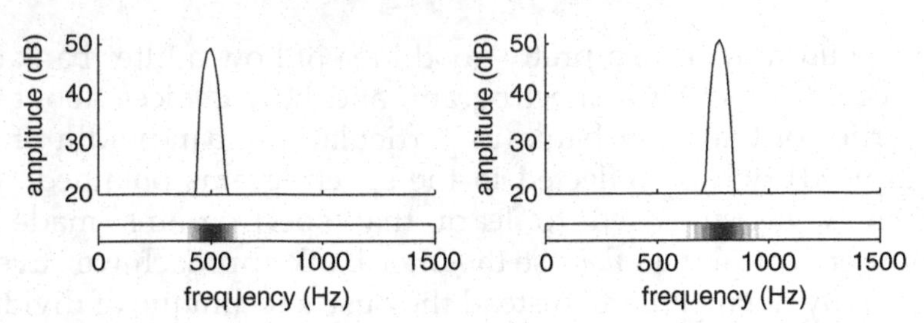

We are now most of the way to a spectrogram! By immediately recoding the amplitude of the spectrum into the darkness of a trace (as done at the bottom of the two spectra above), we now have spectra that look very similar to those we derived from the bar graph of a graphic equalizer (see p. 226). The main difference is that the spectra here, calculated directly from the waveform, have values at many, many frequencies, whereas the bar graphs were limited to a small number of discrete frequencies (for example, 6 in the figure of p. 225). But this difference is really just a cosmetic one. By having many filters in the filter bank of a graphic equalizer, it would be possible to have the spectrum defined at as many frequencies as it is here.

It should now be obvious to you how we can construct a spectrogram from displays like these. First, we need to window the original signal frequently enough (typically, segments overlap by half or less of the window length):

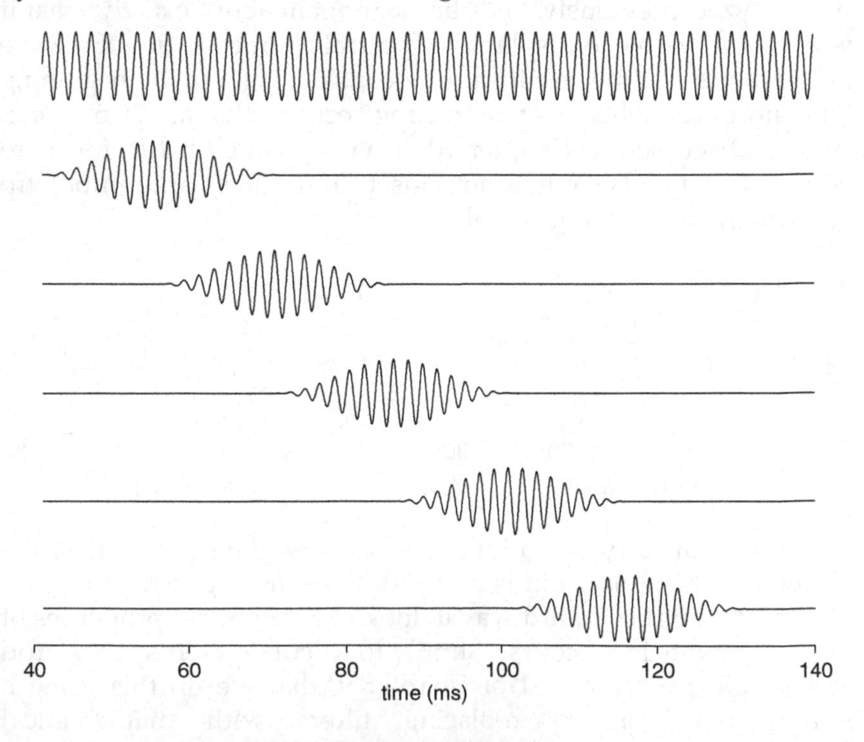

Then, a spectrum is calculated for each of these windowed signals, followed by the recoding of the amplitude dimension of the spectrum into the darkness of a trace. Finally, just as we did for the rows of shadowed cells derived from a bar graph, these horizontal 'shaded' spectra are flipped up and arranged across the page:

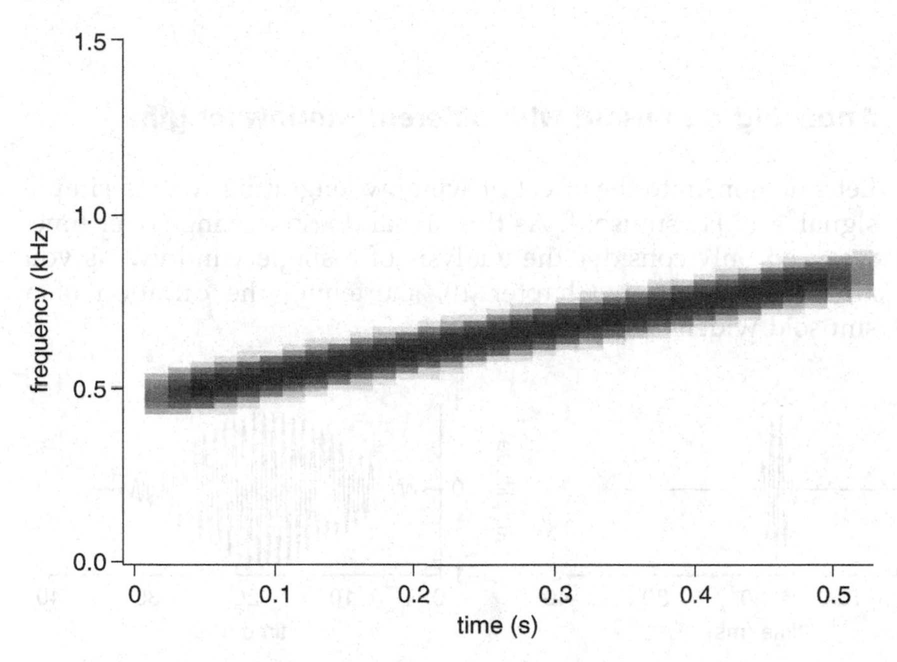

The 'blockiness' across time in this spectrogram results from the fact that the window was moved in 15 ms jumps (half the window length). If shorter time steps had been used, the picture would have been less 'blocky'.

The choice of window length

We have now shown another way of making spectrograms, and as you can see, the spectrograms that arise are more or less the same as ones we constructed with a filter bank. Yet one thing may be puzzling you. Given that filter-bank-derived spectrograms can look so very different, depending upon the choice of filter bandwidth, there should be a parameter in this time-based spectrographic method which also changes the 'look' of the spectrogram. Given the trade-off we have discussed between time and frequency in Chapter 10, it should not be too surprising to learn that time-based spectrograms will look very different depending upon the *duration* of the speech segments analyzed. Roughly speaking, window lengths of about 4–5 ms correspond to a wide-band analysis, whereas window lengths of about 30–40 ms correspond to a narrow-band analysis. It should make some sense to you that the wide-band analysis (smearing in frequency) corresponds to a short-time window (good resolution in time) whereas narrow-band analysis (good resolution in frequency) corresponds to a long-time window (smearing in time).

Analyzing a sinusoid with different window lengths

Let's demonstrate the effect of window length first with a simple signal, a 1 kHz sinusoid. As this signal doesn't change over time, we need only consider the analysis of a single window. As you already know from Chapter 10, shortening the duration of a sinusoid widens its spectrum:

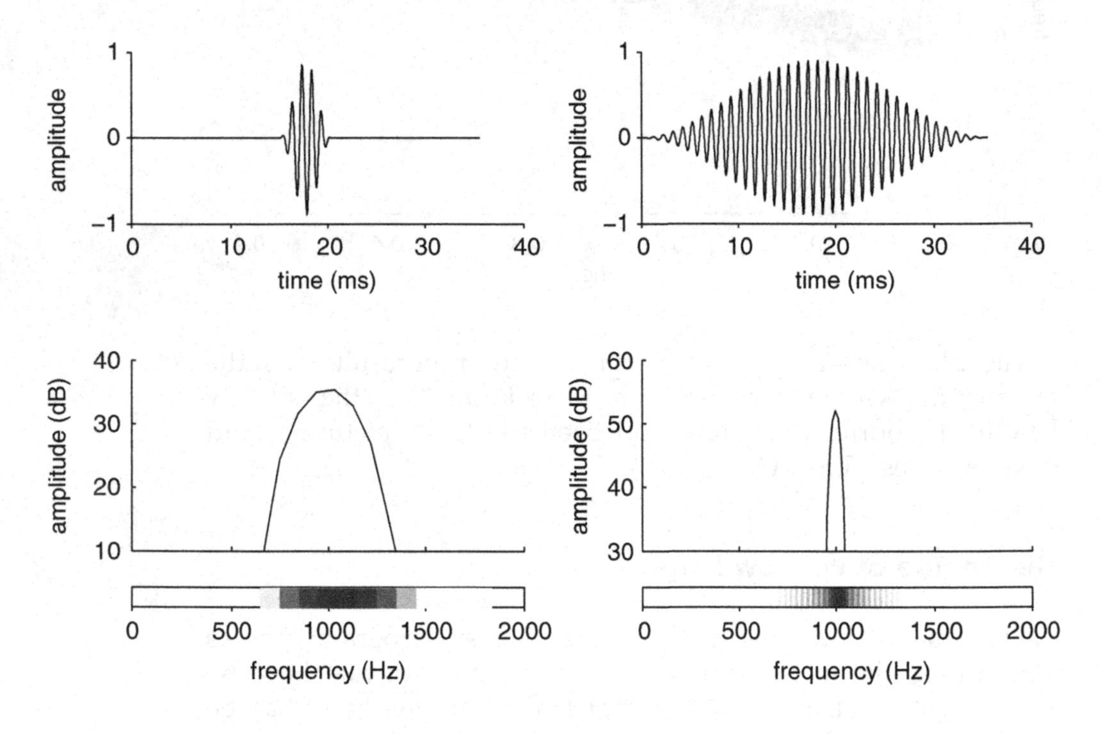

Every window of this signal will look more or less identical, so the spectrogram would simply consist of many of these 'shaded' spectra flipped upright and aligned across the page. Because the shading for the short-time window is considerably wider than the shading for the long-time window, the short-time window will show a much fatter trace than the long-time window, as you have seen before at the top of page 235.

Long- and short-time spectrograms of two closely spaced sinusoids

We have already shown above that a narrow-band analysis does a better job of resolving two sinusoids that are close in frequency than does a wide-band analysis. We would therefore expect a long-time analysis to be better than a short-time analysis in exactly the same way. This can be most clearly seen by comparing the waveforms and spectra obtained by windowing the waveform of the sum of 1000 Hz and 1100 Hz. Let's look first at the long-time analysis:

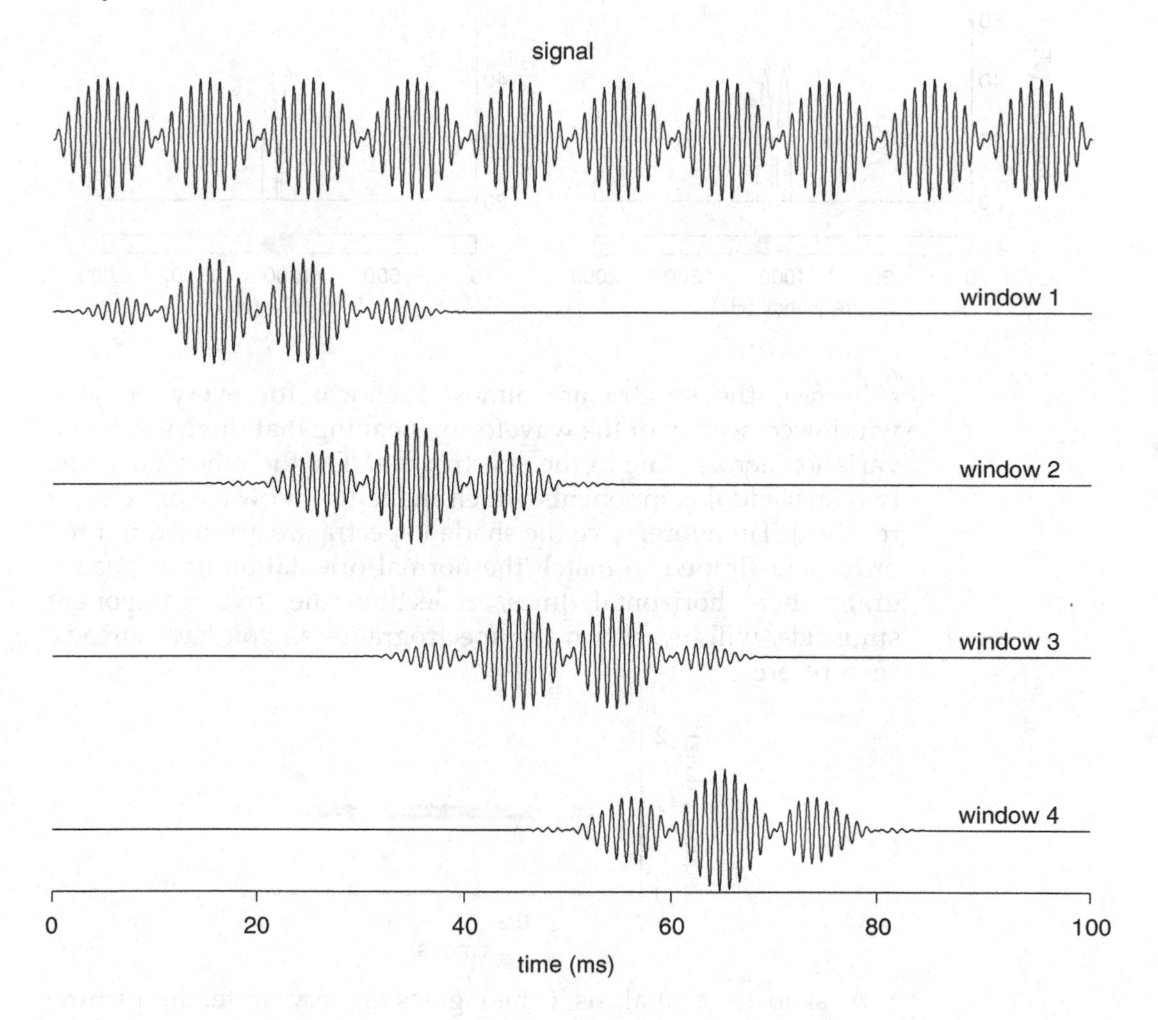

You can see that the long-time window (40 ms) extends over multiple peaks in the waveform, spanning two or three cycles of the 'beat'. Let's look more closely at the first two windowed sections. Although the time waveforms look a bit different in detail, the spectra associated with each are very similar, as you can see here:

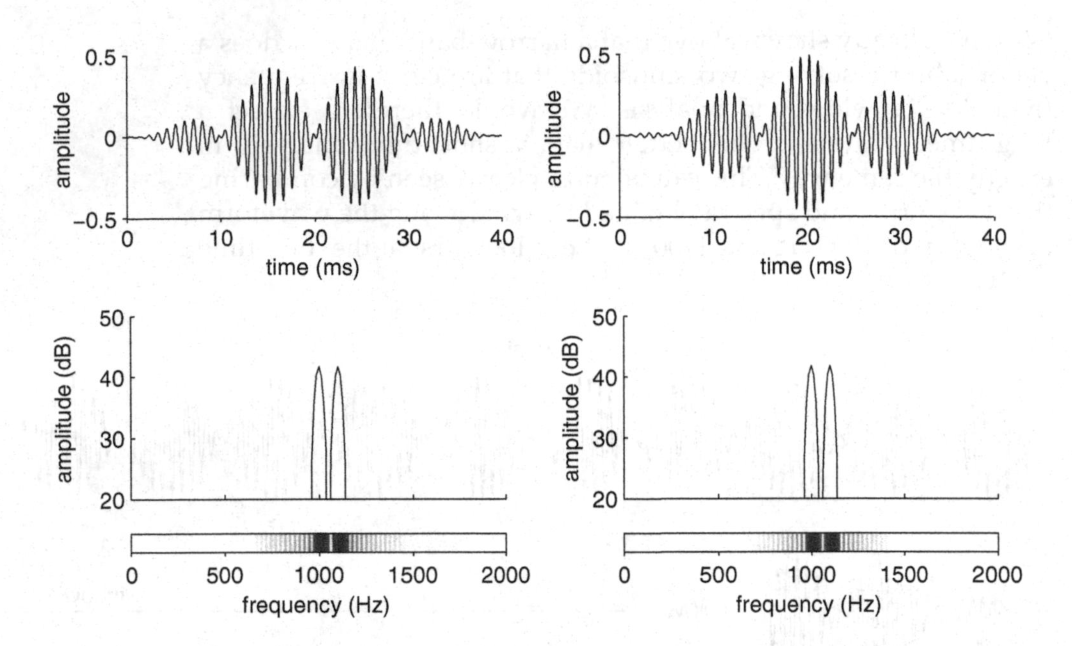

In fact, the spectra are almost identical for every possible windowed section of the waveform, meaning that there will be no variation across time in the spectrogram. On the other hand, the two sinusoidal components which make up the wave are clearly resolved. Therefore, once the shaded spectra are arranged in time order and flipped to match the normal orientation of a spectrogram, two horizontal lines, reflecting the two component sinusoids, will be seen in the spectrogram—as you have already seen before:

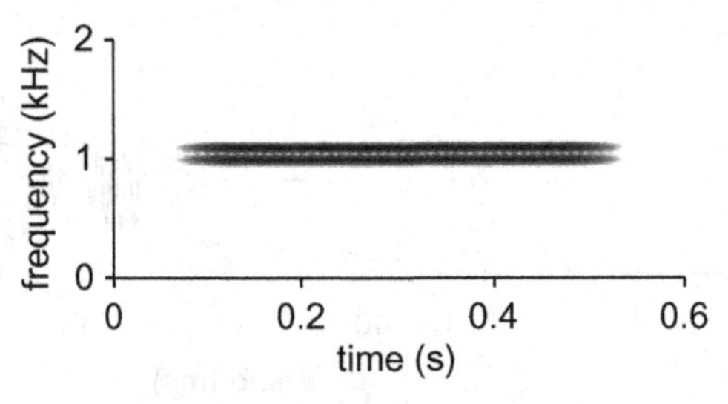

A short-time analysis (4 ms) gives a very different picture. Here the short duration of the window means that the windowed wave varies considerably in overall amplitude throughout the span of the waveform, depending upon whether the window is centred near a maximum or minimum of the beating waveform:

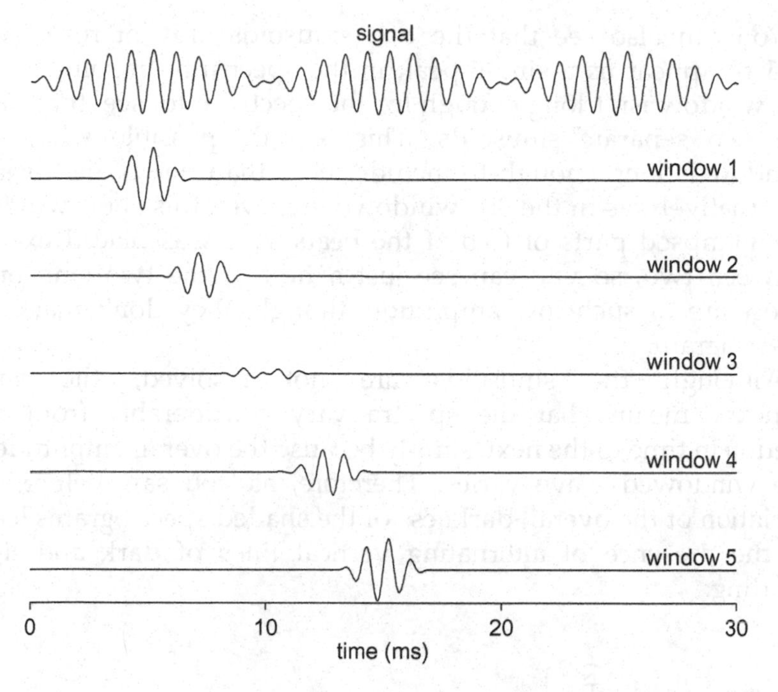

Consider only the 2nd, 3rd and 4th windowed sections of the wave shown above. When we calculate the spectrum associated with the two sections of high amplitude, you can see that high amplitude reflected in the height of the spectrum and also in the dark black of the shaded spectra (at left and right in the figure here):

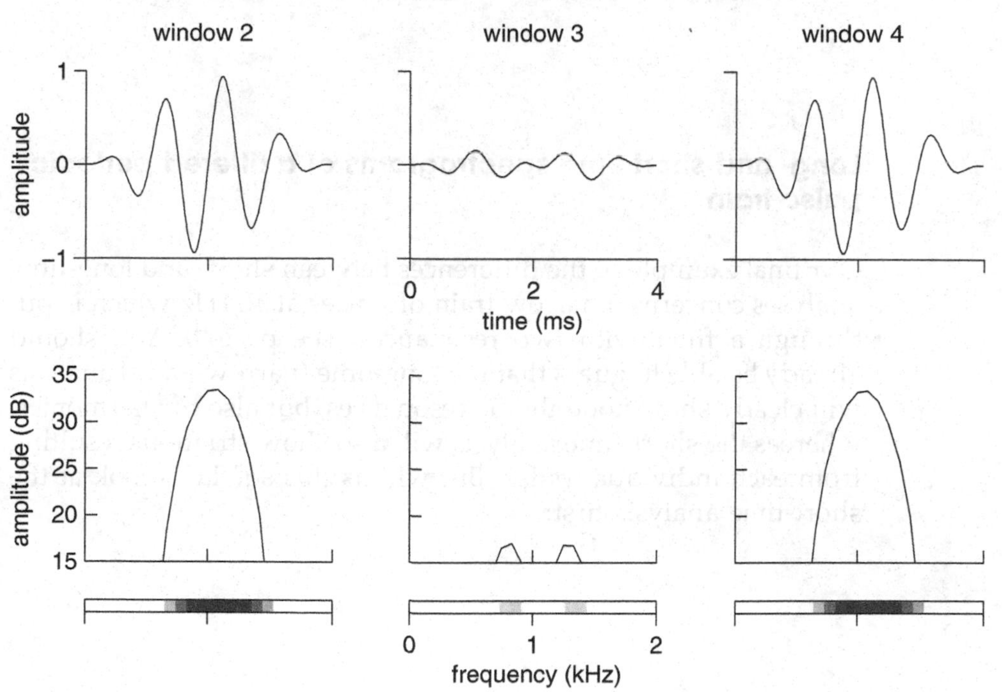

You can also see that the two sinusoids are not resolvable, and so appear as a single peak in the spectrum. In some sense, the window isn't long enough for the spectrum to 'see' that there are two separate sinusoids. This is only possible when the window is long enough to include more than one of the 'beats'. Alternatively, as in the 3rd window here, even this short window has glimpsed parts of two of the beats as it was placed exactly between two, so you can see just a hint of the two sinusoids. These are of such low amplitude, though, they don't mark the spectrogram.

Although the sinusoids are not resolved, the short window means that the spectra vary considerably from one section in time to the next, simply because the overall amplitude of the windowed wave varies. Therefore, as you saw before, the variation of the overall darkness of the shaded spectrograms leads to the presence of alternating vertical lines of dark and light shading:

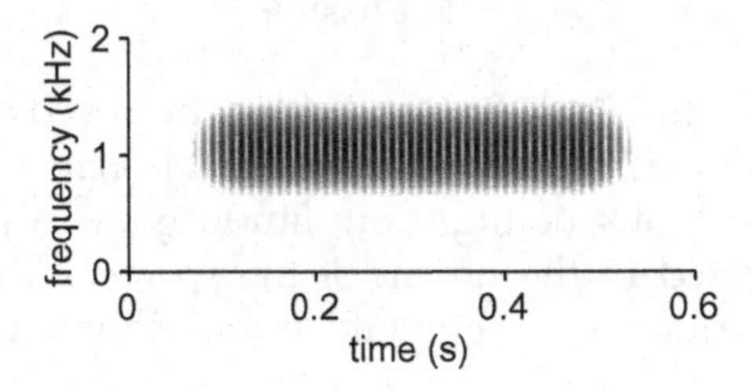

Long- and short-time spectrograms of a filtered periodic pulse train

Our final example of the differences between short- and long-time analyses concerns a narrow train of pulses at 100 Hz which is put through a filter with two resonances (see p. 240). You should already be able to guess that the long-time (narrow-band) analysis will clearly show not only the resonances, but also the harmonics, whereas the short-time analysis will also show striations resulting from each individual pulse. But why is this so? Let's look at the short-time analysis first:

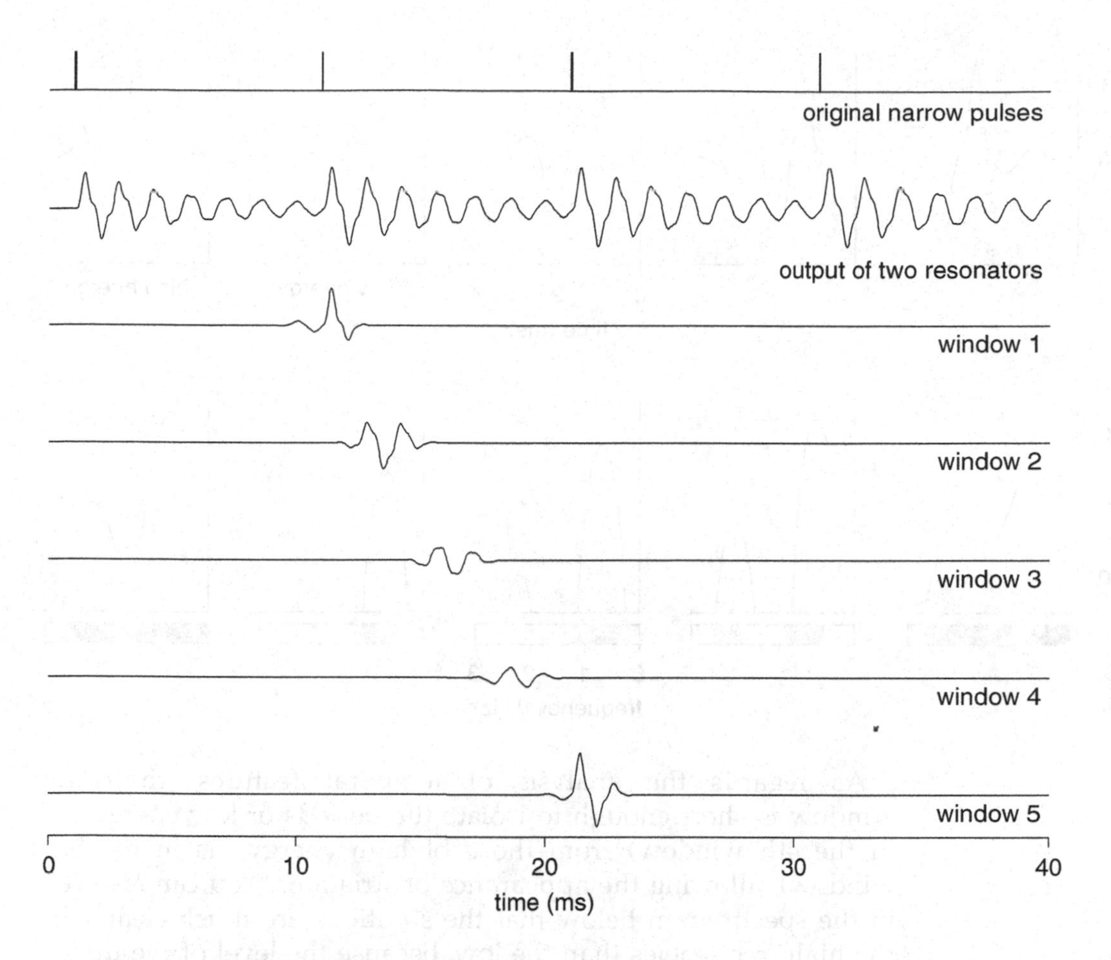

You can see the original narrow pulses at the top of the figure, with the output wave from the resonators just below. This should make the periodicity of the output wave clear. Every pulse results in a cycle of the output wave that is large just after the pulse, and then decays away.

You can also see that the short-time window always includes less than a single cycle of the periodic wave. Therefore, this analysis never 'sees' the periodicity of the wave, so no harmonics will be represented on the spectrogram. Because the window is about the same duration as the main part of the response to a single pulse, the spectrum is being calculated on what is effectively the impulse response of the system used to create the formants (as long as the window is centred on the decaying pulse). Therefore, the spectrum of this pulse is, of course, similar to the amplitude response of the two-resonance filter, as in all but the 4th window on the next page:

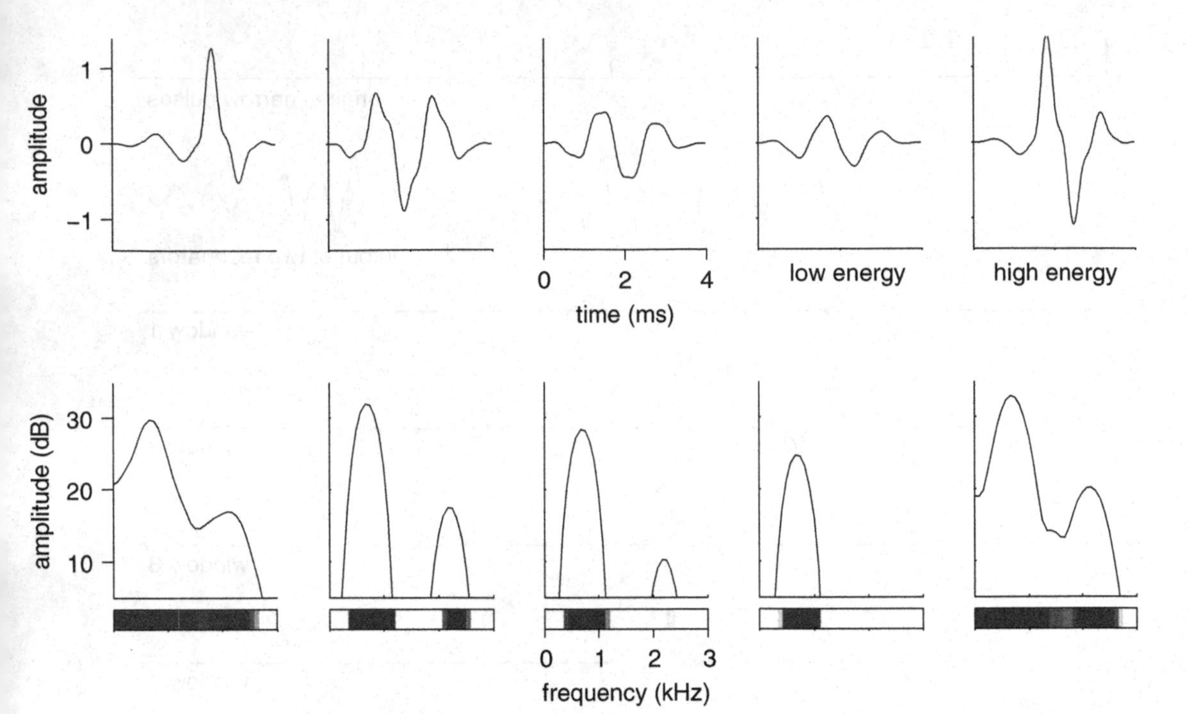

As regards the analysis of temporal features, the time window is short enough to isolate the periods of low energy (as in the 4th window) from those of high energy (as in the 5th window), allowing the appearance of striations. You can also see in the spectrogram below that the striations are much clearer in the high frequencies than the low, because the level of the upper formant is varying more across the windowed sections of the waveform:

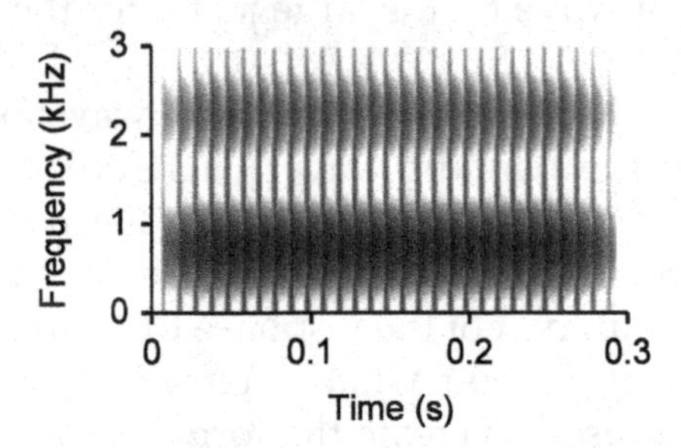

Now let's look at the long-time window:

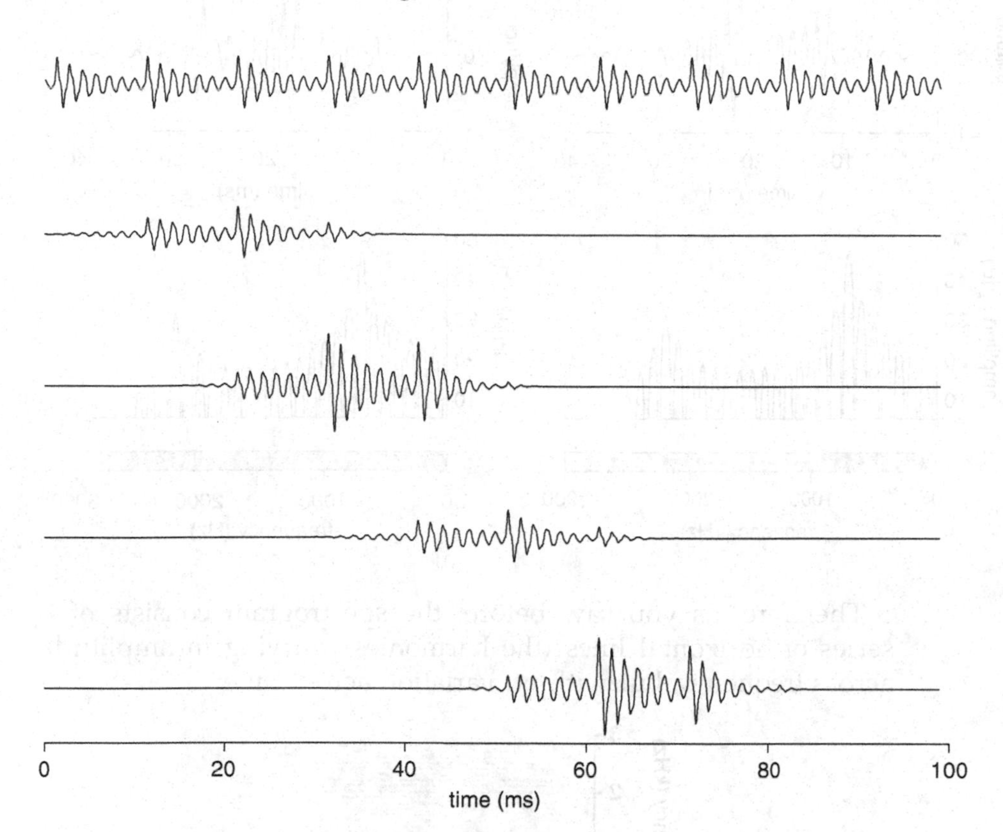

Now the window is long enough to 'see' the clear periodicity of the waveform. Even just the 3 or so cycles included in each window here is enough for the spectrum to reflect the evidence of harmonics (as you saw towards the end of Chapter 10, p. 193). On the other hand, because the long-time window includes multiple peaks and valleys of energy, this variation in energy will not be reflected in the individual spectra for different sections of the windowed waveforms. Every window in the series (with only two shown over leaf) 'sees' a similar waveform:

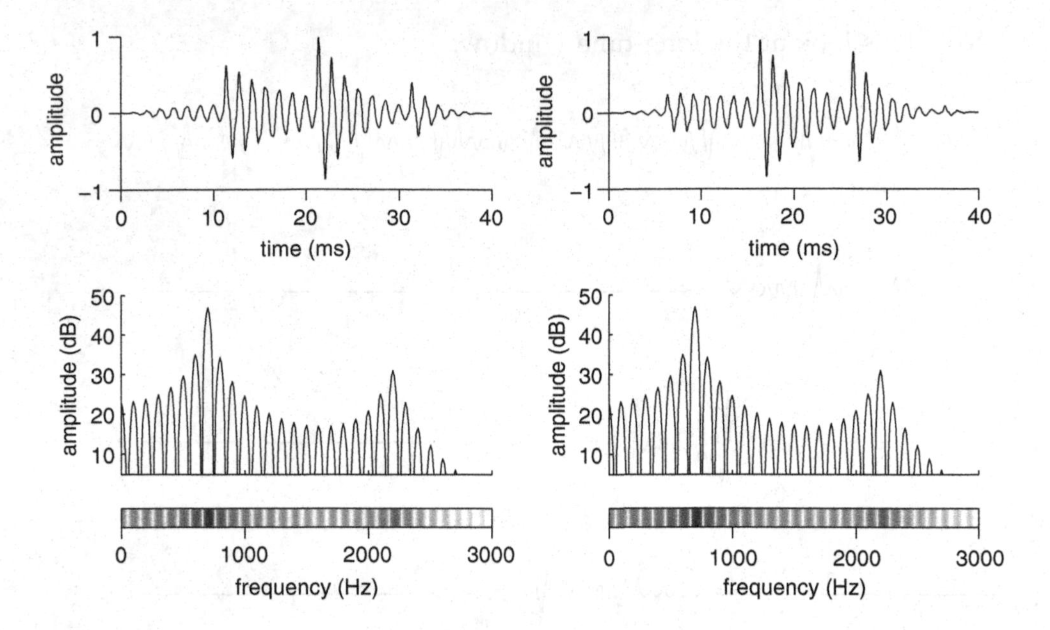

Therefore, as you saw before, the spectrogram consists of a series of horizontal lines (the harmonics), varying in amplitude across frequency, but with no variation across time:

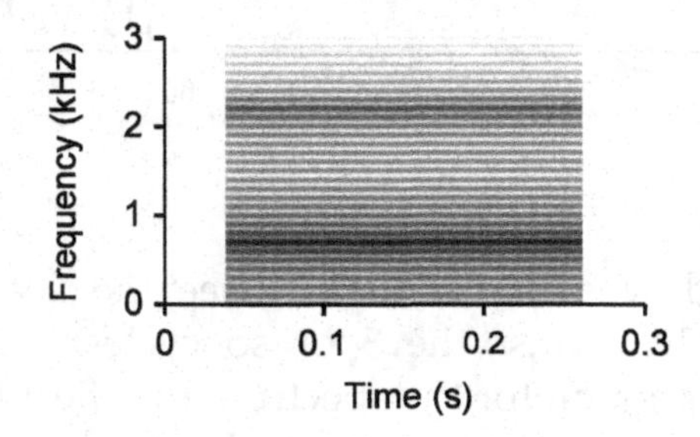

Frequency equalization

This nearly completes our discussion of spectrograms. Before leaving this topic, however, we need to discuss briefly one other possibility in spectrographic analysis that can have important implications for the analysis of real speech—the use of an *equalizing filter.*

Equalizing filters are used, at least in part, to get around a limitation of *dynamic range.* It is only possible to express a small range of darkness by printing onto paper (perhaps 10 dB). In other words, the maximum white of the paper might only reflect about 10 dB more light than the maximum black. Speech sounds have components that often vary in level by 30–40 dB or more. It is, therefore, not possible to have all these components show up on the paper at the same time. Either the lower-level components will not be marked (as seen in the spectrograms of the isolated pulses shown earlier), or a range of higher-level sounds will all be marked at the same maximum black. This limitation only applies, of course, to the time/frequency/amplitude display, not to the sections.

Therefore, it is common to try to correct for the fact that the long-term average spectrum of speech decreases with increasing frequency. In other words, the higher in frequency you go, the less energy you find. To increase these upper frequency components and make their level more comparable to the level of the low frequency components, there is often the option of *equalizing* the signal before it is analyzed. This is really nothing more than a high-pass filter (see 'Exercises'). Commercial spectrograph machines in the past always allowed this possibility.

Such an option is very useful in making spectrograms of speech, and can also be thought of as a rough way of making spectrographic analysis more similar to human auditory analysis. Normal listeners are able to detect sinusoidal components which lie in the range 1–5 kHz better than those below 1 kHz, where the minimal detectable sound pressure level increases as frequency is lowered. The equalizing filter thus makes high-frequency components easier for the spectrographic analysis to 'hear'. It is important to remember, however, that the relative amplitude of frequency components is completely altered. This must be taken into account, especially when making measurements on sections. In fact, for sections, the use of equalizing filters is probably best avoided.

Concluding remarks

This completes our investigation of spectrographic analysis. Although we will generally use spectrograms to display the acoustic properties of human speech, their use need not be restricted to such sounds. *Any* signal could be analyzed in this way, and in fact, much use has been made of spectrograms in studying animal communication sounds. Here, for example, is a spectrogram of a scream from a chimpanzee, along with a section made near 0.25 s:

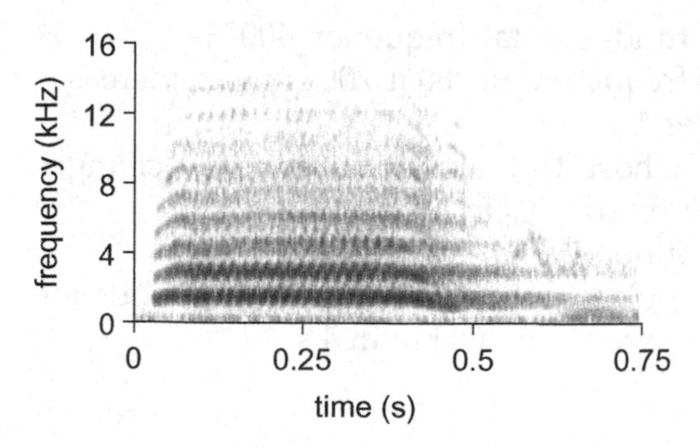

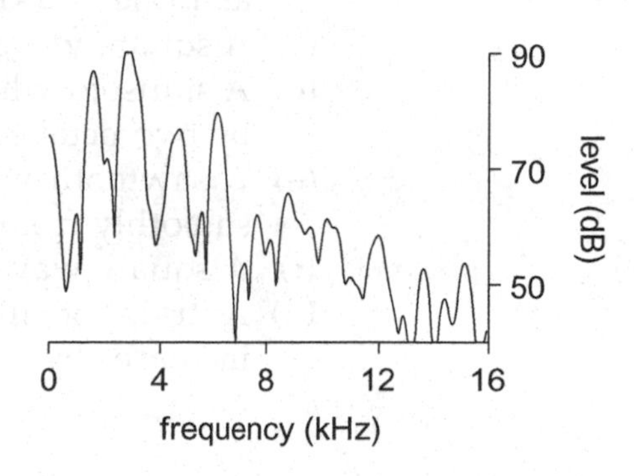

Even though this is a wide-band spectrogram, you can still see the individual harmonics, spaced about 1500 Hz apart, because the fundamental frequency is so high. There also appear to be rapid oscillations in fundamental frequency around its central value, something that in human singers would be called *vibrato*. You may also notice something that looks almost like one cycle of a sinusoidal frequency modulation around 4 kHz and starting about 600 ms into the scream. In fact, this is part of a call from another chimpanzee.

Outside their importance in studying speech, however, spectrograms provide a convenient summing up of many of the principles that we've previously encountered. This is appropriate here because we've now finished our basic exposition of the principles of signal and systems analysis and we'll now see how they apply to some specific problems in the study of speech and hearing.

Exercises

1. Draw half-wave and full-wave rectified versions of sinusoidal, sawtooth and triangle signals with a fundamental frequency of 400 Hz.

2. Show that both half-wave and full-wave rectification are non-linear processes by considering the spectrum of the output of a rectifier which is fed with a sinusoid.

3. Sketch the amplitude response of an equalizing filter for making spectrograms, assuming that it rises at 6 dB/octave from 50 Hz (where it has a gain of −18 dB) to 3.2 kHz, and is then flat to 12.8 kHz.

4. Sketch what you'd expect to obtain for wide- and narrow-band spectrograms (including appropriately sited sections) of the following signals:

   (a) A sawtooth wave with a period of 2 ms.
   (b) A sum of two sinusoids of equal amplitude, one at 200 Hz and one at 2 kHz.
   (c) A square wave of fundamental frequency 400 Hz.
   (d) A sinusoid whose frequency starts at 700 Hz and increases by two octaves over 1 s.
   (e) A sawtooth wave whose fundamental frequency changes smoothly from 100 Hz to 500 Hz over 2 s.
   (f) A square wave of fundamental frequency 100 Hz.
   (g) A train of 10-μs pulses whose fundamental frequency increases by one octave from 100 Hz in 1 s.

(h) A train of 10-μs pulses with a fundamental frequency of 100 Hz passed through a single resonance with a centre frequency of 1 kHz and a bandwidth of 300 Hz.

(i) A train of 10-μs pulses with a fundamental frequency of 100 Hz passed through a single resonance with a constant bandwidth of 300 Hz but whose centre frequency varies from 1 kHz to 2 kHz over 500 ms.

For the waves specified in (b), (f) and (g), also draw the sections that you'd obtain if an equalizing filter with the characteristics described in Exercise 3 were used.

5. Use the following filter shape to do a filter-bank analysis (as in the figures on pp. 231–233) of the first five harmonics of a sawtooth with a fundamental frequency of 100 Hz. What do you notice about the relation of the filter shape to the appearance of a single harmonic in the final spectrum?

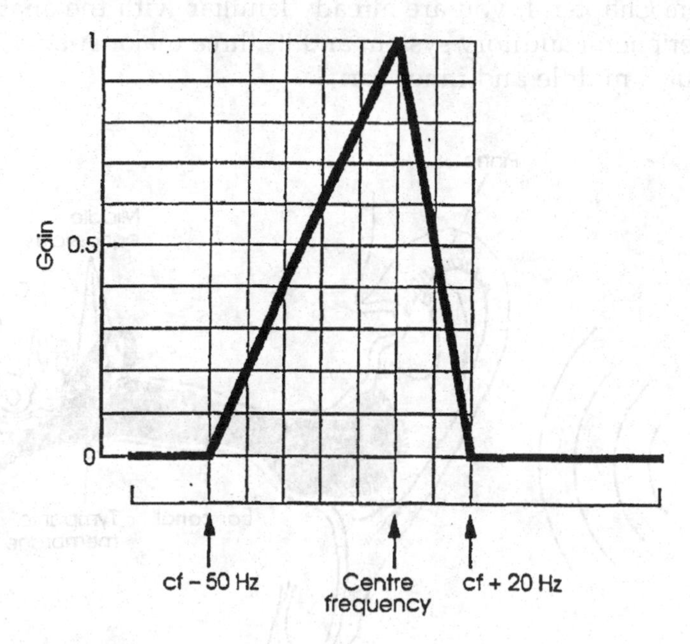

6. Explain (with diagrams) why a short-time analysis would be much better than a long-time analysis for resolving two pulses close in time.

7. With reference to the first section of this chapter (and Chapter 7), propose a practical method of determining the long-term phase spectra of real signals. What properties of the analyzing band-pass filters would be important? What if such filters had a delay? What effect would this have on your method?

CHAPTER 12

Applications to Hearing

So far we've been considering LTI systems in a fairly abstract way, although using specific examples from speech and hearing to illustrate many of the ideas. In this chapter we'll discuss in greater detail how the concepts we've developed can be applied to better understand the functions of the peripheral auditory system.

From Chapter 4, you are already familiar with the anatomy of the peripheral auditory system and its three major subdivisions—the outer, middle and inner ear:

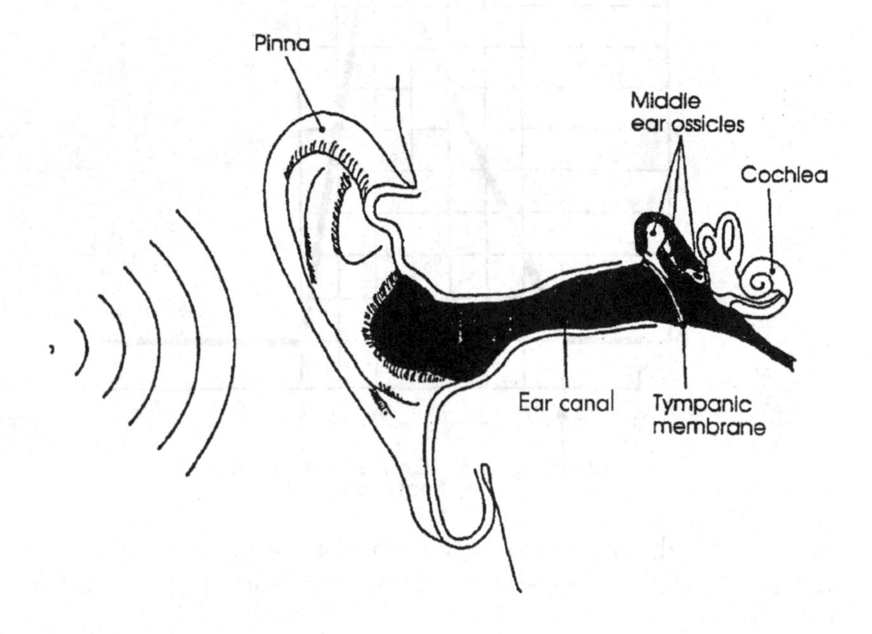

What we will do here is to present an alternative way of thinking about this set of organs—not as 'wet' biological tissue, but as a collection of 'black box' systems. Each of these transforms the signals passing through it in a similar way to the transformations imposed by various structures in the auditory periphery:

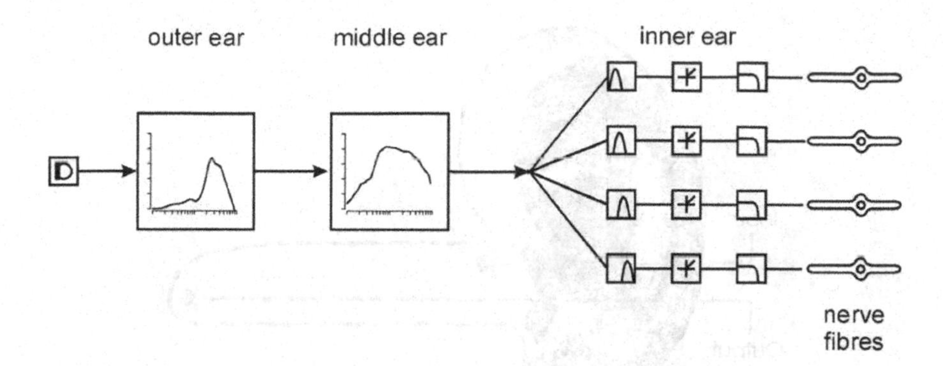

The original acoustic wave in the environment is first transduced by the microphone depicted at the left-hand side of the figure. The resulting electrical wave can then be processed by the rest of the systems in turn, resulting in the representation of the sound that is sent to the brain by the auditory nerve fibres at the far right.

Some of the systems in this chain can be well characterized as LTI, but others cannot. However, even for systems that are not LTI, the concepts of LTI signals and systems analysis can often still be usefully employed. We'll now look at each one of these systems in turn.

Outer ear

We'll begin with the outer ear, which includes the *pinna* and ear canal (also known as the *external auditory meatus*). The head itself also alters the sounds we hear, so we will need to include its effects. As you might expect, the effects of the pinna and head depend upon the direction the sound comes from. Therefore, it will be simpler for us to first describe the acoustic effects of the ear canal, as these do not depend upon sound direction.

The ear canal is a sort of tube stretching from the surface of the head to the *tympanic membrane* (eardrum). Having the tympanic membrane at the bottom of a short canal, rather than at the surface of the skull, significantly changes the sound that acts upon it. More than 60 years ago, Wiener and Ross reported measurements of the amplitude response of the ear canal in several ears (average length 2.3 cm). They delivered sound to the open end of the ear canal and measured the output at the tympanic membrane with a microphone. The input to the system was defined as the sound pressure at the entrance of the ear canal, and its output as the sound pressure at the tympanic membrane (as shown on the next page):

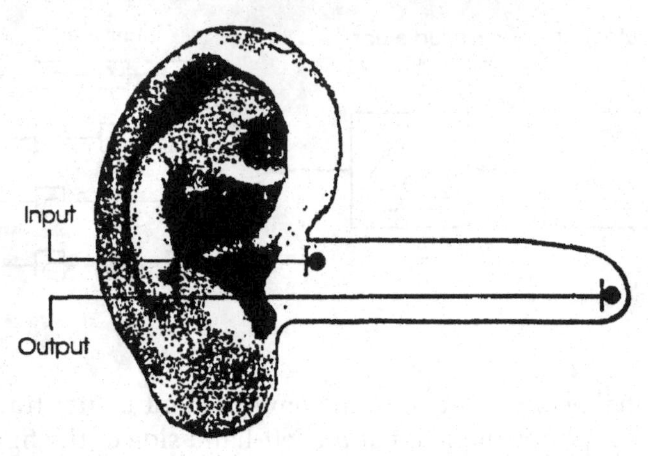

The mean amplitude response that Wiener and Ross obtained was characterized by a single resonant peak near 4 kHz. At low frequencies (below about 1500 Hz), there was little or no effect of the ear canal:

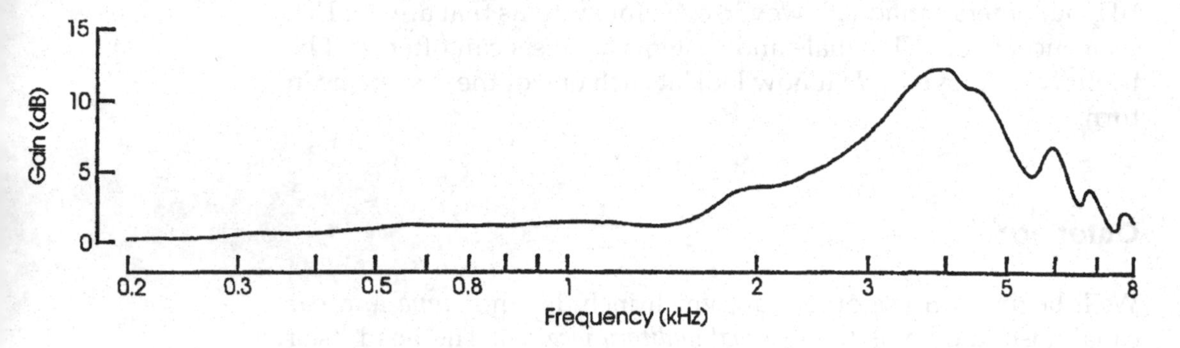

Some understanding of this response can be gained by imagining the ear canal to be a short cylindrical tube closed at one end, as shown below. The input signal is the sound at the opening of the model ear canal, the output is the sound at the closed end of the tube, with the system being the tube itself.

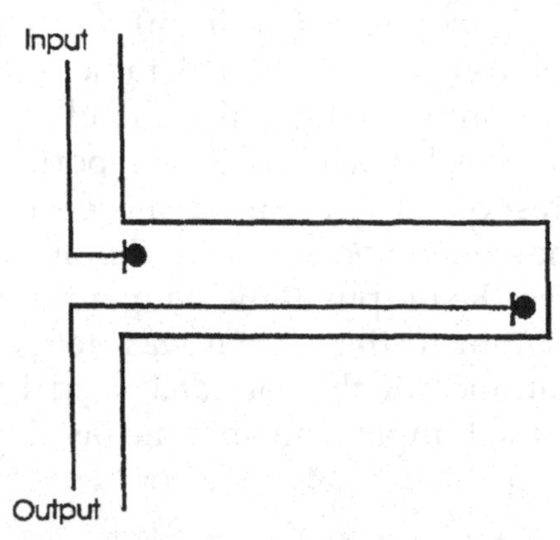

It turns out that the main factor that determines what frequencies are transmitted best in such a system is its *length*.

Before seeing how this is so, we must first introduce a term that you haven't encountered up until now—*wavelength*. The wavelength of a sinusoid is the *distance* the wave travels during one cycle of vibration. This is easiest to understand in a diagram that shows the pattern of acoustic pressure set up by a tuning fork, at some point in time after 'twanging' the fork:

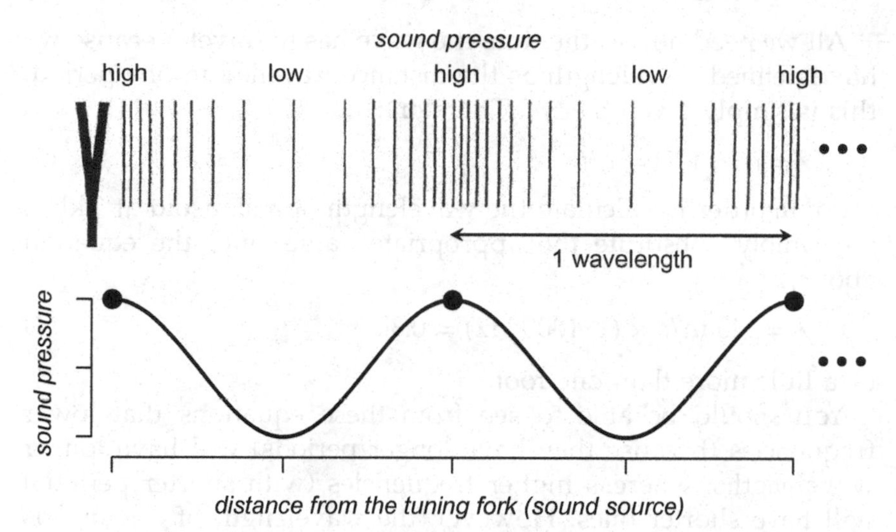

As was explained for the figure on page 8, regions of high pressure are indicated by lines that are close together, while regions of low pressure are indicated by lines spaced more widely apart. The sinusoid at the bottom of the figure is a 'snapshot' of the instantaneous pressure across space at a particular moment of time, with positions of maximum pressure marked with solid circles. This is *not* a waveform, so the x-axis is not time. The stretch between the two places of peak pressure represents a particular distance—one *wavelength* (typically indicated by the Greek letter lambda λ). A wavelength is, thus, equal to the distance between two points in space that are in the same position in their sinusoidal cycle of pressure variations—that is, one period apart.

We want to use a measure of wavelength to relate the dimensions of a system like the ear canal (here its length) to the frequencies that it passes best. It would be useful, therefore, to have a formula that translates between wavelength and frequency. Clearly, the distance a wave travels in a particular time period must depend upon the speed at which it is moving through its medium. For air (the medium), the speed of sound (conventionally symbolized as c) has a value of about 340 m/s (about 770 miles/h).

We can now calculate wavelength (λ) as a function of the frequency (f) of the sinusoidal sound, by considering how far the wave travels in one period. In general, the distance anything travels is simply the product of its speed and the time its journey takes. In a simple formula:

$$\text{distance} = \text{speed} \times \text{time} \quad (1)$$

We can readily rewrite the equation above in symbols as:

$$\lambda = c \times \text{time} \quad (2)$$

All we need now is the time the wave has to travel. Because we have defined wavelength as the distance travelled in one period, this is simply given by $1/f$. Therefore:

$$\lambda = c \times (1/f) = c/f \quad (3)$$

So, in order to calculate the wavelength of a sinusoid at 1 kHz, we simply substitute the appropriate values into the equation above:

$$\lambda = 340 \text{ m/s} \times (1/1000 \text{ Hz}) = 0.34 \text{ m} \quad (4)$$

or a little more than one foot.

You should be able to see from these equations that lower frequencies (because they have longer periods) will have longer wavelengths, whereas higher frequencies (with shorter periods) will have shorter ones. However, the wavelength of a sound is determined not only by its frequency, but also by the speed of propagation of the wave, which in turn depends upon the medium in which the sound is presented. So, for example, if the role of the external ear in the auditory perception of scuba divers was being studied, we would need to know the speed of sound in water—about 1450 m/s. The wavelength of a 1-kHz sinusoid in water would then be:

$$\lambda = 1450 \text{ m/s}/1000 \text{ Hz} = 1.45 \text{ m} \quad (5)$$

or nearly 5 feet.

We can now use the notion of wavelength to characterize features of the amplitude response of our model ear canal. If its walls were infinitely rigid, the response seen below would be obtained. You can see that the amplitude response consists of a series of valleys separated by resonance peaks. Because the system is idealized and has no losses (damping), the amplitude

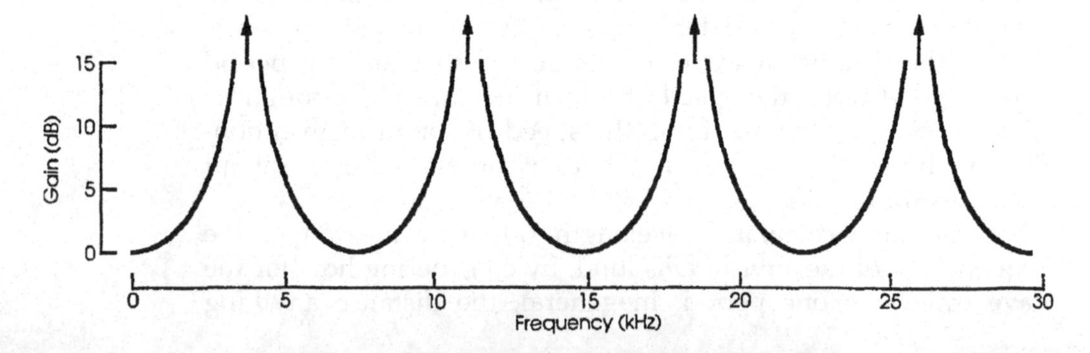

response shoots up to an infinitely-high value. In a real system, such as the ear canal, frictional forces (primarily at the canal walls) damp the amplitude response. This broadens the resonances, and also prevents them getting infinitely high:

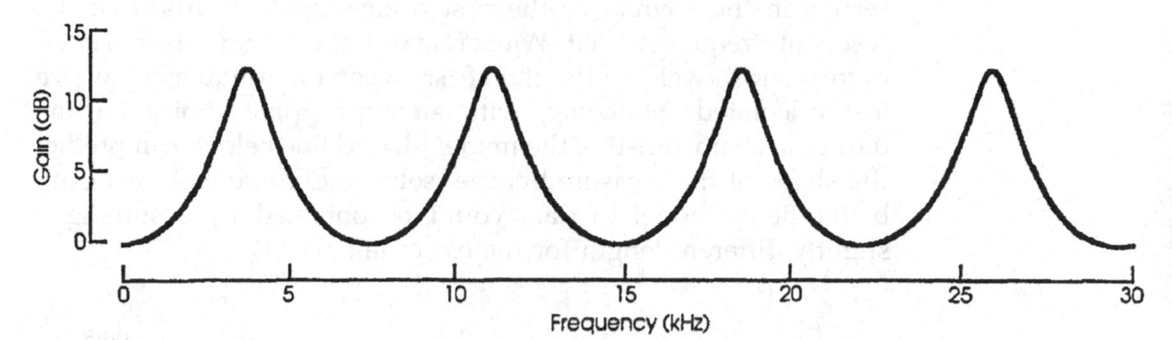

Thus, over the 8-kHz frequency range measured by Weiner and Ross, which includes only one resonance, the amplitude response has a shape like a band-pass filter.

It turns out that the position of these resonances is related in a simple way to the length of the ear canal. The lowest resonant frequency has a wavelength four times the length of the ear canal (called a *quarter-wavelength resonance*). Before working out what frequency this corresponds to, we need to re-arrange equation (3) from above to obtain:

$$f = c/\lambda \tag{6}$$

Now, if the length of the ear canal is L (expressed in metres), its first resonant frequency f_1 has a wavelength of $4L$. Thus, from equation (6), the lowest resonant frequency is:

$$f_1 = c/4L \tag{7}$$

This indicates that ear canals of different length have different lowest resonant frequencies. More specifically, the lowest resonance is inversely proportional to the length of the canal. Thus, longer canals have a lower first resonance, as you might expect from the general physical principle that the bigger something is, the lower its 'frequency'. (Consider the difference between the length of the strings on a violin and those on a double bass). Because frequency is equal to one over the period, this is equivalent to saying that the period of the lowest resonant frequency is directly proportional to the wavelength (again, longer tubes have lower first resonant frequencies). It turns out that the higher resonant frequencies are odd integer multiples (3, 5, 7 and so on) of the lowest resonant frequency.

Let's see how well our simple formula predicts the peak in the amplitude response of the ear canal. The first resonance for an open tube that is 2.3 cm (0.023 m) long should occur at a frequency of $340/(4 \times 0.023) = 3696$ Hz. The next resonant frequency should occur at 3×3696 Hz $= 11{,}088$, the next at 5×3696 Hz $= 18{,}480$ Hz, and so on. As Wiener and Ross only performed measurements up to 8 kHz, we can only compare the model results with the actual results in the vicinity of the first resonance. Note first that the resonant frequency that Wiener and Ross found (near 4 kHz) corresponds well with the first resonant frequency we've just calculated. Moreover, with an appropriate choice for the damping characteristics, the model (dotted line below) can predict the shape of the measured curve (solid line) quite well. An even better fit of model to data could be obtained by assuming a slightly different length for the ear canal:

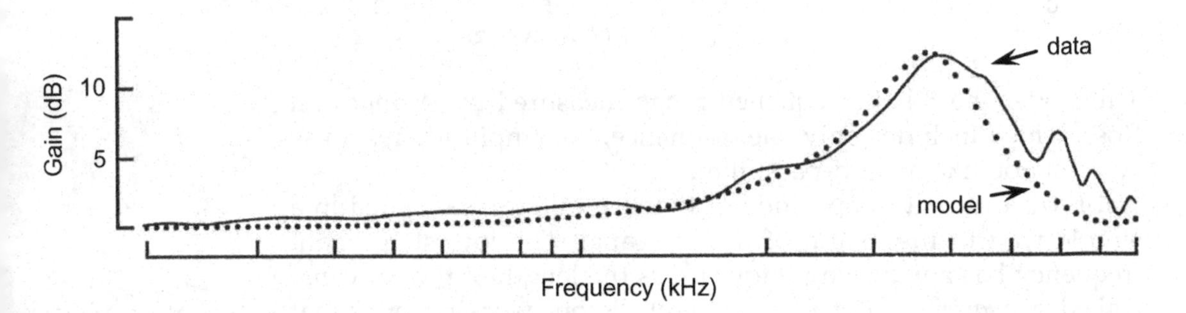

Of course, the acoustic effects of the external ear are not limited to those caused by the ear canal. The head and pinna cause acoustic 'shadows', so that the amplitude response depends upon the orientation of the sound source relative to the ear. In order to get a more complete view of the acoustic effects of the external ear and head, consider the system head-plus-pinna-plus-ear-canal. The input is a sinusoid delivered from different orientations relative to the head (its amplitude determined at the centre of where the listener's head would be) and the output will be taken as the sound pressure level at the eardrum. Defining the input sound pressure level as that arising in the sound field *without* the head present lumps together the effects of the head with those of the pinna and ear canal.

If heads and ears were symmetric, we would get the same result no matter which ear we measured. That is clearly not the case, so a full understanding of the sound field presented to a particular person at both ears would require measurements at *both* eardrums. Here we will only present measurements made at the *left* eardrum.

Shaw has summarized diagrammatically the amplitude responses obtained in several studies. The coordinate system used to denote the angle of the sound source relative to the head (known as the *azimuth*, θ) is represented schematically in the inset. In this set of measurements, the sound source is always presented at the same height, level with the opening of the ear canal (in the so-called *horizontal plane*). So, for example, when $\theta = 0°$, the sound is straight ahead of the listener. When $\theta = 90°$, the sound is directly opposite the left ear and when $\theta = -90°$, the sound is directly opposite the *right* ear. Varying the elevation of the sound would add an extra complicating factor which, although important for a thorough understanding of the effects of the outer ear, will not concern us here:

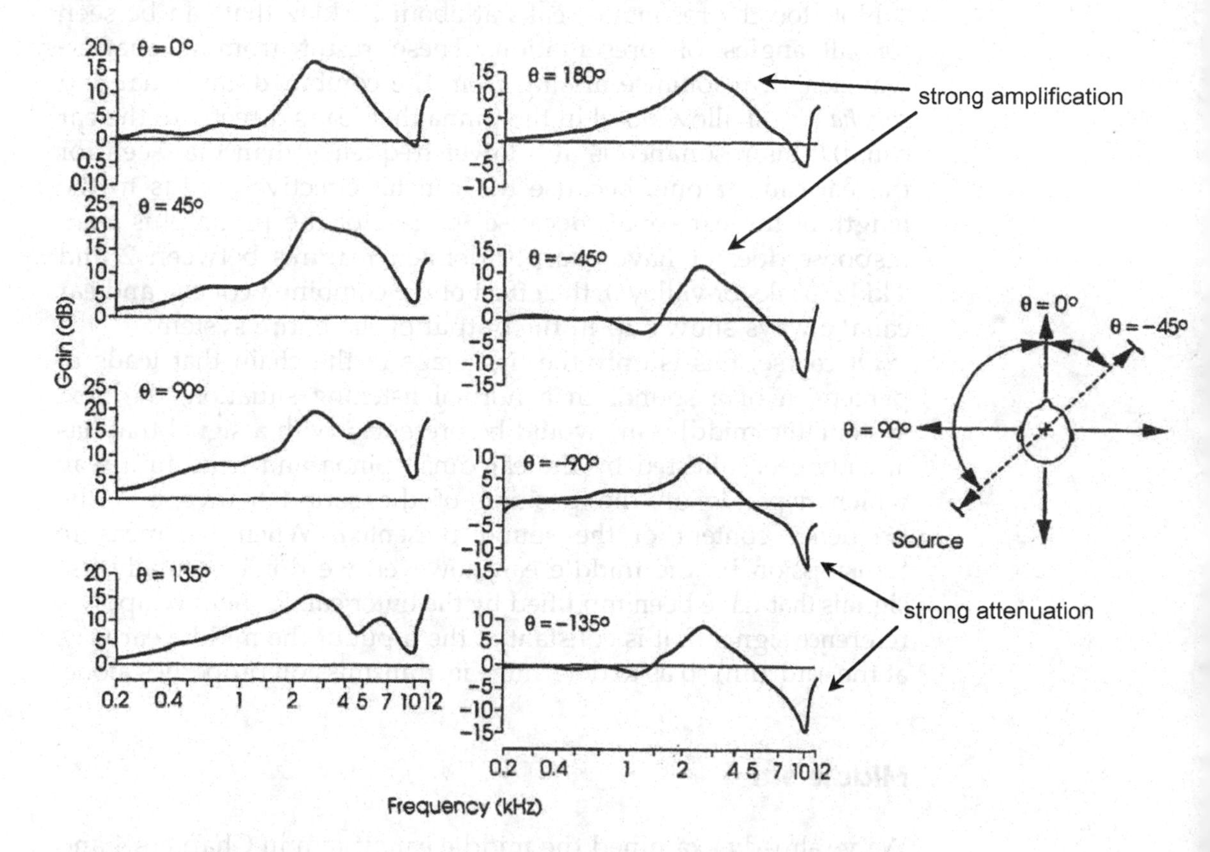

As you can see, the amplitude responses vary greatly depending upon the position of the sound relative to the ear measured. However, these variations only occur when the input sinusoid has a wavelength that is comparable to, or smaller than, the head and pinna. This principle applies generally—the transmission of a sinusoid in a sound field is only affected by objects that are comparable in dimension to, or larger than, the wavelength of the sound. So, the acoustic effects of the head are most marked at frequencies above about 1.5 kHz, equivalent to wavelengths smaller than 22.7 cm (the approximate diameter of an adult male head). These changes in the amplitude response are most dramatic for negative values of θ, where the sound source is on the opposite side of the head from the ear being measured. For these angles, the amplitude response can show a strong attenuation of sound (for frequencies near 10 kHz or so).

Similarly, the measured amplitude responses have gains that are small for low frequencies (near 200 Hz and below), no matter what the value of θ. In other words, the sound field is not affected in this frequency region. This is not surprising, as a sinusoid of 200 Hz has a wavelength of about 1.7 m (about 5½ feet), dimensions not approached by any part of the ear or head, at least of a person!

Note too the resonant peaks at about 2.5 kHz that can be seen for all angles of presentation. These result from a quarter-wavelength resonance arising from the combined ear canal and *concha* (the shallow bowl in the pinna that leads directly to the ear canal). This resonance is at a lower frequency than that seen for the ear canal alone, because the concha effectively adds to the length of the ear canal. Because the rest of the pinna-plus-head response doesn't have sharply defined features between 2 and 4 kHz (peaks or valleys), the effect of the combined concha and ear canal always shows up in the output of the entire system.

Of course, this is only the first stage in the chain that leads to perception of a sound. In a normal listening situation, the next system (the middle ear) would be presented with a signal that has already been affected by the ear canal, pinna and head in a way which depends on the position of the sound source and the frequency content of the sound presented. When we measure transmission by the middle ear, however, we don't normally use signals that have been modified by the outer ear. Rather, we apply a reference signal that is constant at the input of the middle ear (say, at the eardrum) so as to determine its transmission properties alone.

Middle ear

We've already examined the middle ear system in Chapters 4 and 6, describing investigations of the displacement of the stapes for sinusoidal sound pressure variations applied to the tympanic membrane in anaesthesized cats. You'll recall that this system (tympanic membrane-to-ossicles) was found to be LTI. Here, we'll be looking at the amplitude response of the middle ear in humans. These experiments, by Puria, Peake and Rosowski, were performed on temporal bones obtained from cadavers. Although many aspects of peripheral auditory function are very different in living and dead people, it turns out that important aspects of middle ear function are pretty much the same.

The amplitude response was determined on the basis of the same input as was defined for cats (sound pressure near the tympanic membrane) but with a different output. As you know (and which we will discuss more fully in the next section) the stapes moves in and out of the fluid-filled *cochlea* (inner ear), setting up pressure variations which are an essential stage in hearing. Puria and his colleagues decided to use these pressure changes as the output of the middle ear, measured in the cochlear fluids near the stapes footplate using a *hydropressure transducer* (a kind of underwater microphone). Therefore, the 'gain' referred to in the *y*-axis of this figure reflects, on a dB scale, the pressure level in the cochlear fluids relative to the pressure level at the tympanic membrane:

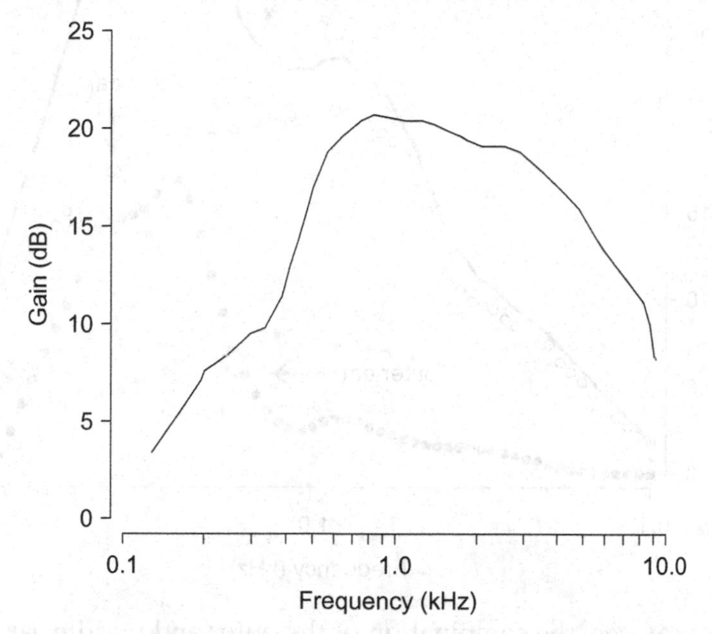

As you can see, the amplitude response is of the form of a band-pass filter centred near 1 kHz, and with quite a broad bandwidth. In other words, spectral components of sounds near 1 kHz are transmitted extremely well through the middle ear, with little or no relative attenuation of spectral components varying from about 500 Hz to 5 kHz.

Let's see how far we have come in our journey through our model of the auditory periphery.

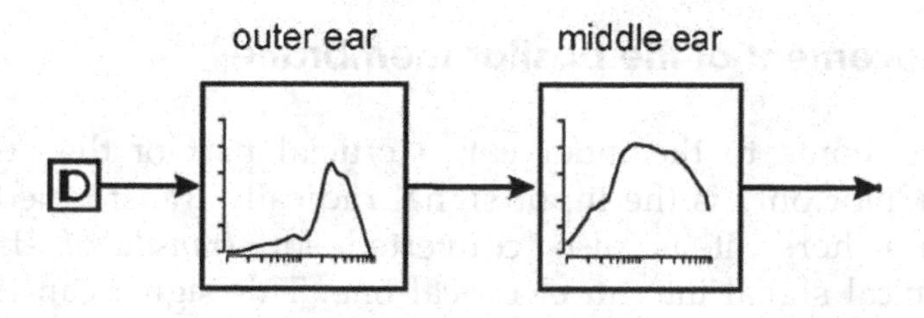

The sound has been picked up by a microphone, transduced into an electrical wave and then filtered by two systems in cascade. You already know from Chapter 6 that it is readily possible to calculate the total amplitude response of these two systems from the individual responses. Since they are expressed on dB scales, it is a simple matter to add together the gains at each frequency:

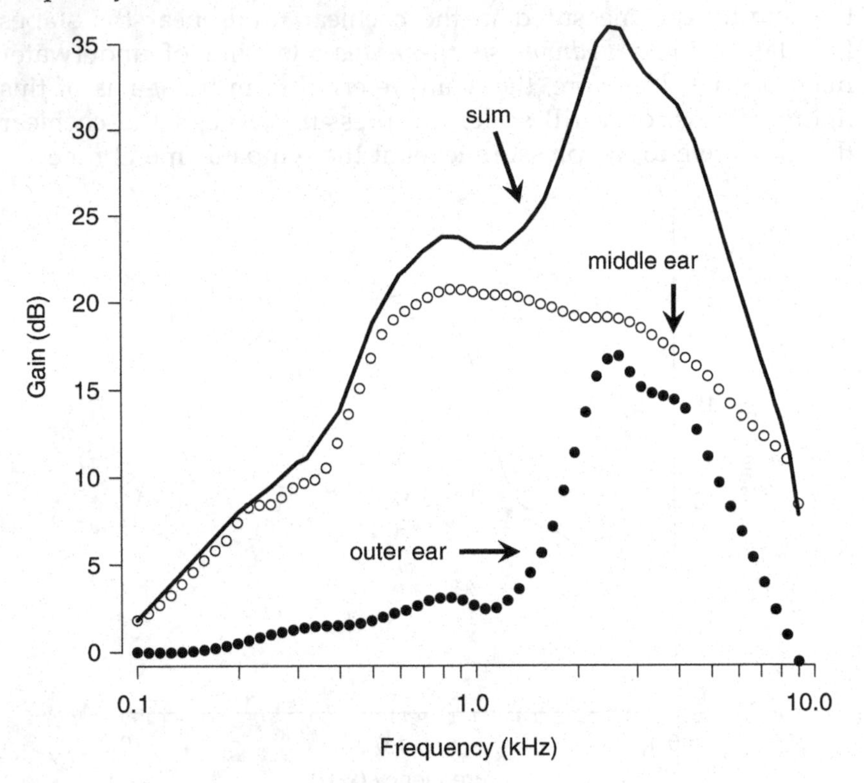

As you can see, the combination of the outer and middle ear leads again to a band-pass response, with its peak dominated by the response of the ear canal plus concha, near 3 kHz. This peak has a direct impact in determining the frequencies of sounds we are most sensitive to. Broadly speaking, it is perhaps not too surprising that sinusoids transmitted effectively through the auditory periphery can be detected at lower levels than those transmitted less effectively.

The movement of the basilar membrane

We now come to the inner ear, a crucial part of the auditory system. Not only is the input signal radically transformed in its structure here, it is also converted—or *transduced*—from a mechanical signal into an electrical one. This signal can then be handled by other parts of the nervous system. Before embarking on a detailed analysis, let's first describe its anatomy and general functioning.

Although the inner ear also includes the organ of balance (*the semicircular canals*) the main structure that will concern us is the *cochlea*. The cochlea is a fluid-filled tube coiled in the shape of a snail. This coiling is pretty much irrelevant to what the cochlea does, so we'll visualize it unrolled to give a clearer picture:

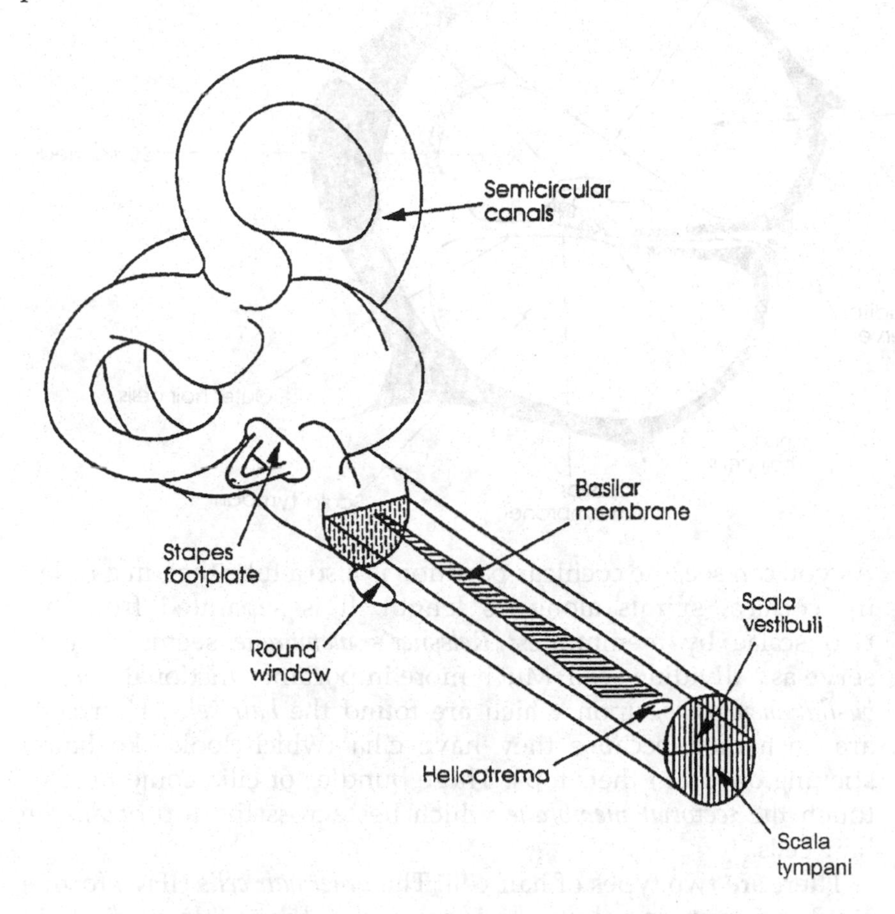

The *cochlear partition* runs down the length the cochlea, dividing it into two chambers (scala vestibuli and scala tympani). Because the partition does not quite reach the end of the cochlea, the two chambers connect through a space known as the *helicotrema*. When the stapes is pushed into the oval window by sound at the tympanic membrane, the incompressible fluid causes the membrane covering the round window to bulge outwards. Similarly, when the stapes is withdrawn from its resting position, the round window membrane moves inward. It, thus, acts as a sort of pressure release, or 'give', to allow the inward and outward movement of the stapes. Because the cochlea is surrounded by rigid bone, the stapes would not be able to move without this 'give'.

The cochlear partition is not simply a barrier but is itself a complex array of structures, as this cross-section shows:

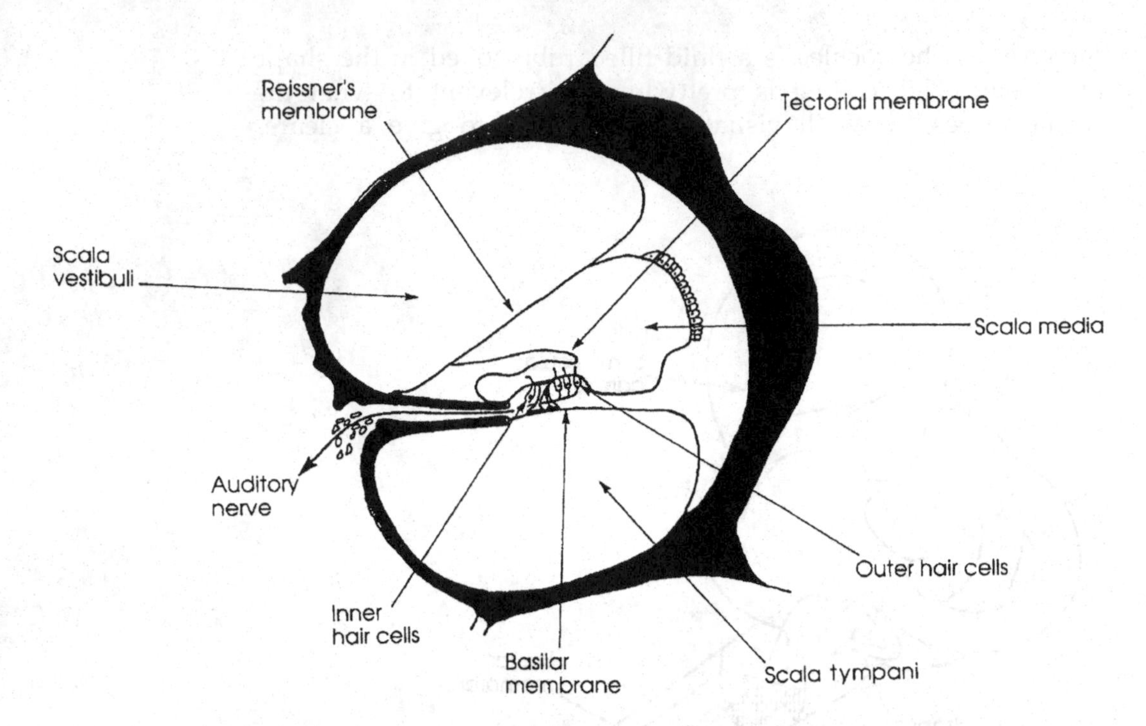

As you can see, the cochlear partition is also a tube that, in a rolled up cochlea, spirals along its length. It is separated from the two scalae by membranes: *Reissner's membrane* seems only to serve as a dividing wall. Much more important functionally is the *basilar membrane*, upon which are found the *hair cells*. Hair cells are so named because they have cilia (which look like hairs) sticking out from their tops. These bundles of cilia come near or touch the *tectorial membrane* which lies across the top of all the hair cells.

There are two types of hair cell. The *inner hair cells* (IHC) form a single row running along the inner part of the cochlear spiral. At the base of the IHCs are the endings of the fibres of the auditory nerve which make synaptic contact with the hair cells. Note that this is the first time in the system that we've encountered any neural elements. Also important to cochlear function are the *outer hair cells*, which are found in three rows, along the outer part of the cochlear spiral.

Although the precise nature of the chain of events that leads from sound to neural firing is still the subject of much controversy, there is general agreement about the major stages. Roughly speaking, this is what happens. The movement of the stapes in and out of the oval window sets up a pressure wave in the cochlea, which in turn causes the basilar membrane to vibrate. As a result of this vibration, the cilia on the IHCs are bent (perhaps due to a sliding motion between the basilar membrane and the tectorial membrane), causing neurotransmitters to be released from the base of the hair cells into the synapse.

The transmitter diffuses across the synaptic gap and causes the nerve to fire. The hair cells, thus, serve as transducers, transforming mechanical vibrations into electrical pulses. The neural 'spikes' are then relayed to other parts of the nervous system.

Because IHCs only cause nerves to fire in those places where the basilar membrane is set in motion (ignoring for the moment the spontaneous firing that goes on even in the absence of sound), the characteristics of this motion are crucial to an understanding of the firing patterns eventually seen on the auditory nerve. Again, there is much controversy, with agreement on certain major principles.

The most important characteristic of the basilar membrane is that it is *selectively resonant*. Not all parts of it vibrate equally well to sinusoidal inputs of a particular frequency. Put the other way round, different frequency sinusoids cause maximum vibration at different places along the membrane. Because the basilar membrane is narrower and stiffer at its basal end (near the stapes) than it is at its apical end (near the helicotrema), the basal end vibrates more to high frequencies than does the apical end. Conversely, a low-frequency movement of the stapes causes the apical end of the basilar membrane to vibrate more than the basal end.

This was first observed directly by von Békésy (pronounced BEH-kuh-shee) using a light microscope. He was able to measure the amplitude of the vibration of the basilar membrane over a significant portion of its length. At right are the results that von Békésy obtained when he presented a sinusoidal input of constant amplitude, at various frequencies, to the stapes of an excised cochlea. You can see that as the sinusoidal input increases in frequency, the peak amplitude of vibration occurs more basally. Note that these curves represent the *maximum* displacement undergone by any particular point on the membrane—the details of the temporal aspects of the vibration have been left out. In fact, every single point on the basilar membrane that moves would be moving in a sinusoidal way at the stimulating frequency.

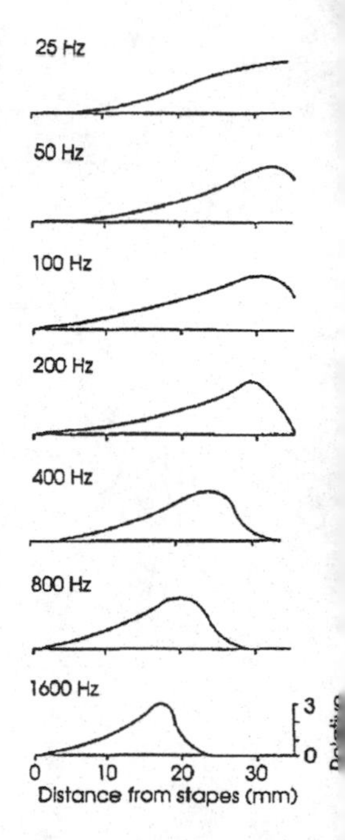

It is important not to mistake these graphs for the amplitude responses we've discussed so frequently. Measuring an amplitude response would necessitate the presentation of a number of sinusoids of different frequency. Because each of these curves is a measure of the motion resulting from a *single* input frequency, they cannot be amplitude responses. They are often known as *excitation patterns* because the response pattern of the entire basilar membrane to a single sound (or excitation) is shown. We can, however, combine together the information from these and a number of other similar curves to obtain the amplitude response of a single point on the basilar membrane. This now familiar way of thinking about a system (here, the basilar membrane) may be schematized as shown here:

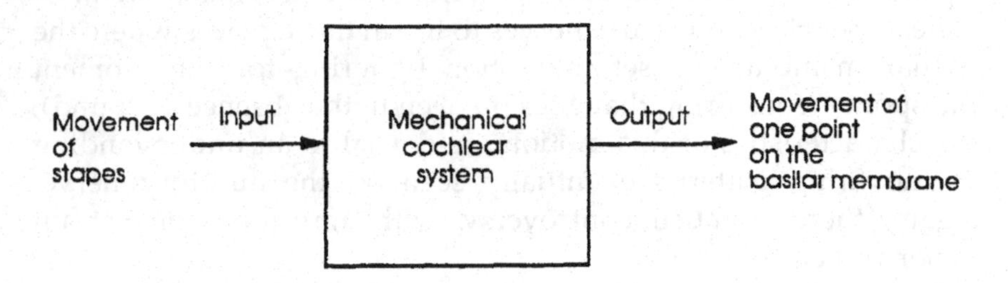

All that needs to be done is to move the stapes sinusoidally at a variety of frequencies and measure the amplitude of the response at a single point on the basilar membrane. This results in the familiar amplitude response. (We'll ignore phase in this discussion although such information can be important.) Of course, for a full understanding of the basilar membrane it would be necessary to measure the frequency response at a number of different places. Here are two such curves that von Békésy measured:

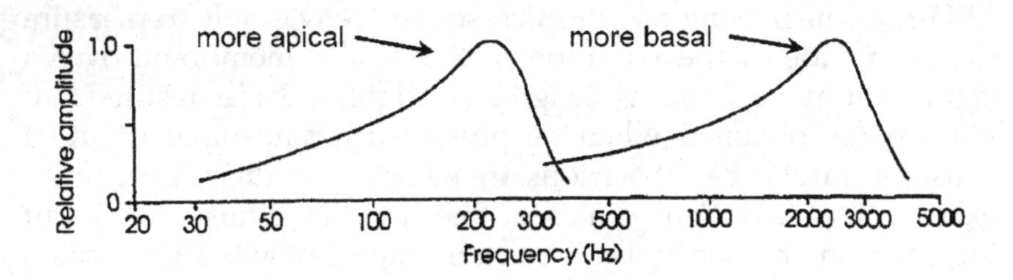

As the curve on the left was obtained from a place on the basilar membrane more apical than that associated with the curve on the right, it is more responsive to low frequencies.

Curves such as these should look very familiar to you—they are nothing more than band-pass filters. One way to think of the basilar membrane, then, is as a sort of filter bank (like those described in Chapter 11). Each point on the basilar membrane corresponds to a band-pass filter with a different centre frequency. As one goes from the base to the apex in the cochlea, the centre frequency of the band-pass filter decreases. This adds a further stage in our model of the auditory periphery:

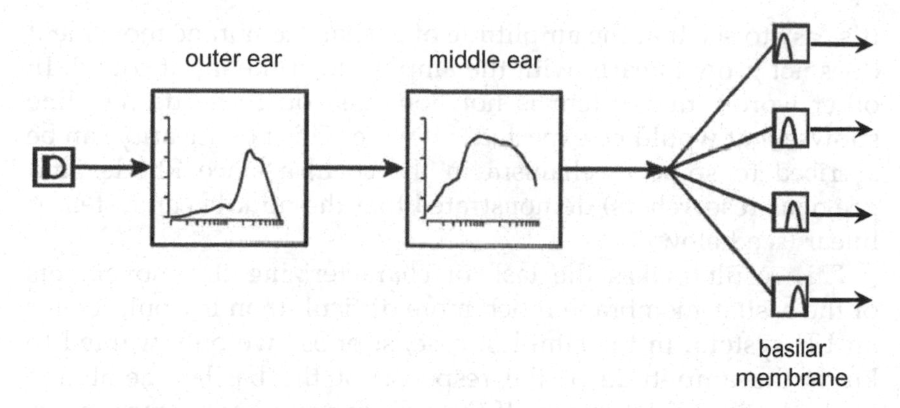

A realistic model would mean an enormous number of filters, of course. Here, for practical purposes, we only show four. Using this model, we can predict the response of any point on the basilar membrane to any input, in the same way we would do for any LTI system. Essential, of course, is the assumption that the transformation between stapes and basilar membrane movement is, in fact, linear. von Békésy claimed that it was in his preparations, and gave as supporting evidence the fact that '... the amplitude of vibration of the cochlear partition, as observed at a particular point, increased exactly in proportion to the amplitude of the vibration of the loudspeaker system ...' which drove the stapes. In other words, he showed that the system was homogeneous.

Some decades after von Békésy's work, however, Rhode showed that the movement of the basilar membrane is highly nonlinear, at least in squirrel monkeys. He also tested homogeneity, in essentially the same way as von Békésy, but with quite different results. Here are Rhode's data for 7.4 kHz, which show the amplitude of the movement of the basilar membrane at a fixed point, as a function of the input sound pressure level. (You have already seen these data in the figure on page 54:

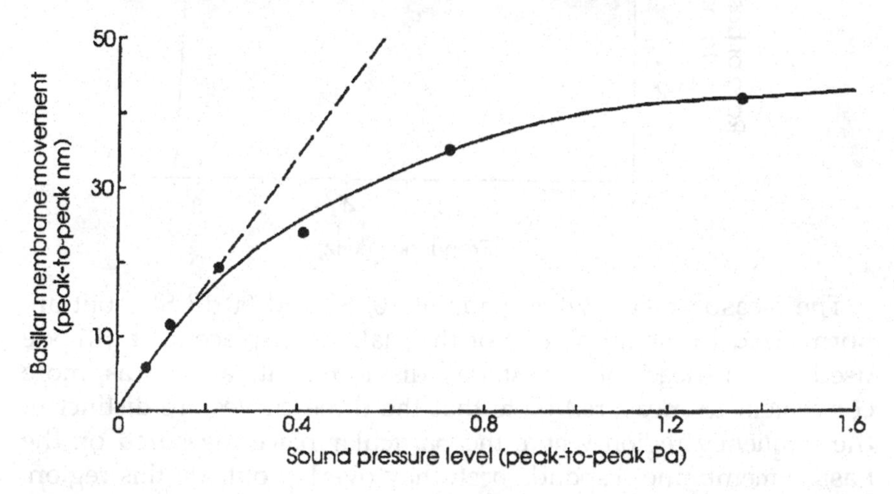

It's easy to see that the amplitude of basilar membrane movement does not grow linearly with the amplitude of the input sound. In other words, the system is not homogeneous. The dashed line shows what would be expected if it were. This nonlinearity can be ascribed to some mechanism in the cochlea since Rhode (and previous researchers) demonstrated that the middle ear system *is* linear (see below).

This result makes the task of characterizing the movements of the basilar membrane much more difficult than it would be for an LTI system. In the simplest case, suppose we only wanted to know the amplitude of the response of the basilar membrane to single sinusoidal inputs. If the system were homogeneous, we would only need one amplitude response, measured at an arbitrary input level for each frequency. We could then use homogeneity to determine the response to any sinusoid. Since the system under consideration here is not homogeneous, we need to look at its amplitude response at a number of levels to know what it will do. Rhode did just this and showed that the shape of the amplitude response did in fact depend on the input level used in the measurement:

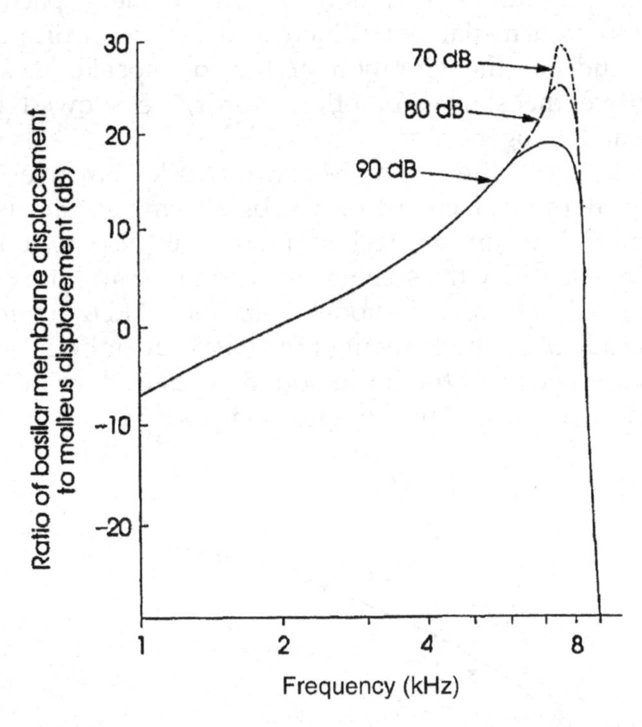

The measurements were made at 70, 80 and 90 dB SPL but are normalized to the amplitude of the malleus displacement. (Rhode used this instead of the stapes displacement, as it was more convenient to measure.) Note that the three curves are distinct in the frequency region where the particular place measured on the basilar membrane responds best; they overlap outside this region.

If the system were linear, all three curves would overlap completely. A nonlinearity like this makes it difficult to apply in a straightforward way many of the concepts we've developed. Take bandwidth, for example. In a linear system it is possible to define a single value for a filter's bandwidth because it does not depend on the level of the input signal. If we tried to estimate the bandwidth of cochlear 'filters' from Rhode's data, however, a single number would not do. Here the bandwidth of the filter increases with increasing level. In such a system, 'bandwidth' would have to be a function of level and not a single value.

It's interesting to note that the cochlear 'filters' seem to operate linearly for frequencies relatively remote from their centre frequencies, but are highly nonlinear for signals near their centre (in the so-called 'pass-band'). This can also be seen in the following diagram where the peak amplitude of vibration for one place on the basilar membrane (most sensitive to 7.4 kHz) is plotted as a function of level for a number of different frequencies:

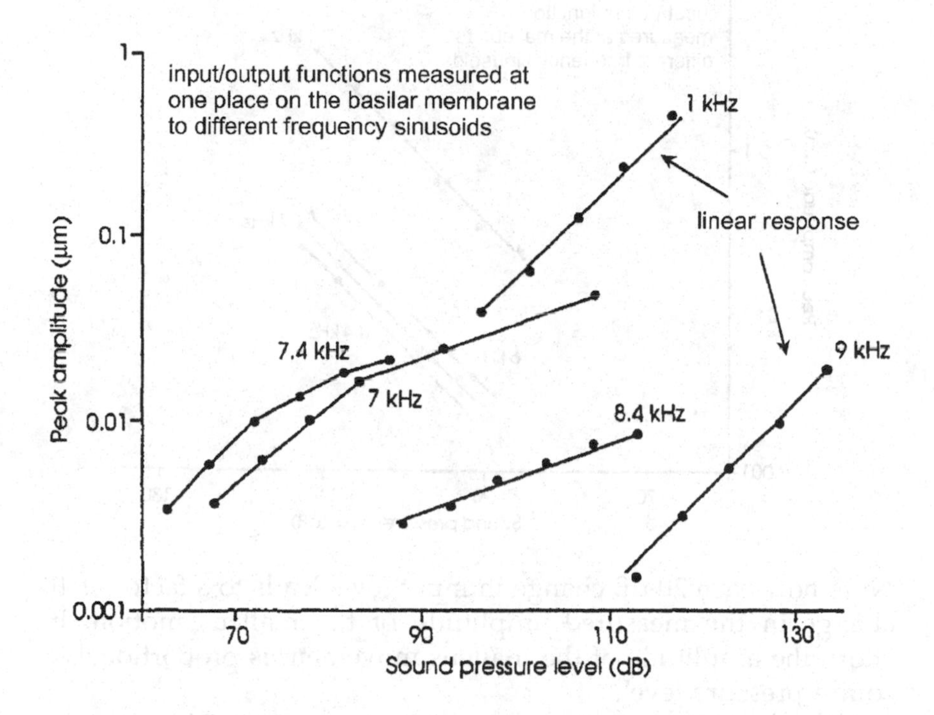

As we saw previously, when the stimulating frequency is near the centre of the pass-band, the amplitude of the basilar membrane vibration grows at a much smaller rate than it would in a linear system. The data for 7.4 kHz are in fact the same used in constructing the figure on page 274, which used linear scales because it is easier to understand homogeneity that way. Here we use logarithmic scales for both axes (log amplitude versus dB SPL).

If this system were linear, the amplitude of movement of the basilar membrane would increase proportionately with sound pressure level for all frequencies. In other words, a factor of 10 increase in the sound pressure level (20 dB) would lead to a factor of 10 increase in the amplitude of movement of the basilar membrane (again 20 dB). Therefore, on an input–output graph like this, all LTI systems would be characterized by a straight line with a slope of 1 (meaning the output grows by 1 dB for each 1 dB increase in the input). Although the system is homogeneous for frequencies of 1 and 9 kHz, it is not homogeneous for the other three frequencies—those in the centre of the pass-band of the 'cochlear filter'. This is another reflection of the finding that the amplitude response curves taken at different levels only overlap outside the 'pass-band'.

Contrast these data with those Rhode obtained when he tested a part of the middle ear for homogeneity. Here he measured the peak amplitude of the movement of the malleus as a function of sound pressure level, and found homogeneity at all frequencies:

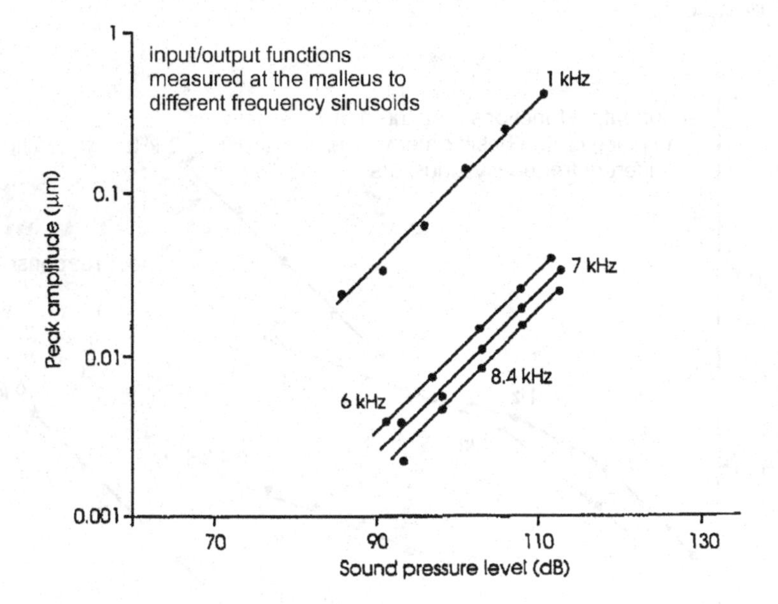

Note how each 20-dB change in input level leads to a factor of 10 change in the measured amplitude of the malleus motion. In short, the amplitude of the malleus movement is proportional to sound pressure level.

The causes of the discrepancies between von Békésy's and Rhode's results on basilar membrane vibration arise from crucial methodological differences between the two studies. Almost certainly the essential one is that von Békésy always used preparations from cadavers. Rhode used live animals and found the nonlinearity to disappear very quickly once the animal had died. Also, von Békésy's technique necessitated the use of extremely high signal levels (up to 140–150 dB SPL) in order to make the movements visible, whereas Rhode used levels that would be encountered in everyday life. Nonlinearities in basilar membrane movements have since been found many times by other groups of experimenters in cats, chinchillas and guinea pigs, and there is now a general agreement that the system is nonlinear.

The details of this controversy are, for our purposes, less important than the way in which the concepts of linear systems analysis pervade the entire discussion. LTI systems serve as a benchmark against which other systems can be compared; hence much effort goes into determining the exact nature of the departures from linearity found in a nonlinear system. Therefore, when Rhode claimed that basilar membrane motion is nonlinear, he did so on the basis of what would be expected of a linear system. This is yet another reason why an understanding of linear systems analysis is crucial to appreciate discussions of even nonlinear systems.

In terms of our model then, we would have to implement nonlinear filters in the filterbank meant to represent basilar membrane movement. In fact, this is not as complicated as it might appear, and many appropriate algorithms for doing nonlinear filtering are available. The details of those won't concern us here. We still need to develop at least one more stage in the model to get to auditory nerve firing.

Transduction by the inner hair cells

Up until this point, we have only been talking about *mechanical* signals, which is to say, those concerning either movement or changes in pressure. In order for any information about sound in the outside world to be relayed to the brain, it needs to be converted into an *electrical* code, as firings on the auditory nerve. This transduction, as mentioned above, is carried out by the IHCs. We will not describe the outer hair cells, although there are three times more of them than the IHCs (three rows compared to one). What has become clear over the last 40 or so years is that the outer hair cells are active and can move, and thus amplify the response of the basilar membrane, especially at low sound levels. In other words, they are responsible for the crucial nonlinearities in basilar membrane movements. However, they play no direct role in transduction, so we will not discuss them further here.

The vast majority of afferent nerve fibres (that is, those carrying information from the ear to the brain) synapse on a single inner hair cell (IHC), with each hair cell having about 10–30 nerve fibres synapsing to it. Therefore, in order to understand the firing on a particular auditory nerve fibre, we only need to consider a single IHC and the movement of the basilar membrane where that IHC lies. Here you can see a schematic of a single IHC, with two auditory nerve fibres synapsing to its base:

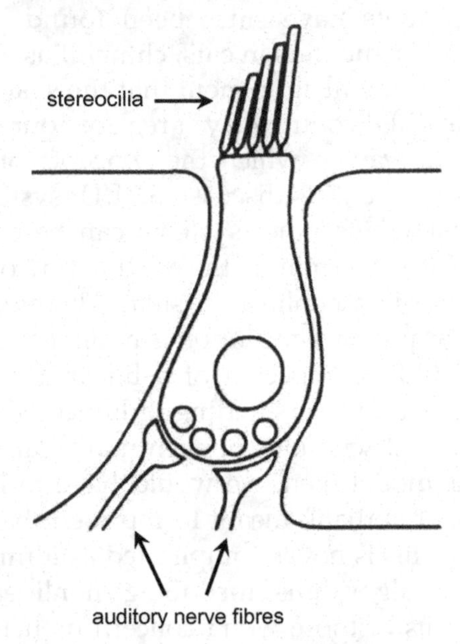

Imagine now presenting a sinusoid at the tympanic membrane. This would create a sinusoidal basilar membrane motion at the frequency of the stimulating sinusoid, with maximum vibration at a particular place on the basilar membrane, as determined by its resonant properties. This, in turn, causes the stereocilia at the top of an IHC in the appropriate place on the basilar membrane to vibrate back-and-forth at the same frequency. When the stereocilia move towards the tallest part of the hair bundle, neurotransmitter is released into the synaptic *cleft* (the tiny gap between the hair cell and auditory nerve fibre ending), making the nerve more likely to fire. When the stereocilia move the other way, neurotransmitter is taken up out of the cleft, making the nerve less likely to fire. Therefore, as long as there is time for the neurotransmitter to be injected and removed from the synaptic cleft, the nerve will tend to fire in synchrony with the stereocilia movements—at the same phase of the stimulating sinusoid. Here, for example, is the genuine firing pattern of an auditory nerve fibre to a section of a 300 Hz sinusoid:

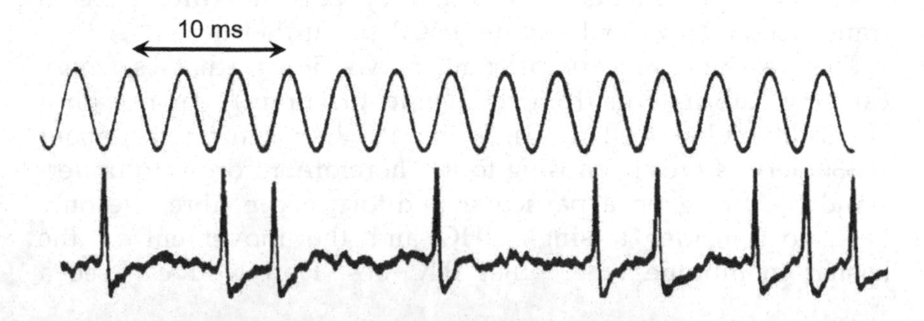

Nerves don't fire on every cycle of the stimulating wave, but when they do fire, it is at a similar point in the cycle of stereocilia movement (here when the sinusoid is at its lowest value). Because the movement of neurotransmitter in and out of the synaptic cleft takes some time, this synchrony is only present for sinusoids up to certain frequencies. Roughly speaking, synchrony is strong up to about 1.5 kHz, and then tails off, becoming undetectable for frequencies above about 5 kHz. At that point, the nerve fires without regard for the phase of the stimulating waveform, because the neurotransmitter does not have time to be released and cleared. Therefore, there is a more or less constant amount of it in the synaptic cleft through the time corresponding to one period.

We somehow need to account for these processes (the synchrony of nerve firing and its dependence on frequency) in the IHC portion of our model of the auditory periphery. It turns out to be easy to do this with a combination of rectification and a low-pass 'smoothing' filter, a concept you have met before when discussing the construction of spectrograms. For making spectrograms, we used full-wave rectification, because we wanted to account for *all* the energy in the wave. In order to model the way the IHC only releases neurotransmitter when the stereocilia bend in one particular direction, we use *half-wave* rectification. Following the rectification with a low-pass filter with a cut-off of about 1.5 kHz will filter out any fluctuations that would lead to synchrony at high frequencies. Our model is now complete, and looks like this:

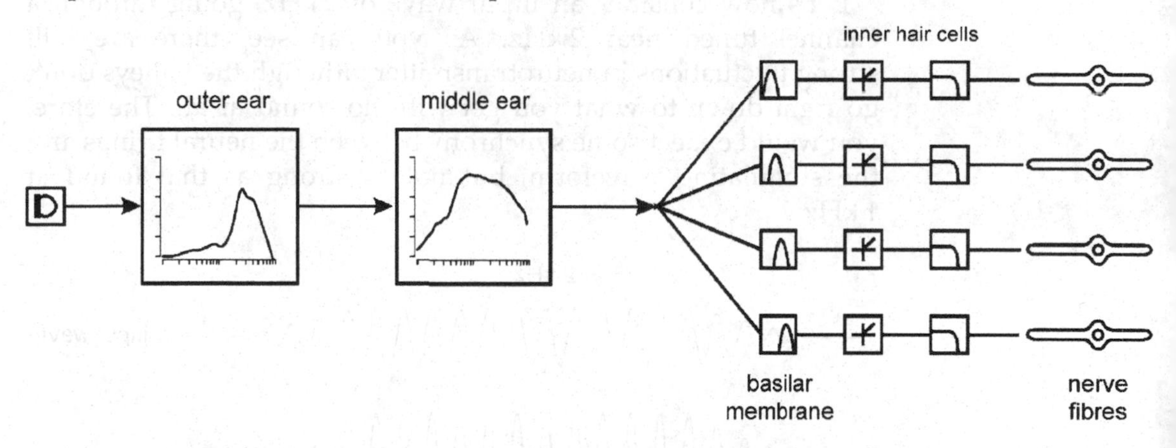

Before discussing how this model could be used, let's try to get a better feel for how the rectification and smoothing would simulate what happens in the IHC. Consider putting a 1 kHz sinusoid into the model. Assuming it is sufficiently intense, this will appear strongly as a sinusoid at the output of one of the band-pass filters simulating basilar membrane filtering with a centre frequency near 1 kHz. Half-wave rectification and smoothing would lead to the waveforms here:

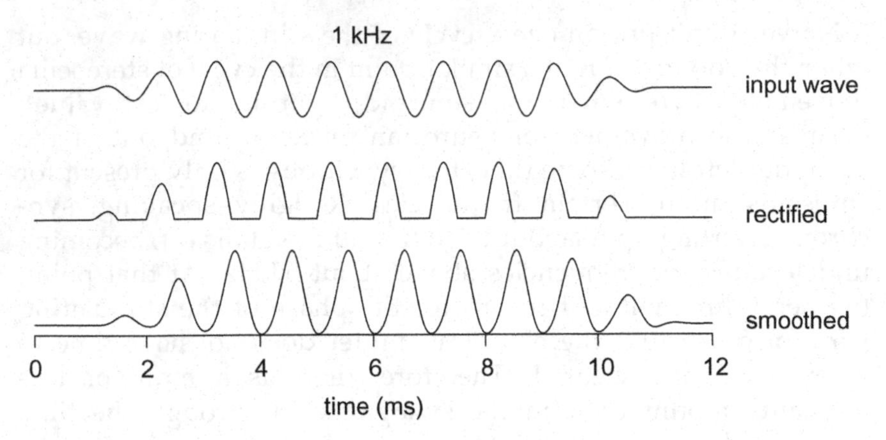

You can think of the final waveform at bottom as representing the amount of neurotransmitter in the synaptic cleft, or the probability that the nerve will fire at any particular moment. When the wave representing the amount of neurotransmitter is high, the nerve is likely to fire. When it is low, it is very unlikely to fire. At the 1-kHz frequency used here, the strongly fluctuating amount of neurotransmitter means that the nerve firings would be highly synchronized to the input wave. Perhaps the best measure of synchrony is the extent to which the peaks and valleys in the amounts of neurotransmitter vary. Here, the valleys go right down to the amount of neurotransmitter present in the absence of any sound, so this represents the maximal degree of synchrony.

Let's now consider an input wave of 2 kHz, going through a channel tuned near 2 kHz. As you can see, there are still strong fluctuations in neurotransmitter although the valleys don't go right down to what you get with no sound at all. Therefore, you would expect some synchrony between the neural firings and the stimulating waveform, but not as strong as that found at 1 kHz:

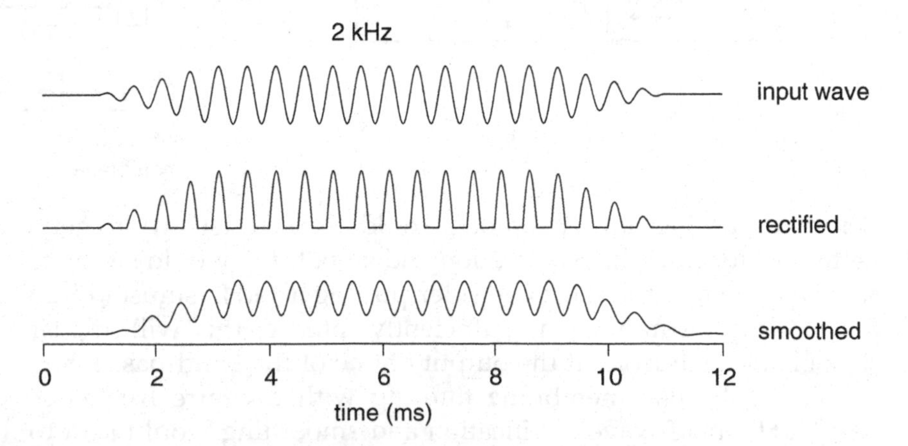

For an input wave at 3 kHz, there is even less evidence of fluctuations in the amount of neurotransmitter. We would, therefore, expect little synchrony of nerve firing with the stimulating wave:

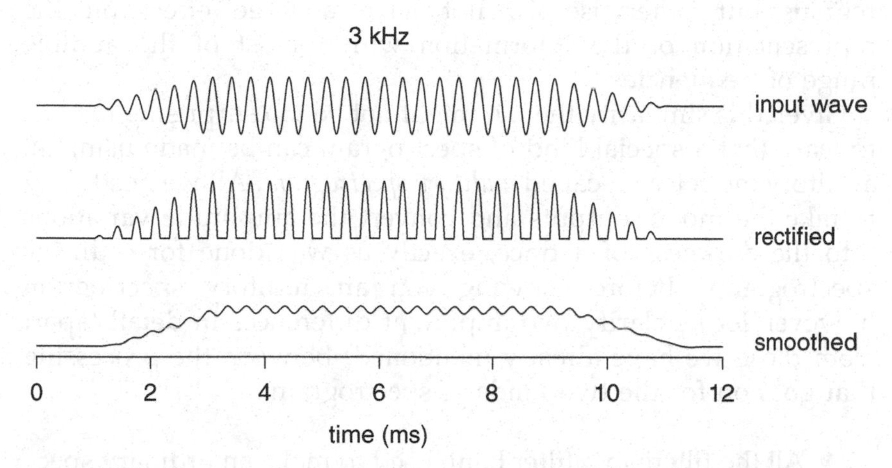

Finally, at 6 kHz, the amount of neurotransmitter in the synaptic cleft increases as the sound is turned on, but there are no fluctuations related to sound frequency at all. Note again that it is not the half-wave rectification that is responsible for this loss of synchrony—it is the low-pass filtering that matters.

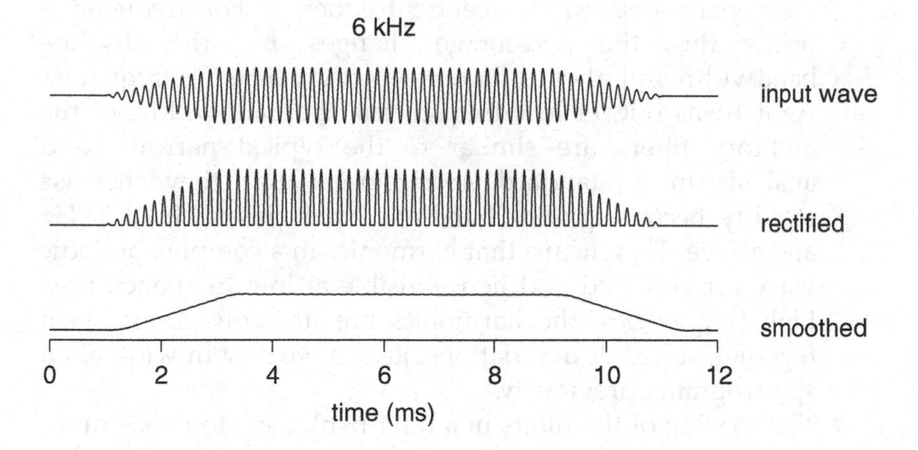

Making an auditory spectrogram

It probably has not escaped your attention that the model of the auditory periphery that we have developed in this chapter is structurally very similar to the collection of systems we described in Chapter 11 for making spectrograms (p. 225). In both cases, an input signal is fed to a filter bank, with each channel output being rectified and smoothed. The only significant difference in overall structure is that the auditory model has band-pass filters to represent the effects of the outer and middle ear. These serve only to amplify or attenuate various frequency regions but otherwise do not have a large effect on the representation of the information within most of the audible range of frequencies.

Given this similarity then, it might not be too surprising for you to learn that a special kind of spectrogram can be made using an auditory model, a so-called *auditory spectrogram*. All we need do is to take the model outputs and convert the amplitude variations into the darkness of a trace, exactly as was done for ordinary spectrograms. Before showing you an auditory spectrogram however, let us clarify two important differences in detail (apart from those we have already mentioned) between the processing that goes on for the two kinds of spectrogram.

- All the filters in a filter bank used to make an ordinary spectrogram have the same bandwidth, whether that is a wide or narrow band. In the auditory periphery, bandwidths increase as we move from the apex to the base of the cochlea—in other words, they increase with increasing centre frequency. We need to include this aspect in our auditory spectrograph. For frequencies of about 1 kHz and above, the bandwidth of an auditory filter is approximately a fixed percentage of its centre frequency. For frequencies below this, the percentage changes, but the absolute bandwidth still always increases with increasing frequency. As it turns out, at low frequencies, the bandwidths of the auditory filters are similar to the typical narrow band analysis in a standard spectrogram. But they increase steadily, becoming wide band at frequencies of about 2 kHz and above. This means that harmonics in a complex periodic wave are resolved and hence visible at low frequencies. At high frequencies, the harmonics are unresolved, and beat together, so result in striations, just as we saw in wide-band spectrograms previously.
- The spacing of the filters in a filter bank used to make an ordinary spectrogram is linear, whereas an auditory filter bank has a spacing that corresponds to the way in which sinusoidal frequency maps onto place on the basilar membrane, a so-called *tono-topic* map. For frequencies of about 1 kHz and above, this mapping is logarithmic. For frequencies below 1 kHz, the mapping is somewhere between linear and logarithmic. So, for example, if you make a spectrogram over the frequency range from 20 Hz to 20 kHz (the audible range of frequencies), half of an ordinary

spectrogram is taken up by frequencies above 10 kHz, which hardly matter at all to us. On an auditory spectrogram, however, the midpoint would be at about 1.8 kHz, which is a reasonable reflection of the relative importance of these two bands.

We'll only look at the auditory spectrogram for one particular wave, one for which you have already seen ordinary spectrograms (p. 241). This is a periodic train of narrow pulses with a fundamental frequency of 100 Hz, which has been put through a cascade of two resonators, one at 700 Hz and one at 2200 Hz:

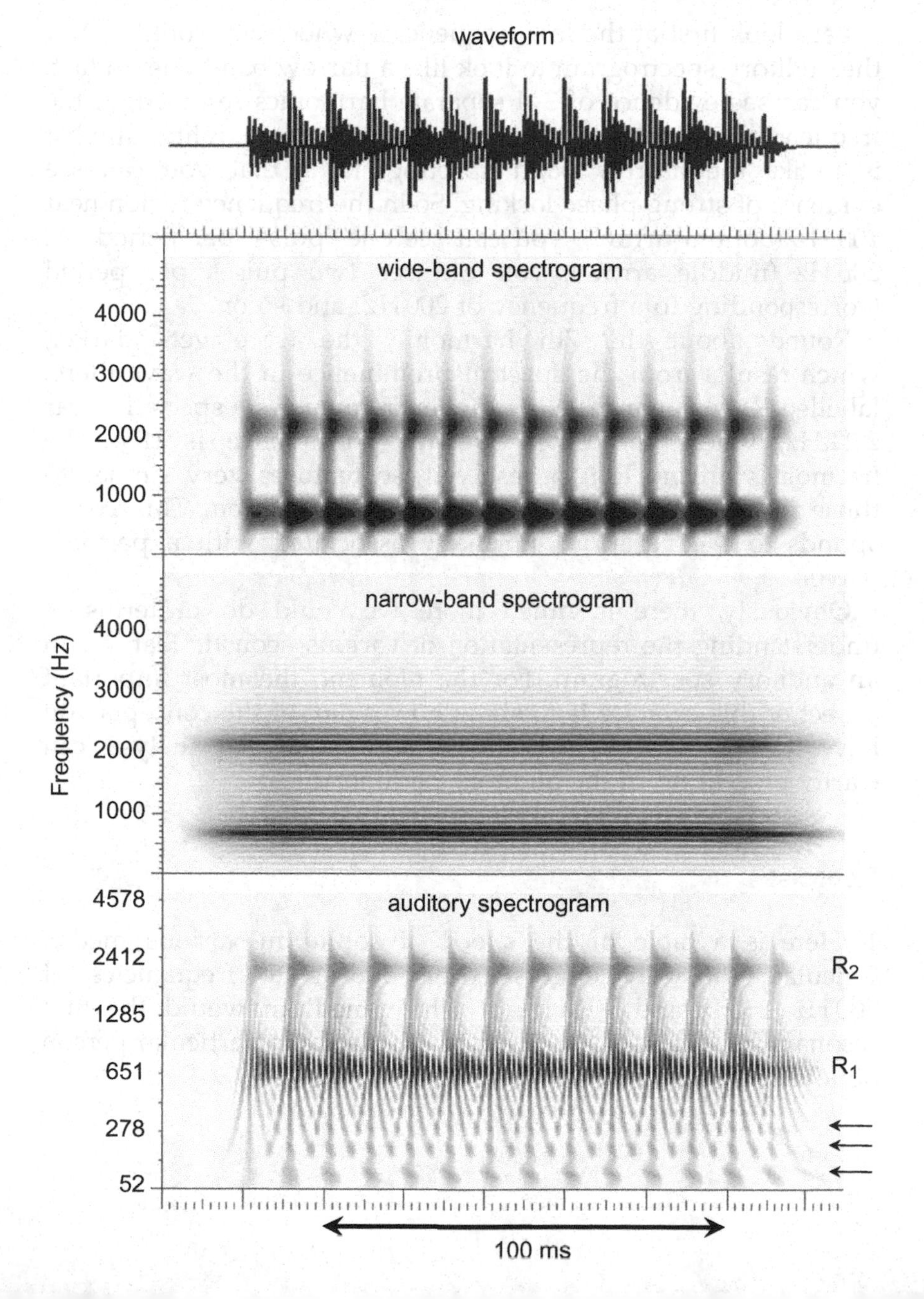

Going from top to bottom, you can find the waveform, and three different kinds of spectrogram. The time axis is the same for all the four panels, but look at the frequency axes on the spectrograms. The ordinary spectrograms have, of course, a linear scale, but the auditory spectrogram is more or less logarithmic, at least for the frequency range above about 1 kHz. You can confirm this by taking the ratios of successive numbers on the axis, which are about equally spaced in distance. So, $4578/2412 = 1.90$ and $2412/1285 = 1.88$. But the scale is *not* logarithmic at low frequencies, because $278/52 = 5.35$, which is not close to 1.90 or 1.88.

Let's look first at the low frequencies, where we would expect the auditory spectrogram to look like a narrow band one. In fact, you can see evidence of 3–4 separate harmonics resolved, at the frequencies indicated by the arrows at the bottom right. But what is unlike the narrow band spectrogram is that you can see evidence of strong phase locking. So in the frequency region near 100 Hz (bottom arrow), you can see one 'pulse' per period. At 200 Hz (middle arrow), you can see two pulses per period (corresponding to a frequency of 200 Hz) and so on.

Round about the 7th harmonic, the trace gets darker, which results from the spectral prominence in the wave there, labelled R_1 (resonance 1). At higher frequencies, especially near 2.2 kHz, where the other spectral prominence is (R_2), the harmonics are no longer resolved, so features very similar to those in the wide-band spectrogram can be seen. This corresponds to the striations normally associated with a periodic wave.

Obviously, there is much more we could do in terms of understanding the representation of various acoustic features in an auditory spectrogram. For the moment, the most important aspect of this exercise is to show how many of the concepts you have learned with regards to systems and signal analysis can clarify processing in the auditory periphery.

Exercises

1. Here is a table of the speed of sound in various media. Calculate the wavelengths corresponding to frequencies of 500 Hz, 2 kHz and 10 kHz. In which medium would the first resonant frequency of the auditory ear canal of a particular person be lowest and highest?

Medium (m/s)	Speed of sound
Air	340
Water	1450
Oxygen	317
Hydrogen	1286

2. Calculate the first resonant frequency of the auditory ear canal (length = 2.3 cm) of an adult female scuba diver while in air and under water. Compare the two values.

3. How much would the frequency of the first resonance of the ear canal change over the course of life if it was 2 cm long at birth, and reached 2.4 cm in adulthood?

4. Discuss all the reasons you can think of for the advantages of saturating nonlinearities in natural systems.

5. On page 264, in attempting to model the amplitude response of the ear canal, we noted that an even better fit of model to data could be obtained by assuming a slightly different length for the ear canal. Would a better fit be obtained if the assumed length of the meatus was longer or shorter, and why?

6. Sound waves are affected by the head when they have a wavelength that is comparable to, or smaller than, the head and pinna. For an adult, the acoustic effects of the head are most marked at frequencies above about 1.5 kHz, equivalent to wavelengths smaller than 22.7 cm (the approximate diameter of an adult male head). What frequencies would be most affected for a child with a head size of 15 cm?

7. The motion of the middle ear ossicles can be modelled approximately as a pendulum (a weight suspended from a pivot so it can swing freely). The period of swing of a pendulum (T) is given by the formula:

$$T = 2\pi\sqrt{L/g}$$

where L is the length of the pendulum and g is the local acceleration due to gravity (9.8 m/s^2). What is the length of suspension of the middle ear ossicles for a frequency of 1 kHz (roughly the centre frequency of the middle ear system)?

CHAPTER 13

Applications to Speech Production

Just as the concepts of LTI systems analysis can be used to characterize many aspects of hearing, so too are they relevant to speech production, as the articulators form a system that is LTI (at least to a first approximation). An LTI model can thus be used as a way to understand the structure of speech sounds, as a basis for speech signal analysis, and for the electronic synthesis of speech. Here we will be concerned with the first of these areas.

Source-filter theory of speech production

Whenever speech is produced, an excitatory signal must be produced, which is then modified by the vocal tract in much the same way that other LTI systems modify the signals that pass through them:

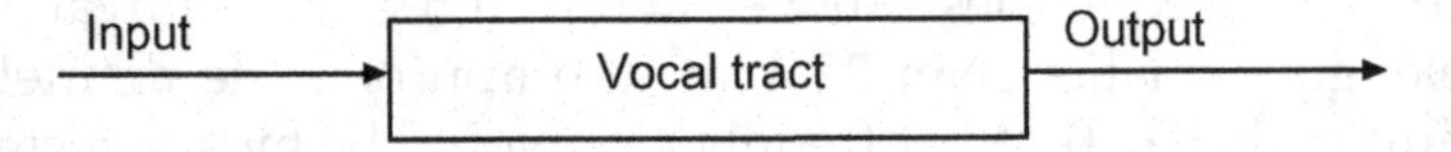

As the input excitatory signal is often known as the *source*, and the vocal tract acts as a *filter* (really just another name for a linear system), this approach is known as the source-filter theory of speech production. Before going on to explore the source signals commonly encountered in speech production, and how the vocal tract filter alters these sources to produce speech, we'll briefly review the relevant functional anatomy. Here's an X-ray cross-section of the vocal tract with a line drawing alongside it indicating the important structures:

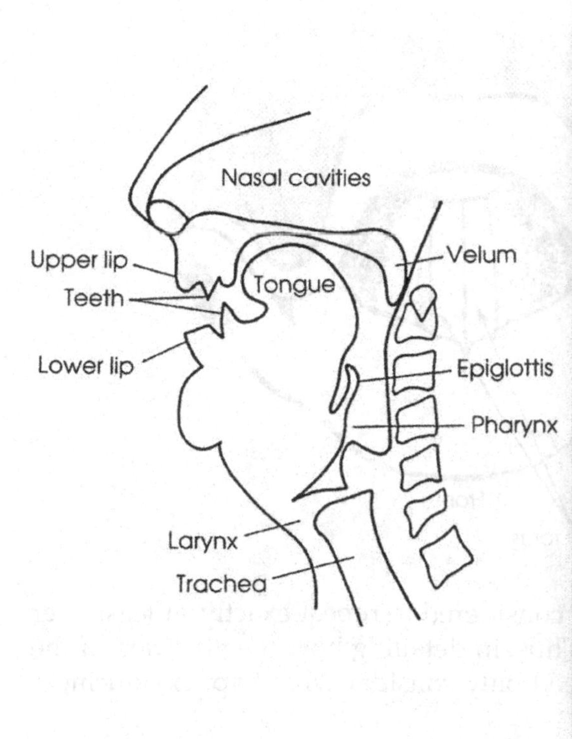

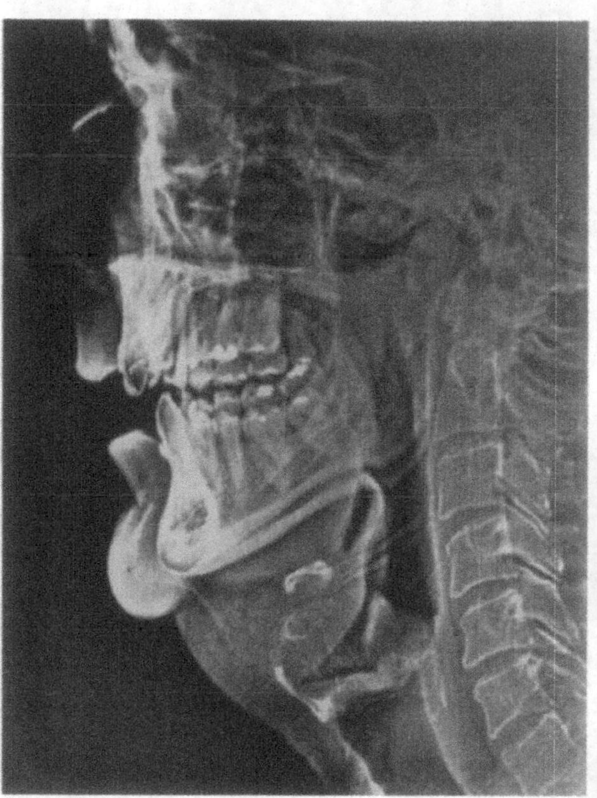

There are three basic divisions of the articulatory system: the part below the larynx (not much of which is seen in the pictures above), the part above and the larynx itself.

Below the larynx are the lungs and muscles that allow the lungs to inflate or deflate. The lungs operate like bellows, forcing air through the trachea and into the vocal tract, where it can be converted into sound. Although sound can be generated in other ways, this mechanism provides the vast bulk of speech energy.

The air that is forced into the vocal tract enters the *larynx*, a structure made up of cartilage and muscles. The *vocal folds* (or 'cords') run back to front between two of the cartilages. The way they would look if seen from above and within the vocal tract is shown at the top of the next page.

When the source arises from the vibration of the vocal folds (the primary source of sound energy for speech), there is said to be *voiced* excitation. This vibration arises from a combination of aerodynamic forces caused by air flow from the lungs and appropriate positioning and tensioning of the folds. These two factors interact to produce movements of the vocal folds that are nearly periodic—hence, you will often see voiced excitation referred to as being *quasi-periodic*. For many purposes, the

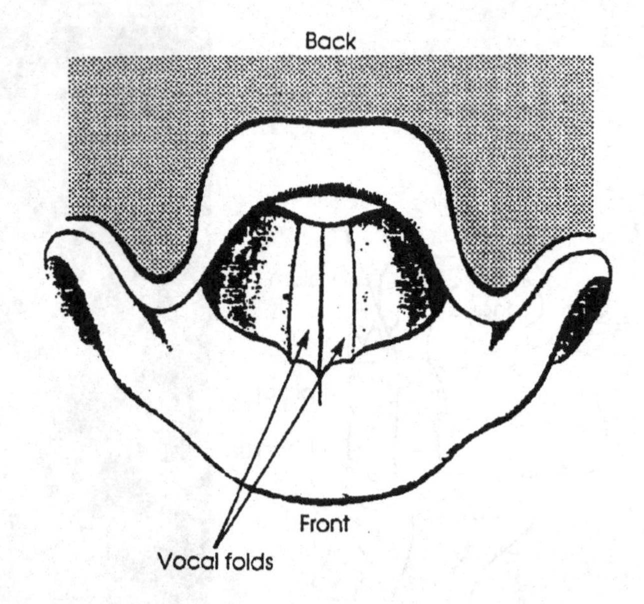

vibration pattern can be considered to repeat exactly, at least over short stretches of time. Thus, in detailing how the vibration of the vocal folds arises, we need only consider what happens during a single cycle.

Assume that we're at a point in the cycle where the folds are closed. As air flows from the lungs, pressure below the folds increases, eventually pushing them apart, allowing free flow of air into the vocal tract through the *glottis* (the space between the folds). Because air is flowing, the pressure between the folds drops, creating a suctional force (the *Bernoulli* force) which, along with the natural springiness of the folds, brings them together again. The pressure underneath the folds then builds up again, until they are forced apart and the whole cycle repeats. The stages in one such vibration cycle can be seen here in this view of the folds from above:

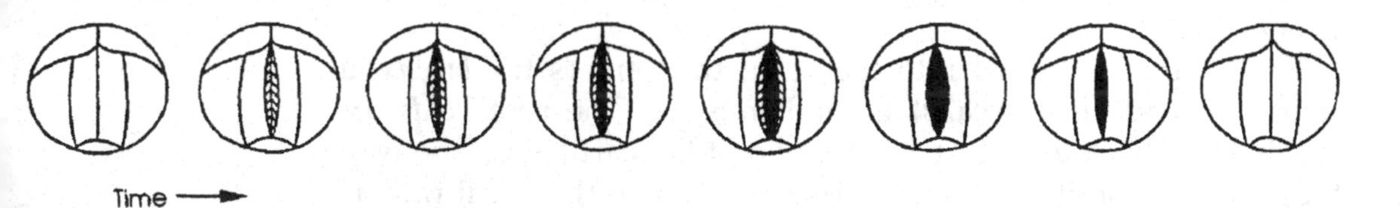

The net effect of the opening and closing of the vocal folds is to 'chop up' what would otherwise be a smooth flow of air from the lungs, thus providing a source of sound energy. We'll consider the acoustic properties of voiced excitation later. For now, we'll just note that the rate at which the vocal folds vibrate can be varied by muscles within them which cause the stiffness to vary, or by changes in subglottal pressure. We hear changes in vocal fold vibration rate as changes in *voice pitch* or melody.

Of course, the sound from the larynx must pass through, and hence be modified by, the vocal tract before it is heard. This is what is implied when the vocal tract is referred to as a filter. The principal factor that determines the amplitude response of the vocal tract filter is its shape, which can be varied by moving certain of the articulators (primarily the tongue, lips, jaw and/or velum).

Before we go on to the application of the LTI source–filter model to speech production, we must add one caveat. In connected speech, the anatomical structures move continuously. Consequently, the vocal tract system does not, strictly speaking, have one of the crucial properties of LTI systems—it is not time invariant. Its shape, and hence its amplitude (and phase) response, varies over time. Since many of the concepts that we have developed apply only to LTI systems, some care is necessary. On the whole, as long as we look at the response of the vocal tract only over periods of time where it does not significantly change shape, it can be treated as if it is time invariant.

We'll now consider the amplitude spectra of some speech sounds in detail (phase spectra play little or no role in the perception of speech). It's perhaps worth restating that this isn't a book about acoustic phonetics, so we won't be describing the properties of all sounds—only some important examples that are good illustrations of the ideas we have presented earlier.

Amplitude spectrum of /ə/ at a constant voice fundamental frequency

We'll start with the vowel /ə/ (called 'schwa')—the vowel used in the pronunciation of the word 'a' in the phrase 'Have a nice day'.

/ə/ is a good example for our purposes for several reasons. Firstly (like all pure vowels), it can be sustained. In other words, the vocal tract can be held in a constant shape during its production. As the shape of the vocal tract determines its filtering properties, when the shape is fixed, so is the filtering that occurs. In short, the system can be *time invariant* during the production of /ə/. Secondly, voiced sounds like /ə/ can be spoken with a constant rate of vocal fold vibration. Thus, the source can be treated as purely periodic. The final factor that makes /ə/easy to deal with is that the shape of the vocal tract appropriate to produce it has relatively simple transmission properties.

In order to determine the spectrum of /ə/corresponding to a particular source and filter, we can employ the principles developed in Chapter 8. There, you'll remember, we illustrated a frequency domain method of establishing the output spectrum of *any* signal passed through *any* LTI system. So, paralleling the procedure in the earlier chapter, we'll:

(1) Specify the amplitude spectrum of the excitatory source signal that enters the vocal tract.
(2) Obtain the amplitude response of the vocal tract when it's set for producing /ə/.
(3) Determine the spectrum of the output speech sound either by multiplying the amplitude spectrum of the source signal by the amplitude response of the vocal tract (if the amplitude scales are linear) or by adding them together (if the amplitude scales are logarithmic). In fact, because dB scales are almost always used for representing the spectra of speech sounds, the two relevant curves will be *added*.

Step 1: Determine the amplitude spectrum of the source

The acoustic excitatory signal that results from the action of the vocal folds can be idealized as a sawtooth waveform. Since the folds peel open gradually and snap together fast, they have a sharp trailing edge. Here is the excitatory signal we will assume (a sawtooth with a fundamental frequency of 100 Hz) and its spectrum:

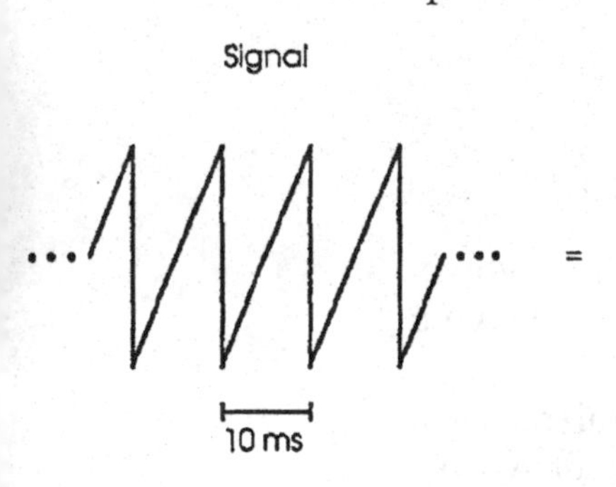

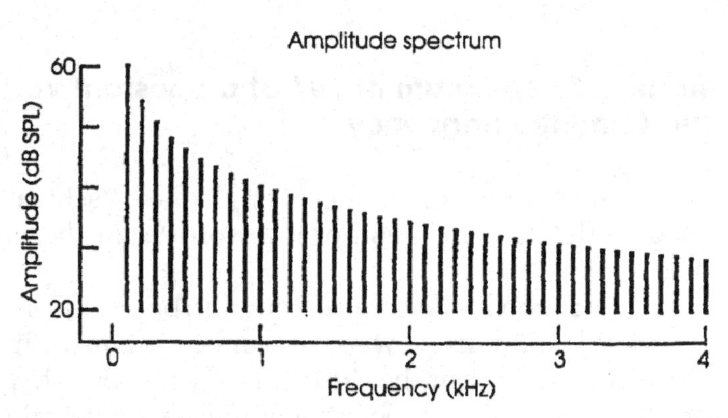

Step 2: Determine the amplitude response of the vocal tract

Next, we need to determine the filtering caused by the vocal tract. The vocal tract shape during production of /ə/ is usually represented in simplified form as a tube of uniform cross-sectional area. Moreover, it can be treated as if it is closed off at one end (the end where the vocal folds are) and open at the other (the mouth). Therefore, it is similar in important respects to the model of the auditory meatus that was discussed in Chapter 12, although obviously much larger! There, we showed that the resonant frequencies were at $c/4L$ and odd-integer multiples ($3x$, $5x$ and so on) of this frequency (c being the speed of sound in air, 340 m/s, with L the length of the tube). Since a male adult's vocal tract is about 17 cm long, you should be able to confirm that the resonant frequencies occur at 500 Hz, 1500 Hz, 2500 Hz, 3500 Hz and so on. Therefore, the amplitude response of a 17-cm tube open at one end and closed at the other looks like this (assuming some reasonable degree of damping):

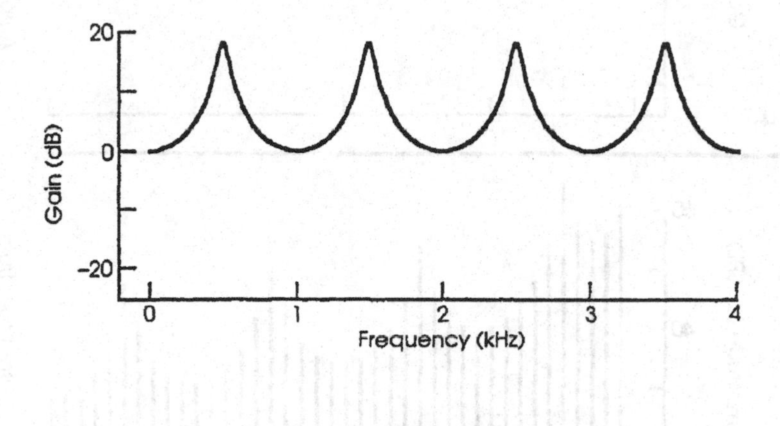

Step 3: Determine the output amplitude spectrum

As shown in Chapter 8, the spectrum of the signal at the output of the vocal tract can now be calculated by adding together the spectrum of the source and the response of the filter (because both are on dB scales). This process, and its result, is depicted on the opposite page.

The peaks in the output speech spectrum are referred to as *formants* and are numbered, starting at 1, from the low-frequency end. Formants are a reflection of the resonances of the vocal tract. It is a general property of speech sounds that formant peaks are almost always caused by the vocal tract filter even though the overall speech spectrum results from the interaction of the spectral shapes of both the source and the filter. Here, for instance, it is clear that the overall decrease in amplitude across frequency results not from the vocal tract filter, but from the spectral shape of the source.

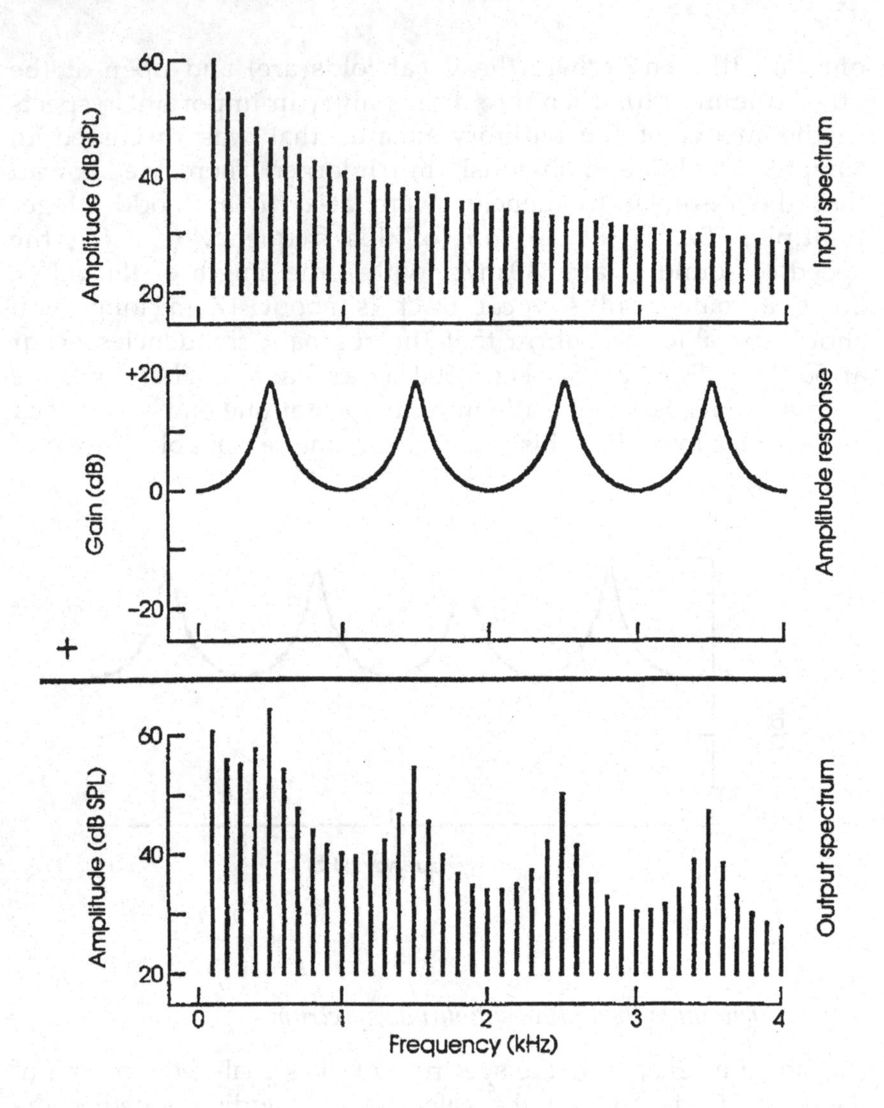

Amplitude spectrum of /ə/ at a different voice fundamental frequency

It is now relatively straightforward for us to determine what happens if a source signal of a different frequency, say with twice the voice fundamental frequency, excites this same vocal tract.

As before, we'll approximate the excitatory signal with a sawtooth with a sharp falling edge. The only difference between the previous source and this one is that the source now has a period half as long (5 ms rather than 10 ms). Thus, the fundamental frequency will be higher and the harmonics spaced at 200 Hz rather than 100 Hz. As the signal is still a sawtooth, its amplitude spectrum still falls off at 6 dB/octave. So, the waveform and spectrum of the source with a fundamental at 200 Hz looks like this (those of the 100-Hz sawtooth are shown again below for comparison):

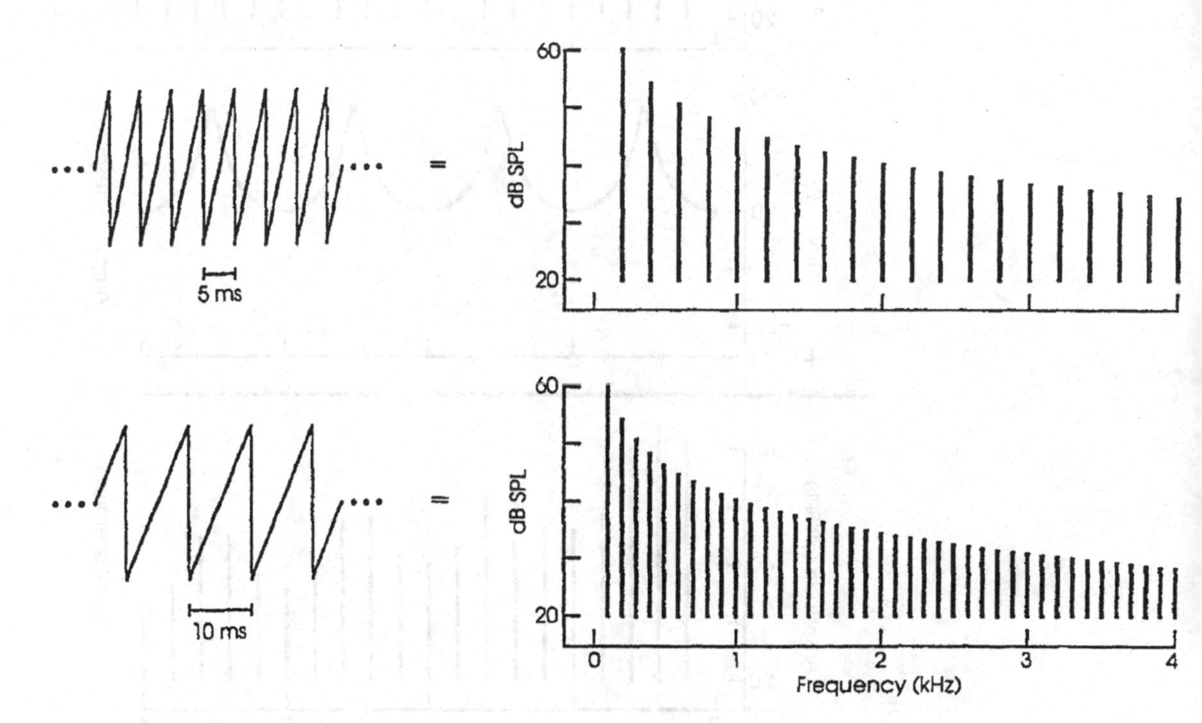

As the vocal tract filter has not been altered, its amplitude response remains as before. Therefore, to obtain the spectrum of the output, we add the new source spectrum and previous filter amplitude response together (as shown on the next page).

The relatively wide spacing of the harmonics in the output spectrum arises from the relatively wide spacing of the harmonics in the source (that is, its higher fundamental frequency). As before, the formants can be easily seen. However, because the harmonic spacing is relatively wide, and the resonance bandwidths have not changed, there are fewer harmonics within each formant peak. Thus, the shape of the formants is less clearly defined for the higher fundamental frequency source. As women have higher-pitched voices than men (that is, higher fundamentals) and so greater interharmonic spacing without concomitant increases in formant bandwidth, the definition of the formants tends to be less clear than in men. This can make it difficult to identify the position of formants in female speech.

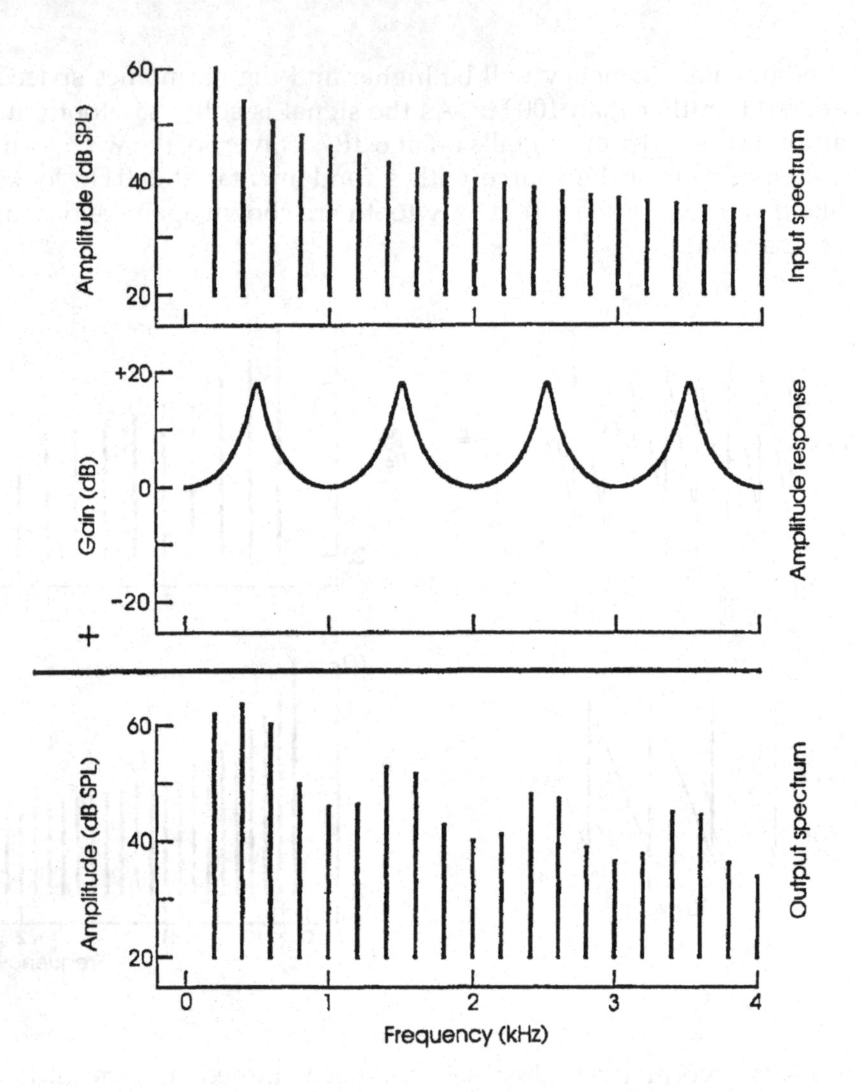

Spectrographic representation of /ə/ at different voice fundamental frequencies

Although single spectra can be important in analyzing speech sounds, spectrograms too are of great utility. Therefore, we'll show how these synthetic /ə/ sounds appear on spectrograms. It might be a good idea to review the section towards the end of Chapter 11 in which we examined spectrograms of pulse trains which varied in fundamental frequency, or were passed through a filter which emphasized selected frequency regions. Many of the features exhibited by those non-speech sounds are present in speech sounds as well.

Over the page, then, are wide-band spectrograms and sections of the /ə/ vowel with the two different fundamental frequencies. Recall from Chapter 11 that spectrograms are plots of the distribution of energy over frequency as a function of time. Sections, on the other hand, show the distribution of energy at a particular moment of time— a *short-term spectrum*

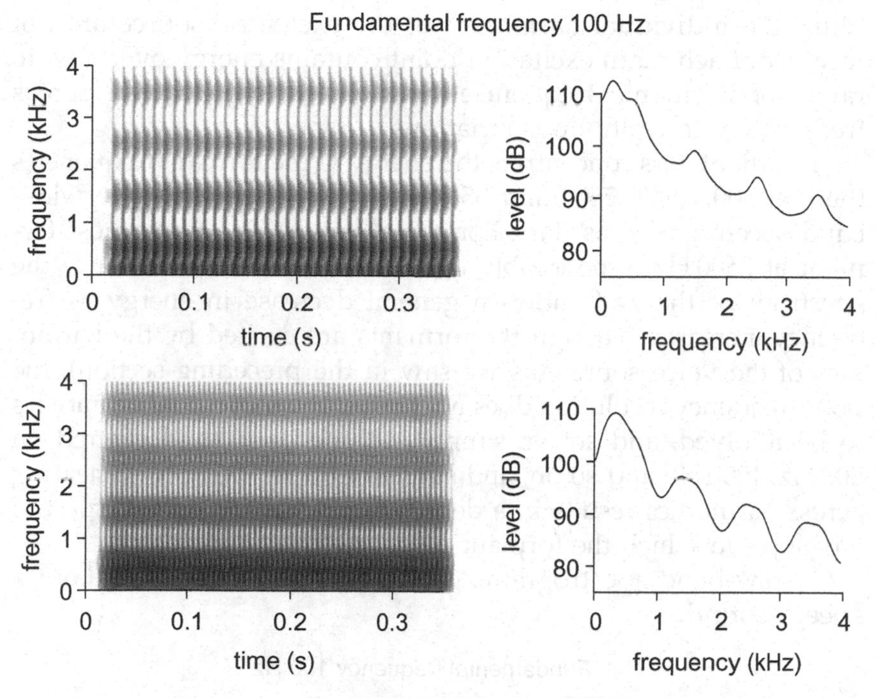

You'll remember that a wide-band analysis gives good resolution of the temporal aspects of a sound but that some frequency information is lost. This is most clearly seen in the way that voice fundamental frequency is reflected in a wide-band spectrogram. The properties of the voiced source arising from the vibration of the vocal folds are such that the amount of excitation they provide is not constant over the period of a single cycle. It appears that the amount of excitation is determined not by the amount of air flowing, but rather by the degree to which it is *changing* over time (called the *derivative* of the flow). Since the vocal folds come together much faster than they peel apart, the airflow is changing fastest there, leading to maximum excitation at that point in time. Good temporal resolution allows the time of the main excitation of the vocal tract to be seen. At each point of main excitation, a vertical stripe (a *striation*) is seen on the spectrogram. Since there is only one striation per cycle, the time between them can be used to measure the excitation frequency—the separation between the vertical lines gives the period of each cycle of vocal fold vibration. The separation between the striations is less (shorter period) when the vowel is spoken on the high voice fundamental frequency than when it is spoken on the low one.

The poor representation of spectral information results from the 300-Hz bandwidth of the analyzing filters. Since the harmonic spacing for both fundamentals used is narrower than this bandwidth, each filter spans across more than one harmonic. Thus, the individual harmonics of the excitation source are not resolved. Each main excitation point contains energy over a wide range of harmonic frequencies, and these are smeared across frequency into continuous striations.

The articulators concentrate the energy at the formant frequencies (here at 500, 1500, 2500 and 3500 Hz) which appear on the wide-band spectrograms as dark horizontal bands. The uppermost formant at 3500 Hz is noticeably less dark (less intense) because the spectrum of the /ə/ shows a general decrease in energy as frequency increases. Though the formants are excited by the harmonics of the voice source (as we saw in the preceding section), the poor frequency resolution does not permit the individual harmonics to be resolved, and so we simply see one dark band centred on 500 Hz, 1500 Hz and so on, and similarly for the sections. Smearing across harmonics results in a depiction only of the overall spectral envelope, in which the formant peaks are clearly delineated.

Narrow-band spectrograms give different information about a speech sound:

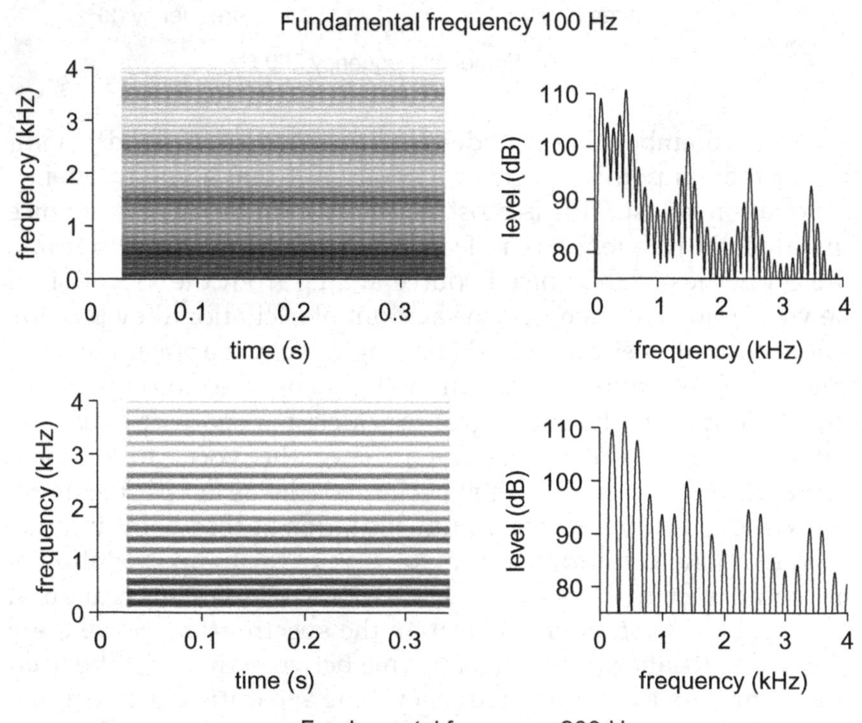

For both fundamentals, the harmonics of the voice source are resolved separately, because the analyzing filters have a bandwidth (at 45 Hz) smaller than the interharmonic spacing. Thus, the spectrograms have the appearance of a set of horizontal lines, each of which corresponds to a single harmonic. Poor temporal resolution means that the point of maximum excitation is not indicated—energy is smeared across time periods at least as long as the fundamental period. Therefore, the fundamental must be calculated from spectral, and not temporal, features. This can be done by tracing the frequency of any particular harmonic over time (and dividing by the harmonic number), or simply by determining the spacing between the harmonics. Practically speaking, it is usually easiest to determine the frequency of the 10th harmonic, which is then divided by 10 by moving the decimal point one place to the left.

You can also see that energy is concentrated at the formant frequencies because the harmonics are more intense there, and similarly for the sections. Each harmonic of the voice source is clearly seen, with the spectral envelope being indicated by their peaks. They thus look nearly identical to the theoretical amplitude spectra shown above—the only difference being that the earlier amplitude spectra are considered to be analyses over an infinitely long period of time, while these are short-term spectra. There is little difference here because the signal being analyzed is steady state.

This description of the difference between narrow- and wide-band analyses, although fairly typical, doesn't necessarily hold for all voice fundamental frequencies. If the rate of vocal fold vibration and, therefore, the voice fundamental frequency, is high enough (as is more likely to happen with children, or adult female speakers), the harmonics might be resolved even by wide-band filters (at the same time eliminating striations). This can be seen in the following spectrograms (with equalization to emphasize higher frequencies) of a female soprano who is singing the vowel /ə/ at a fundamental varying around 640 Hz):

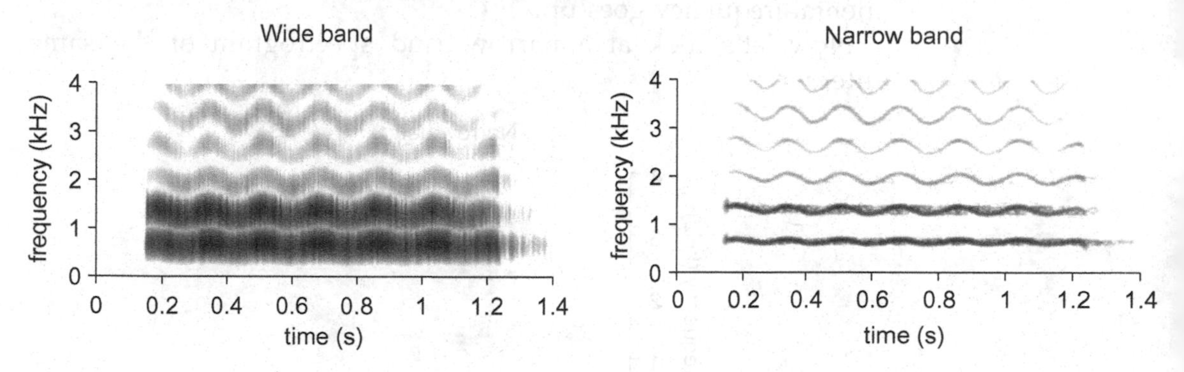

Both the wide- and narrow-band analyses resolve the harmonics, although of course the harmonics *appear* broader in the wide-band analysis (look again at the analysis of the 1-kHz tone in Chapter 11, p. 235). The 'wavyness' of the traces is due to *vibrato*, a cyclical variation of fundamental frequency typical of singers trained in the Western art tradition (but not necessarily of singers trained in other traditions).

Spectrographic analysis of vowels with a changing fundamental

You might have noticed that, as long as we were dealing with synthetic speech, voice fundamental frequency could be absolutely (inhumanly) constant. In real speech, of course, and exhibited plainly in the previous spectrograms of sung speech, the voice fundamental is never constant for long. These changes can be readily seen in spectrograms (although not always so clearly as for the vibrato above). Let's now examine a wide-band spectrogram of a 'real' vowel, spoken with a rising intonation—equivalent to an increase in fundamental frequency:

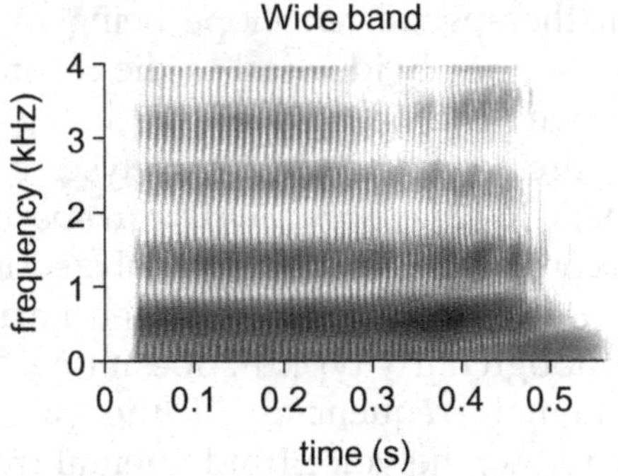

Here, the speaker is an adult male with fundamental frequencies typical of such, so the analyzing filters do not resolve the harmonics of the voice source at any point in the utterance. If you look carefully, you might be able to see the striations coming closer together over the time course of the spectrogram. This shows that the fundamental period decreases as the voice fundamental frequency goes up.

Now let's look at a narrow-band spectrogram of the same utterance:

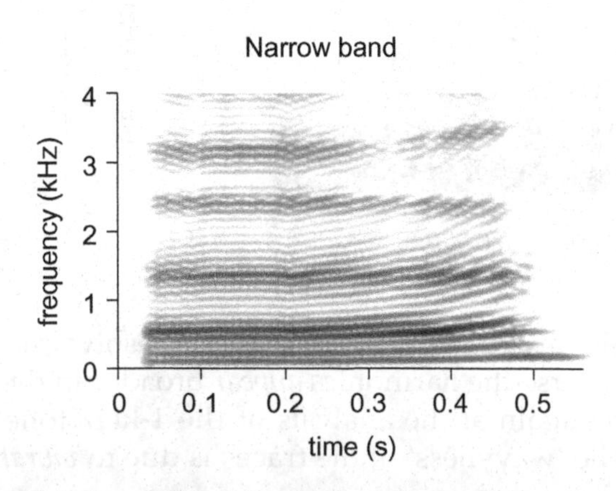

Right through the sound, the harmonics are seen, again most intensely at the formant frequencies. Of course, the harmonics get further apart as the voice fundamental frequency increases.

Properties of other vowel sounds

Other pure vowels can be produced in a similar way to /ə/, differing only in the position in which the vocal tract is held. This can be seen from the following tracings from X-ray photographs taken while two different steady-state vowels were being spoken. Each is labelled with a phonetic symbol appropriate for the vowel being produced and a word that includes that vowel too. Beware! The vowels in the specified words are pronounced differently in different accents of English. What we've given is right at least for Southern British English and all American accents:

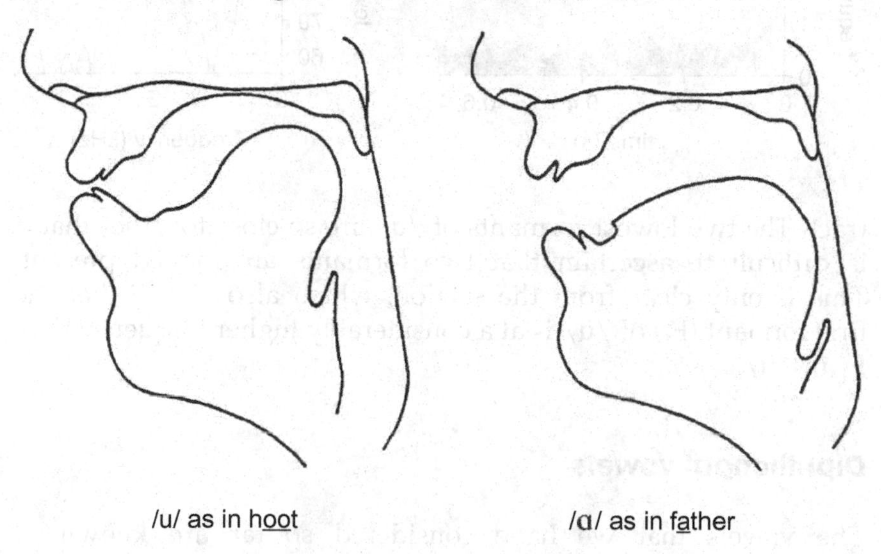

/u/ as in hoot   /ɑ/ as in father

The vocal tract shapes are somewhat more complex than that for /ə/. For example, in /ɑ/, the vocal tract is wider at the front (towards the mouth) than it is at the back. In /u/, the vocal tract is narrow at the lips, wider further in and then narrower again at the velum.

It's possible to determine the amplitude response of these (more complex) vocal tract shapes in a number of ways, both theoretically and empirically. For our present purposes, however, it will suffice to examine some wide-band spectrograms and sections of these two vowels, as uttered by an adult male speaker (seen on the next page).

The dark bands in the spectrograms corresponding to the formants are more or less constant over the full duration of the vowel, since the speaker did not alter the shape of his vocal

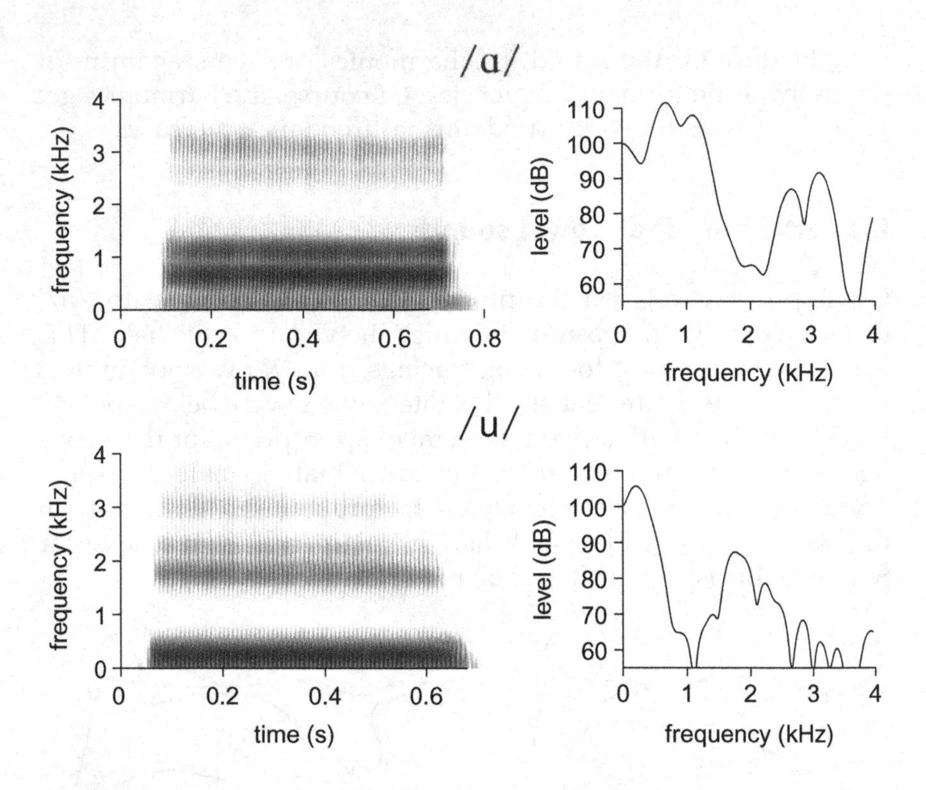

tract. The two lowest formants of /ɑ/ are so close together that it is difficult to ascertain that two formants are indeed present. This is only clear from the section, which also shows that the first formant (F_1) of /ɑ/ is at a considerably higher frequency than F_1 in /u/.

Diphthongal vowels

The vowels that we have considered so far are known as *monophthongs*. Monophthongs are produced, at least in isolation, with a fixed vocal tract shape throughout the time that they are spoken. A second class of vowel-like sounds are the *diphthongs*. Just as monophthongs are single vowels, diphthongs are a combination of *two* vowels. Examples of diphthongs would be /eɪ/ as in 'hay' or /au/ as in 'how'. If you say /eɪ/, you'll notice that your articulators move from the position about like that used to produce /e/ (similar to the vowel in 'bed') to one for /ɪ/ (the vowel in 'bid'). The property that monophthongal vowels and diphthongs share is that they are produced with a relatively open vocal tract. So, if you say /eɪ/ or /au/ you'll notice that at no point do the articulators come very close together. This is a general characteristic of vowels and vowel-like sounds.

Here are spectrograms of /e/, /eɪ/ and /ɪ/ spoken with an approximately constant voice fundamental frequency:

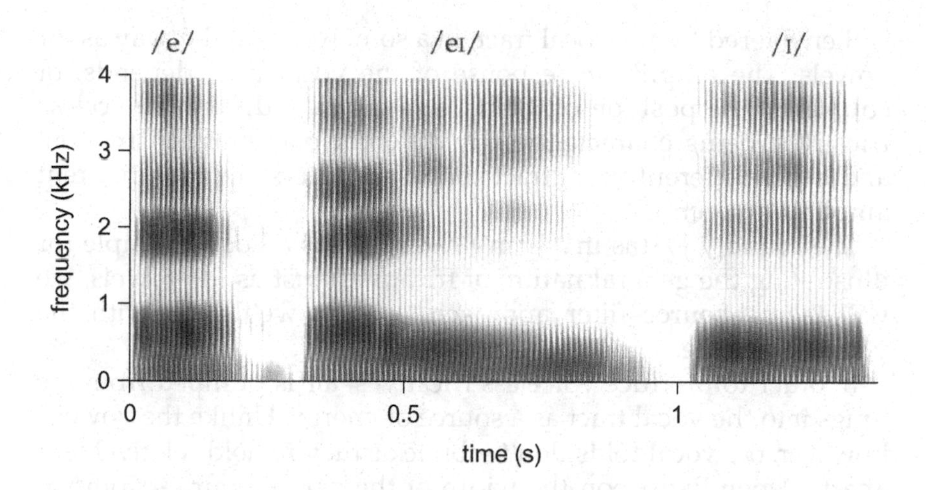

At the start of the diphthong the formant frequencies are appropriate for /e/, and at its end for /ɪ/. Between these points, the formants move smoothly, reflecting the fact that the articulators are moving smoothly from one vowel configuration to the other.

Voiceless fricatives

So far, all the sounds that we've discussed have employed the same (voiced) source of excitation. We did mention that other sources of excitation were possible, and we'll now consider one class of sounds to illustrate this—the voiceless fricatives. *Fin, thin, sin, shin* and *him* all begin with a voiceless fricative, and these are the only five to occur in English. If you say each one of these sounds in isolation, prolonging them to be easier to hear, you will note that they all sound 'hissy' or 'noisy', quite unlike the strongly pitched 'buzz' that vowels have. This 'noisiness' arises because the source of excitation for voiceless fricatives is itself aperiodic (hence 'noisy'), quite unlike the strongly periodic voiced source.

As you say these sounds, you may note that there is some point in the vocal tract at which the articulators are held close together, creating a *constriction*. In /f/, for example, the constriction is at the upper teeth and lower lip, while for /s/ it is between the tongue and the roof of the mouth. When air from the lungs flows sufficiently fast through a constriction, it becomes *turbulent*, and generates a band of noise that is usually wide in its spectrum. Furthermore, in many fricatives, the turbulent stream strikes an obstacle—in /s/, for example, the teeth—and it appears from work by Shadle that the noise that is thus generated is of considerably greater amplitude than the turbulence noise generated at the constriction. Regardless of the details, this noisy source is then filtered by the vocal tract, in a somewhat similar way as for vowels. The amplitude response of the vocal tract depends, of course, on the position of the articulators. The different voiceless fricatives, being characterized by different places of constriction, and hence different vocal tract shapes (or filters), exhibit different amplitude spectra.

The sound /ʃ/ (as in 'sham') can serve as a good example for illustrating the general nature of fricatives. Just as for vowels, we will take a source–filter approach, so first we'll deal with the voiceless source.

In order to produce voiceless fricatives, air is pumped from the lungs into the vocal tract as a source of energy. Unlike the vowels, however, the vocal folds don't vibrate as they're held relatively far apart. Depending upon the width of the gap, the air escaping at the vocal folds may or may not cause excitation of the vocal tract at this point. Excitation by air escaping through this gap (the *glottis*) is an important source for /h/ but is less important for the other voiceless fricatives.

For /ʃ/ in particular, the vocal folds are sufficiently far apart to have little effect. Therefore, the air flows relatively unimpeded until it comes to the narrow constriction formed by the tongue and the roof of the mouth:

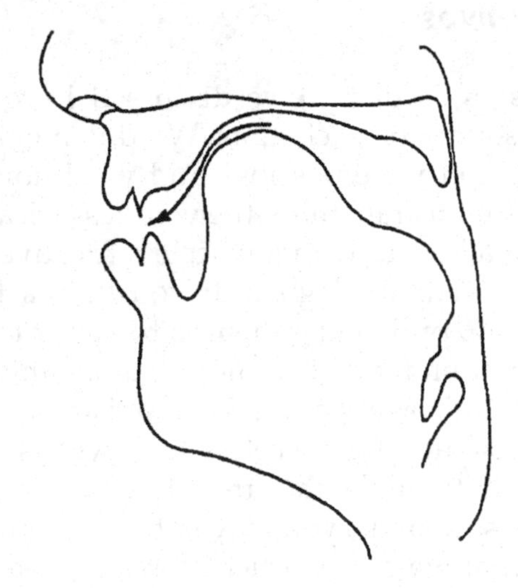

The narrowness of the constriction means that the speed of the flow increases significantly through the constriction. When its speed becomes sufficiently high, the pattern of flow changes from a smooth streaming into a random and turbulent one. This turbulent flow then strikes the teeth, the result being a random series of changes in air pressure—noise. All turbulent airflows are aperiodic, and so the amplitude spectrum for this excitation does not contain energy at discrete frequencies—it has a *continuous spectrum*. We cannot, of course, tell from this alone what spectral composition the sound has. Studies employing models of the vocal tract suggest, however, that it is quite broad, with a shallowly falling spectral envelope. The next stage, as with vowels, is to characterize the effect of the vocal tract filter (the LTI system). For the production of /ʃ/, we can think of the vocal tract as being made up of four main features—the constriction itself, the resonators in front of the constriction (formed by the front of the mouth, lips and teeth), an obstacle in front of the air jet as it leaves the constriction (the teeth) and the resonators behind the constriction (including the lungs):

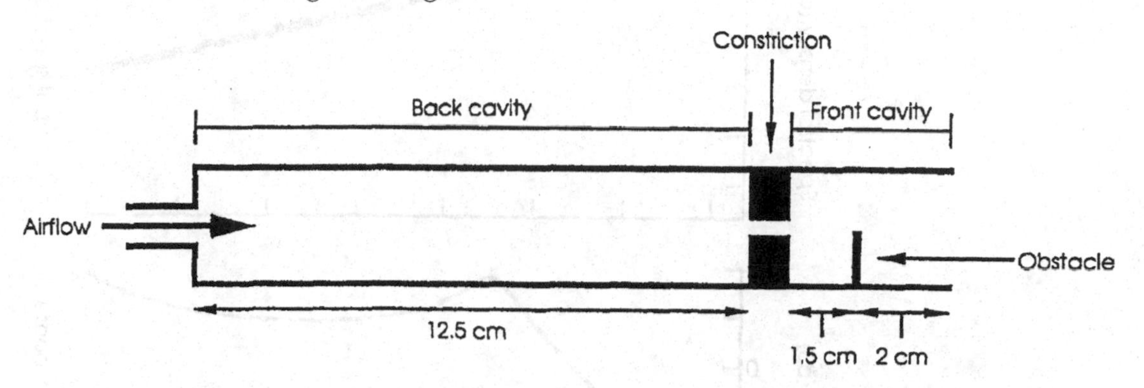

As long as the constriction is fairly narrow and long enough (as it is here), the back cavities and the constriction itself exert little or no influence, so we need not consider them any longer. The remaining resonances of the system (due to the front cavity) can be calculated in just the way we've done for the auditory meatus and neutral vocal tract. The front cavity can be considered as a tube open at one end (the mouth) and closed at the other (because the constriction is small). Thus, we already know that it will have a quarter-wavelength resonance, and odd-integer multiples thereof. As the front cavity has a length of 3.5 cm, its resonances will be at about 2429 Hz, 7286 Hz, 12.14 kHz and so on. Over the frequency range investigated here, only the first two resonances matter.

There is one further complication. Because the main energy of the source appears to arise at the obstacle (representing the teeth), there is a cavity *behind* the source, bounded by the obstacle and the constriction. This cavity also has resonances, but because of the position of the source, the cavity *absorbs* sound at frequencies near its resonance peaks (as measured at the mouth) instead of amplifying it. These regions of energy absorption are known as *anti-resonances* and produce dips or valleys in the amplitude response. It turns out that this cavity can be roughly modelled as a tube that is closed at *both* ends. Theoretical calculations show that this leads to an anti-resonance at low frequencies (below 200–300 Hz) and a first higher-frequency resonance which is half-wavelength (that is, given by $c/2L$, where $L = 1.5\,\text{cm}$). Even this first resonance, at about 11.33 kHz, is out of the range of frequencies that we're investigating.

To obtain the final predicted output spectrum of /ʃ/ then, we add the source spectrum to the amplitude response obtained from the combination of two resonances and one anti-resonance:

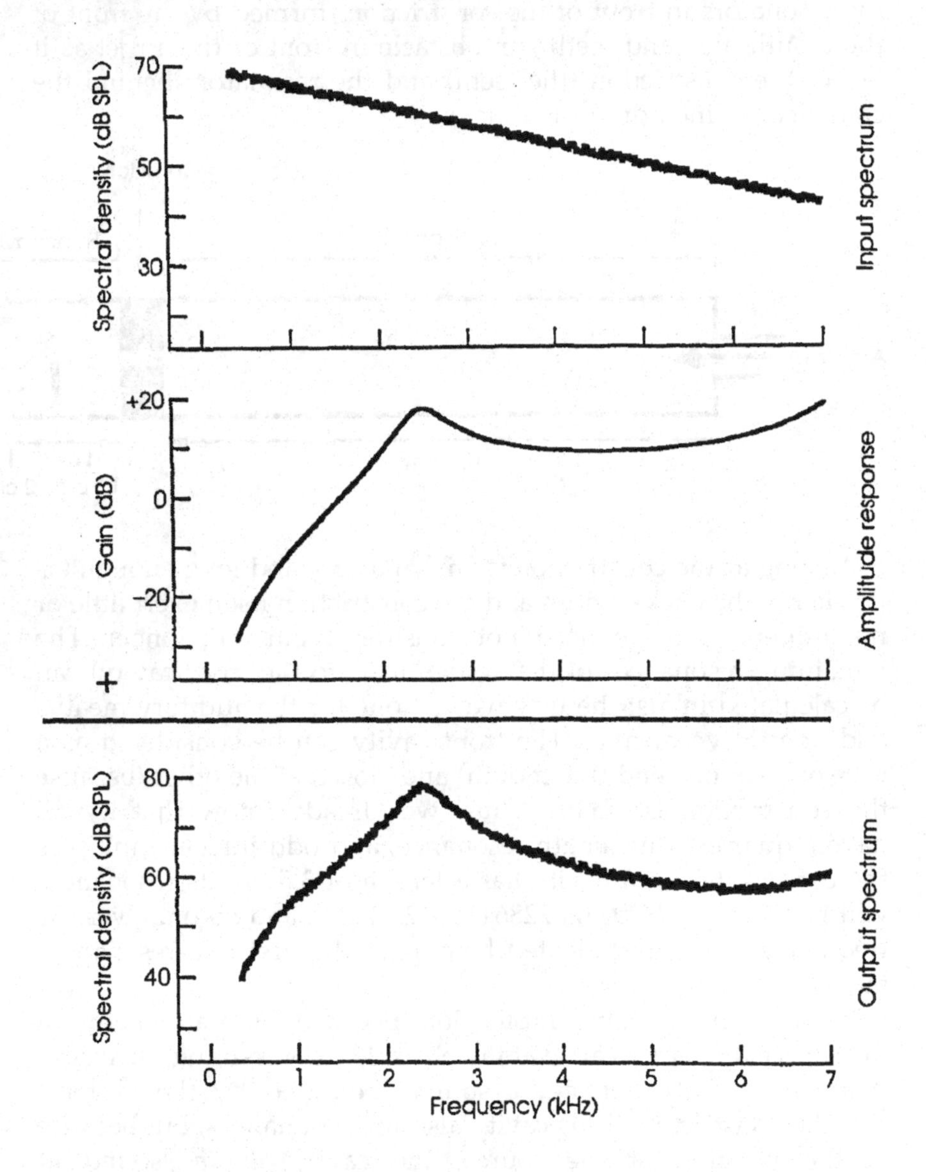

Note in the output spectrum the concentration of energy at high frequencies, with a single broad formant peak. The frequency of this peak is determined primarily by the resonance of the front cavity. As the front cavity will be shorter for fricatives whose constriction is made further forward in the mouth, such fricatives will have higher-frequency formants. To give a specific example, an /s/, with its constriction quite close to the front of the mouth, should have a main concentration of energy higher in frequency than that of the /ʃ/. This is clearly seen in wide-band sections and spectrograms of the two sounds, as uttered by an adult male:

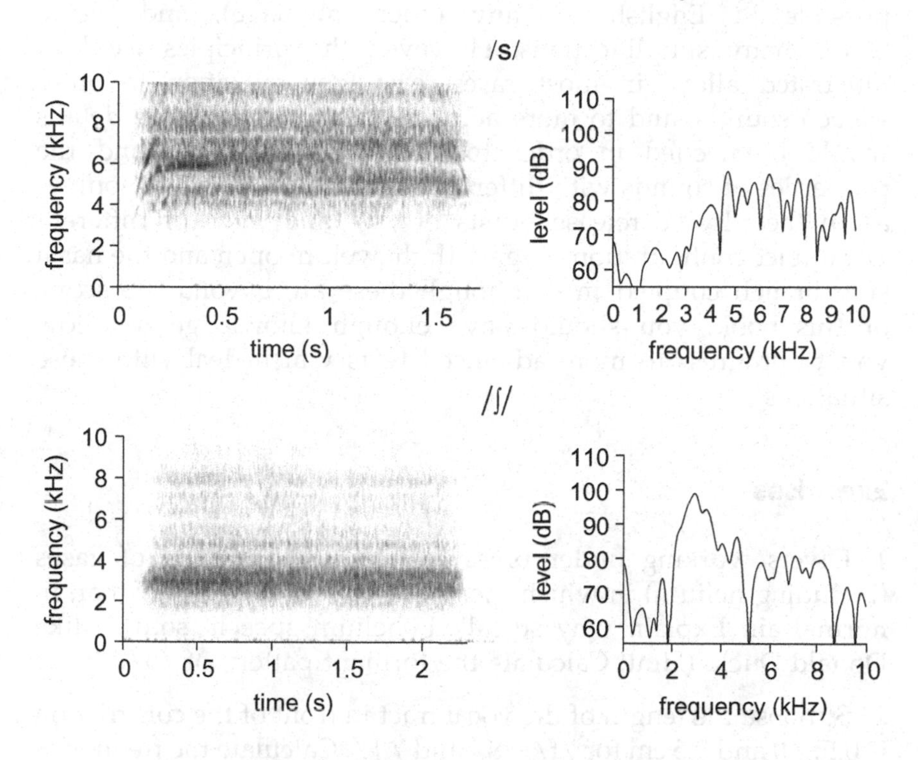

You should be able to give a thorough interpretation of this figure without any help from us by now! But to get you started, compare the positions (in frequency) of the main concentrations of energy in the two sounds. Why is there a lack of harmonic structure? Why don't the two spectrograms change much over time?

Summary and comments

In this chapter, we have shown the application of linear systems techniques to speech. For each class of sound, we described the *source* of excitation involved and the *filter* action of the vocal tract. Moreover, in the case of diphthongs, the influence of a change in vocal tract shape across time was described. Each class of sound was selected for discussion because it illustrated points of importance that can be generalized to other classes of speech sounds. Thus, with vowels, we used a voiced source of excitation and a stationary filter. Diphthongs were again excited by a voiced source, but the vocal tract filter varied over time. Finally, the fricatives exemplified the action of a voiceless source, and the acoustic effects produced by a source which is positioned somewhere along the vocal tract (not at its bottom).

Clearly, these three classes of sound do not exhaust all those possible in English (or any other language), and we've made many simplifications. However, the principles we *have* illustrated allow, in most cases, easy generalization to other speech sounds, and to more accurate descriptions. More details would be needed in order to allow you to understand the properties of sounds with different sources (e.g. transient sources as in the plosive release bursts of /k/ and /p/) or different vocal tract configurations (e.g. with the velum open and the nasal side branch coupled in). Although these are beyond the scope of this book, you should have enough knowledge to allow you to progress to more advanced texts which deal with these situations.

Exercises

1. Divers working under pressure breathe a mixture of gases (including helium) in which the speed of sound is higher than in normal air. Explain why so-called 'helium speech' sounds like Donald Duck. (Hint: Calculate the formant pattern of /ə/.)

2. Suppose the length of the vocal tract in front of the constriction is 0.5, 1.0 and 2.5 cm for /f/, /s/ and /ʃ/. Calculate the frequency of the front cavity resonance for each fricative.

3. Explain why any harmonic can be used to calculate the fundamental frequency of voiced speech.

4. In the introduction, we mentioned that the concepts of linear systems analysis are also relevant to speech synthesis. Illustrate how you would employ a signal generator and (a) a serial and (b) parallel combination of band-pass filters to produce two vowels of your choice. How can you make a vowel with a falling intonation contour? A diphthong?

5. Write a letter to a friend explaining that formants are not harmonics in any sense. Why would the notation Fx for fundamental frequency be preferable to F0 (for one thing, 'x' could stand for 'x'-citation)?

6. Estimate (using a ruler and calculator) the minimum and maximum frequencies reached during the vibrato in the figure on page 297. Do some outside reading and calculate the musical interval this corresponds to. How does the size of this compare with the intervals normally used in Western scales? Why are the vibrato excursions bigger for the 10th harmonic than for the first

harmonic? Would this be true if frequency were displayed on a log scale? Why or why not?

7. Using a ruler and a calculator, estimate the formant frequencies for the two vowel sounds /ɑ/ and /u/ independently from the spectrograms and sections in the figure on page 300. How do these numbers tally with published data? (See, for example, Ladefoged, P. (1975). *A Course in Phonetics*, Harcourt Brace Jovanovich: New York.)

8. In Chapter 1, we described the use of an artificial larynx for oesophageal speech. Explain how linear systems analysis helps in understanding what is going on.

9. The spectrograms of nasal vowels contain formants associated with the nasal tract. An experimenter wanted to check this for herself. She made a spectrogram of herself speaking a nasal vowel. Then she made a tube to lengthen her nasal cavities. She then made a recording of herself speaking the same nasal vowel with the contraption fitted to her nose. Finally, a spectrogram was made. The pair of spectrograms were inspected to see which formants had shifted in frequency. Explain why you think she thought the formants associated with the nasal cavities would move if the extension was fitted. Which way would the formants move?

10. The speech of a woman is different to that of a man. Write an essay describing why this is so.

11. You are given spectrograms of the three fricatives /f/, /s/ and /ʃ/, which show a first peak in the spectrum at 7900, 5700 and 2500 Hz, respectively. What front cavity lengths would account for these values? Do they seem reasonable for what you know about these three fricatives?

12. Speech can be spoken at different rates. Describe what happens to the formant patterns seen in spectrograms. (Read: Lindblom, B.E.F. (1963). Spectrographic study of vowel reduction. *Journal of the Acoustical Society of America 35*, 1773–1781.)

13. It has been argued that Neanderthal people could not produce human speech. Write an essay describing why this is so. (Read: Lieberman, P., Crelin, E.S. and Klatt, D.H. (1972). Phonetic ability and related anatomy of the newborn, adult human, Neanderthal man, and chimpanzee. *American Anthropologist 74*, 287–307.)

14. Describe what you would do to analyze the speech of two speakers speaking two different vowels simultaneously. One of the speakers is whispering, and the other is speaking on a constant fundamental frequency. What might you expect to see on a spectrogram of such a combination of sounds?

CHAPTER 14

An Introduction to Digital Signals and Systems

All of the systems we have discussed so far have involved real-world signals involving pressure changes (like speech sounds) or movements (like those of the middle ear bones) or changes in voltage (like the wave fed to a loudspeaker). As we will discuss further below, such signals are, as are all signals we directly experience, *analogue* signals. Perhaps surprisingly then, almost all the equipment we use to record, store, analyze and reproduce these signals is no longer analogue, but *digital*. In fact, as a result of technological advances in the last 20 or so years, digital signals and systems have almost entirely superseded analogue systems in most applications. Perhaps the best-known example of this is the fact that almost all music is now recorded and reproduced digitally.

As a result of these developments, anyone in the fields of speech and hearing (even those not involved in research) is guaranteed to meet digitally based equipment in the course of their work. For example, any work done with a computer involves digital signal processing although this may not always be apparent to the user. The intelligent use of such equipment necessitates at least a rudimentary understanding of its working principles. Unfortunately, it would take at least another whole book for a proper exploration of the properties of digital signals and systems! Our goals here are more limited—to acquaint you with the basics of this pervasive and rapidly expanding field, without going into too much detail. In any case, as you'll see, many of the principles that we have already developed apply in a straightforward way to digital signals and systems, although often in a slightly modified form.

Pros and cons of digital techniques

Before beginning, however, it is well to keep in mind the main advantages of digital techniques. First, regarding signals, is the possibility of making essentially perfect copies of an original.

As you may know, if you take an ordinary audio-tape recording (as made, for example, using the cassette recorder described in Chapter 4), and make a copy of it on another tape recorder, the copy, no matter how good the equipment, isn't quite the same as the original. If you copied that copy, things would be even worse. Copying the previous copy in a chain in this way will eventually lead to an absolutely useless reproduction. This is not so with digital signals—it is possible to make each copy completely equivalent to the original. This is a major attraction of compact discs which store signals digitally—the signals coming out of your CD system should be as good as those on the studio master. A related property is that digital signals are much more resistant to noise and degradation when transmitted (as in, for example, telephones and televisions).

The second primary advantage concerns systems. Digital systems can be extremely flexible and allow the routine use of processing schemes that would be impossible to perform with analogue systems.

What is an analogue signal?

We've been bandying this word 'digital' about without much care, counting on you having a vague understanding of it. It's time now to define this term more precisely. The best way we can do that is by first giving a name to the kinds of signals that are *not* digital—*analogue* signals.

In fact, analogue signals are the only ones dealt with in this book so far. We didn't need to make this explicit, because there were no other kinds of signals to contrast them with. So, for example, speech sounds or musical tones, or the displacements of the middle ear bones, are all examples of analogue signals.

There are two crucial aspects of analogue signals which distinguish them from digital signals. First, they are continuous in time. This is another way of saying that at any particular moment of time that you tried to measure a signal, it would have *some* value. Between any two given moments, then, there are always an infinite number of instants at which the signal exists. Thus, you cannot write down all the values of analogue signals in a table (no matter how big the table or short the signal) because an infinite number of entries would be required. You should already be familiar with this limitation on the use of tables to express essentially continuous information from Chapter 6. There we showed why a graph could be better than a table in representing an amplitude response appropriately.

Not only are analogue signals continuous in *time* (normally plotted as the x-axis), but they are also continuous in *amplitude* (usually on the y-axis). This is another way of saying that, at a particular moment in time, the signal can have any amplitude—it is not restricted to some set of values. So, for example, an electrical signal coming from a microphone at a particular moment could have the value of 6.232071 V, or 1.21440633 V, or 0.1016173263926 V. The point is that whenever we make a measurement (here with a voltmeter), we round off to however many digits our equipment handles. But the signal itself could have any value, and so, in general, we would need an infinite number of digits to represent its true value.

Let's illustrate these two points by thinking about the difficulties of writing down (in a table) the values making up a single period of an analogue sine wave of frequency 100 Hz and peak amplitude 1 V:

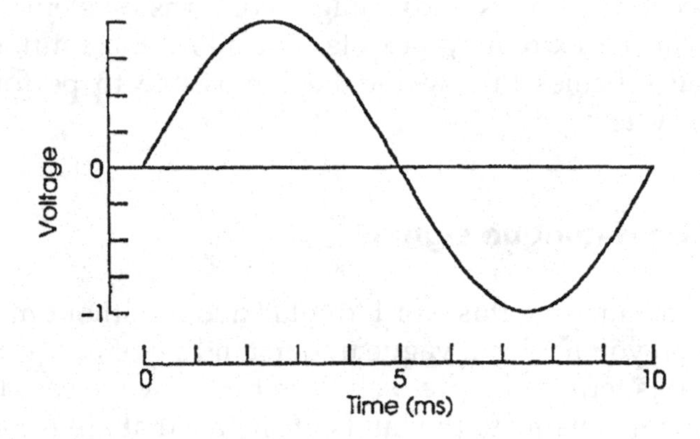

Consider first the problem of writing down any single value with a fixed number of digits. Although this is easily done for, say, time values of 0, 5 and 10 ms, where we know that the amplitude of the sine wave is 0, what about at a time of 1.25 ms? As 1.25 ms is 1/8th of the total period of 10 ms, and there are 360° in each period, we need to take the sine of 360/8. This is equivalent to the sine of 45° ($=360° \times [1.25\,\text{ms}/10\,\text{ms}]$) which a calculator says is 0.7071068. As good an approximation as this might be, it is not the exact value. The sine of 45° is actually $\sqrt{2}/2$, an irrational number which, as with all such numbers, takes an infinite number of digits to write down. Nor is this an unusual case. Many (in fact, most!) of the values of the sine wave are irrational.

Now, let's address the problem of the number of entries we would need to specify this waveform (ignoring the problem of the number of digits that we'd need). We'll take a sort of 'Zeno's paradox' approach. Certainly, we'd want to specify the endpoints of the waveform, 0 and 10 ms. Then we'd specify a point halfway between these two, at 5 ms. We could then go on to write down the value of the sinusoid at a time halfway between 0 and the last value picked. This would give us a series of values at 2.5 ms,

1.25 ms, 0.625 ms, 0.3125 ms, 0.15625 ms, and so on. The problem is that this series never ends, so no finite length table could describe the sinusoid completely.

To summarize, analogue signals are continuous in time and amplitude. They exist at every moment, and their amplitude at a particular moment can take on any value. This is why we normally use other kinds of analogue representations to represent them adequately—in our case, graphs. A graph of a signal is also continuous in time and amplitude.

What is a digital signal?

Although a graph can be a complete representation of an analogue *or* digital signal, a table can only be a complete representation of a digital signal. Here, for example, is a table that represents a particular digital signal:

Time (ms)	Amplitude (V)
0.0	0
0.5	2
1.0	3
1.5	−3
2.0	2
2.5	−1
3.0	1
3.5	0
4.0	−2
4.5	0
5.0	−1
5.5	1
6.0	3
6.5	0
7.0	3
7.5	−2
8.0	0
8.5	−1
9.0	−2
9.5	−1
10.0	1

There are two properties of digital signals that make it possible to write them down in a table. First, they exist only at discrete, equally spaced moments of time. So, as here, we might have a value for a signal every ½ ms. *Between* these sample points the signal doesn't really exist—it only has values *at* the sample points. Any signal that has this property is known as a *discrete-time signal*, since it only exists at specific, discrete moments in time.

Second, the amplitude values that the signal can take on are limited to a set of discrete values, here a whole number of volts. This notion should not be new to you—we often restrict the possible values that numbers can take. For example, the interest payable on a loan in France would be rounded to the nearest Euro cent; in the United States or the United Kingdom to the nearest penny. When we measure a space for a shelf, we round to the nearest millimetre or 1/16th of an inch. A signal whose possible values are restricted in this way is said to be *quantized*, as two unequal values must differ by at least a fixed quantum (a single Euro cent, penny or millimetre in the examples above). This is quite different from analogue signals, for which there is no minimum difference between two unequal values.

To summarize, there are two reasons why digital signals can be written down in a table and analogue signals cannot. In a given time period, there is only a finite number of sample points in a digital signal; in an analogue signal, there is an infinite number of values that the signal takes on. For a given amplitude range, there is only a finite number of digits that are needed to write down any particular value of a digital signal; the value of an analogue signal needs, in general, an infinite number.

Of course, we could also draw a graph of the digital signal tabulated above:

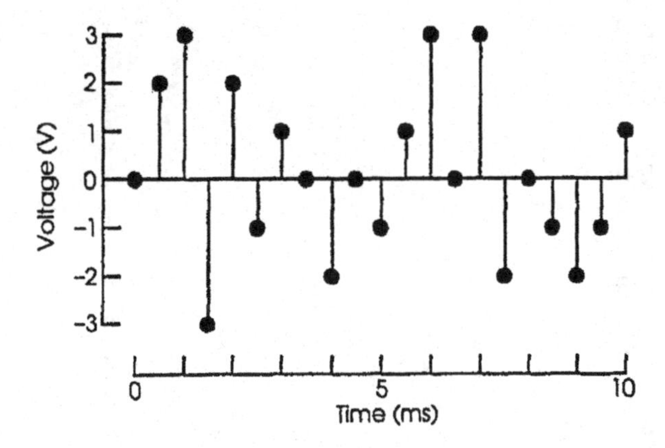

Note the distinct way in which a digital signal is represented here. In order to emphasize that it exists only at discrete moments, a small circle is drawn for each defined value of the waveform. From this circle is drawn a vertical line to the horizontal axis.

This is only one example of an arbitrary waveform, so let's give a further illustration with a digital version of a signal you are more familiar with—a single period of a digital 100-Hz sine wave. We'll assume an inter-sample time (or *sampling period*) of 0.5 ms and rounding of each amplitude value to three significant digits:

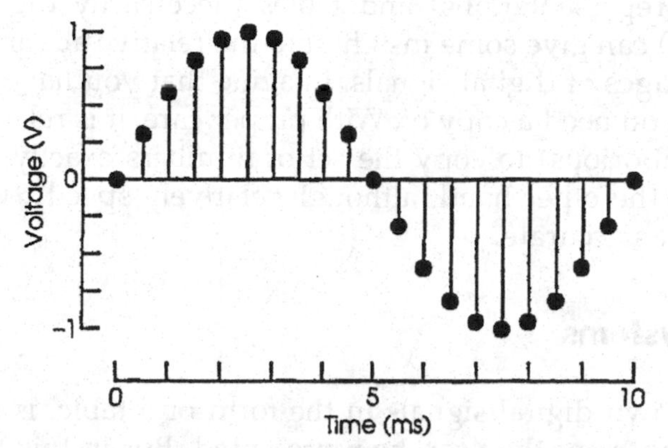

Although we've drawn this as a graph for you to see that there is still a sinusoidal wave shape, the information can also be represented in a table:

Time (ms)	Amplitude (V)
0.0	0.000
0.5	0.309
1.0	0.588
1.5	0.809
2.0	0.951
2.5	1.000
3.0	0.951
3.5	0.809
4.0	0.588
4.5	0.309
5.0	0.000
5.5	−0.309
6.0	−0.588
6.5	−0.809
7.0	−0.951
7.5	−1.000
8.0	−0.951
8.5	−0.809
9.0	−0.588
9.5	−0.309
10.0	−0.000

Because the signal only exists at discrete moments of time, there isn't the problem of infinitely long tables that occurs with analogue signals. Because the values are quantized to a certain number of digits, there isn't the problem of amplitude values that need an infinite number of digits to be written down.

Thinking about the differences between graphs (normally analogue representations) and tables (necessarily digital representations) can give some insight into the relative advantages and disadvantages of digital signals. Imagine that you have a table of numbers you need a copy of. With proper care, it is relatively easy (though laborious) to copy the set of numbers exactly. Tracing a graph, on the other hand, although relatively speedy, would not be nearly as accurate.

Digital systems

Writing down digital signals in the form of a table, is of course, not the only way they can be represented. But in this form, as a simple list of numbers, they can be easily entered into and hence stored and operated on by a digital computer. Much of the development of digital signal techniques, in fact, was due to the desire to apply the flexibility of general-purpose computers to signal-processing problems.

In this mode, then, computers are used as *digital systems*, the name given to any system that has digital signals for inputs and outputs. In the same way, all the systems we have investigated until now, with analogue signals as input and output, are known as *analogue systems*.

Not all digital systems are general purpose, however, although they are all computers of a sort. For reasons of speed and economy, more and more use is being made of hardware specially constructed for a specific task—a type of computer with a fixed program.

No matter what particular form the digital system takes, special or general purpose, it's relatively rare for a complete system to have digital signals both at the input and output. In other words, we rarely encounter completely digital systems, especially for any task concerned with speech and hearing. We can neither speak digitally nor listen to digital sounds directly. At least one end of the process (and very often both) must be an analogue signal. Typically, practical systems are a series of systems, some digital, some analogue and some which are the go-betweens.

As an example, let's look in a very broad way at the steps involved between a recording of a solo piano player and your listening to a compact disc of it at home as shown on the next page.

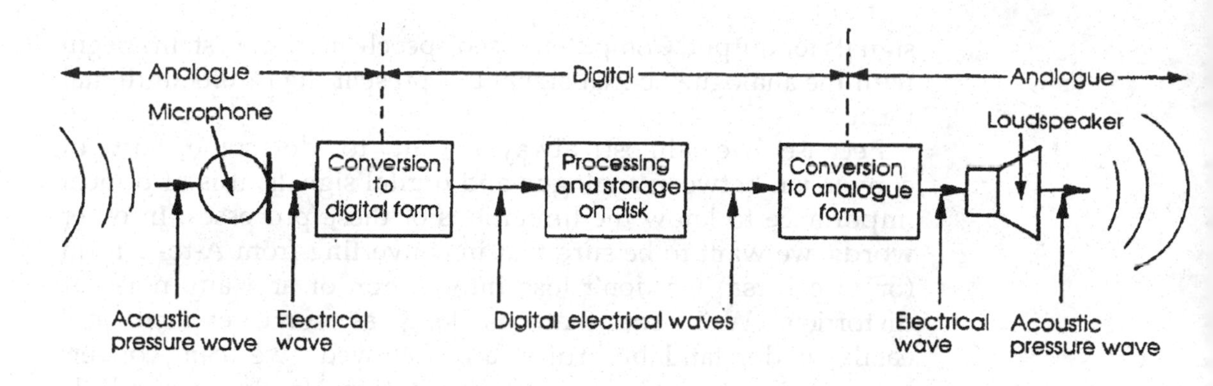

The starting point is, of course, the sounds created by the pianist as she or he strikes the keyboard. This analogue acoustic signal is first transduced into an electrical signal by a microphone. Although the information is now in a different form (electrical instead of acoustic), it is still analogue. In order to make use of digital techniques of storage and processing, it's necessary to transform the analogue electrical wave into digital form—known as performing an *analogue-to-digital* (usually abbreviated as A-to-D) conversion. Now the information is stored as a series of numbers, and so can be copied and stored with little or no degradation. The acoustic information, represented as a long list of numbers, is coded and pressed into the surface of the compact disc.

In order to listen to the disc, all these steps must be performed in reverse. The compact disc player reads the series of numbers from the disc, but in digital form they are of little use. The digital electrical signal must first be transformed into analogue form by *digital-to-analogue* (usually abbreviated as D-to-A) conversion, before being amplified and then transduced, 'for your listening pleasure', into sound again by a loudspeaker.

This chain of events is fairly common in many systems which are based on digital techniques: (1) the transduction of information into analogue electrical form followed by A-to-D conversion, (2) the processing and/or storage of that information by a digital system and (3) the reconversion into analogue form by D-to-A conversion and final transduction to sound.

Although purely digital systems are fairly rare in speech and hearing, there are many instances in which only D-to-A or A-to-D are performed. For example, in a digital speech synthesizer, the signals originate in a digital form but are converted to analogue signals for output. Computer-based speech analysis systems begin with the analogue speech signal but present the results in digital form.

Because we almost always want to do some sort of conversion between analogue and digital signals, it is of utmost importance to know the limitations of these processes. In other words, we want to be sure that in converting from A-to-D form (or vice versa) we don't lose information or add any noise or distortion. We'll see that, as long as some explicit and easily understandable rules are followed, we can convert between the two kinds of representations freely, losing little information. Let's begin with the process of converting an analogue signal into digital form. As digital signals differ from analogue signals in two ways, we consider the process of transformation to consist of two stages: *quantization* followed by *sampling*.

Quantization

Let's suppose that we want to use a digital system to process the 100-Hz analogue sinusoid we looked at above. For convenience, we'll make the peak amplitude 5.5 V instead of 1 V.

The first step that's necessary is to quantize the signal so that it can only take on a fixed set of amplitude values. In order to do this, we must define two parameters: (1) the maximum voltage levels of the signal to be represented and (2) the number of different levels that are possible. Since the sinusoid goes between +5.5 V and −5.5 V, this will be our defined range. To show the effects of quantization clearly, we'll begin with a small number of possible levels—say, 11 running from −5.5 to +5.5.

How, then, do we transform the amplitude values of the analogue signal, which are continuous, into this digital discrete scale? One way is to use a rule based on rounding to the nearest digit. We simply take the analogue amplitude value and round it to the nearest number of volts. The easiest way to represent this information is in the form of an *input–output* function, similar to those we used in Chapter 11 to portray rectification. The x-axis is the instantaneous amplitude value of the input analogue signal, while the y-axis gives the instantaneous value of the quantized output as shown on the next page.

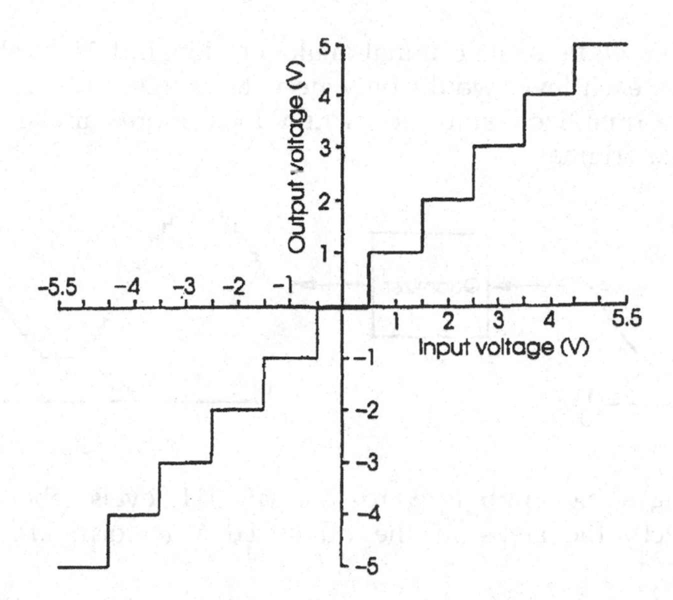

From this figure, for example, you can see that input voltages between −0.5 and 0.5 V are quantized as 0, while input voltages between −4.5 and −5.5 V are quantized as −5.

Let's put our 100-Hz sinusoid through this quantizer, and see what comes out:

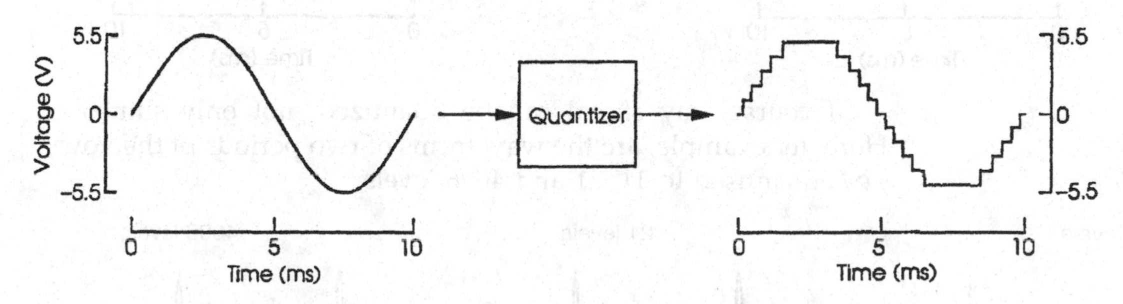

The quantized signal is a sort of sine wave with steps. The amplitude values it takes are, of course, limited to the 11 possible that we defined.

Rarely does a quantizer use as few steps as this, however. If we think of the quantized signal as an approximation to the original analogue signal, then we can get better approximations by increasing the number of available levels.

This should be apparent by considering how the resolution (in amplitude) of the quantizer varies with the number of possible levels. Here, 11 levels were used to represent a total voltage range of 11 V (−5.5 to 5.5 V); hence, each level represented a range of levels 1-V wide in the original analogue signal. If 21 levels were possible, each level would only need to represent about 0.52 V, and the quantized sinusoid would look more similar to the analogue original:

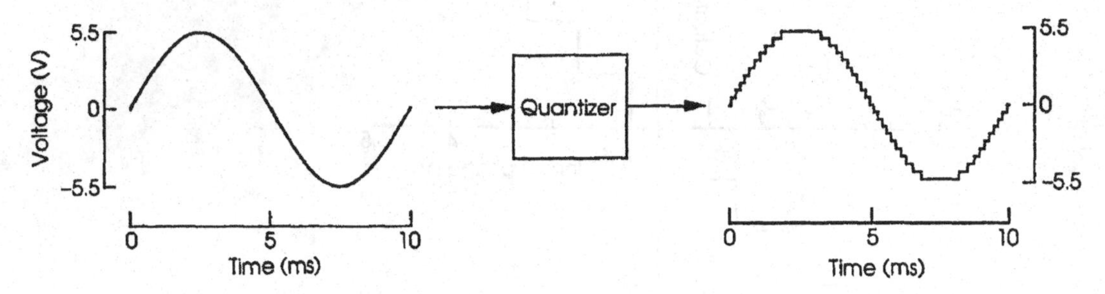

Taking a big jump upwards to, say, 111 levels (about 0.1 V per level), the steps in the quantized waveform are barely visible:

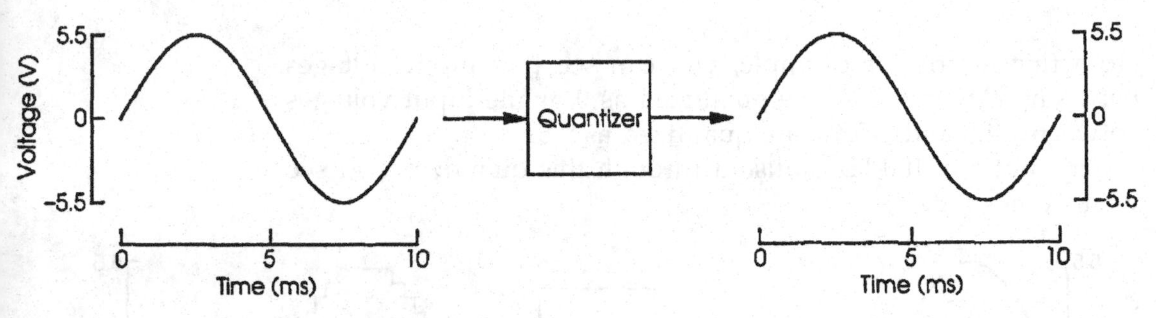

Of course, any signal can be quantized, not only sinusoids. Here, for example, are the waveforms of two periods of the vowel /ɑ/, quantised to 11, 21 and 4096 levels:

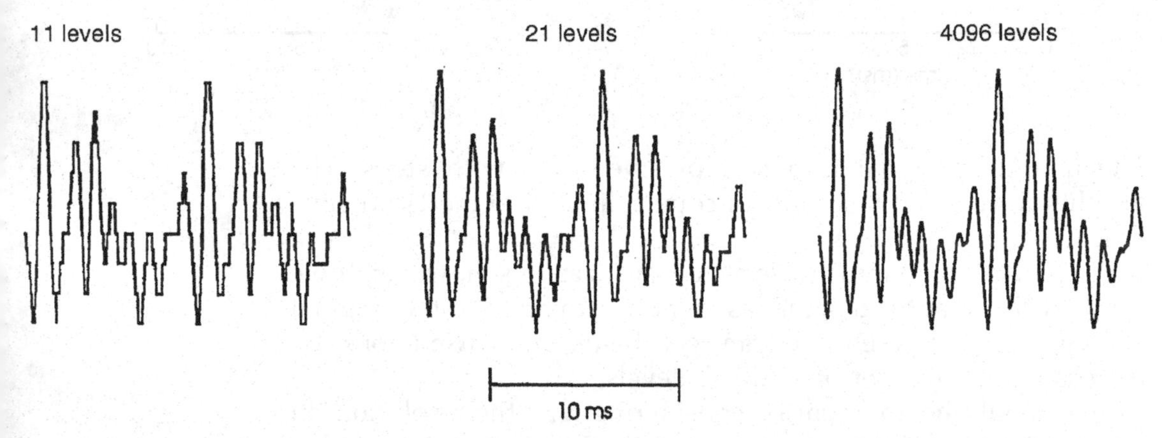

Real-life digital systems almost always use a number of levels that can be expressed as an integer power of two (that is, as 2^n, where n is a whole number) because almost all digital hardware (computers included) express numbers internally in the base 2 or *binary* system. In such a system, it's necessary to specify the maximum number of *bits* or binary digits that are available to represent each amplitude value. A 1-bit system can express two levels, and each extra 'bit' doubles the number of available levels. (In a similar way, a base 10 system has 10 available levels for each digit, and each added digit increases by a factor of 10 the number of available levels.) So, an 8-bit system has 256 levels and a 12-bit system 4096. In fact, this is the way digital systems are normally described—as an *8-bit system,* not a system with 256 quantization levels (although the two mean the same thing).

For high-quality audio applications (as in the compact disc), 14–16 bits are used, but many tasks can be done with as few as 8 or 10 bits. As always, different situations demand different degrees of accuracy. The crucial point is that signals can often be quantized finely enough (that is, with enough levels) to be treated as if *any* amplitude value were possible. In other words, the quantized signal can be considered such a good approximation to the original that the effects of quantization can be ignored. Thus, most of the theoretical developments in digital signal processing (though by no means all) assume an unquantized signal. In the rest of our development, we too shall ignore quantization effects.

Sampling

We're now halfway to a digital signal. Although we can restrict the amplitude values that an analogue signal takes on with a quantizer, the quantized signal still exists at every moment in time. What we need to do now is restrict to discrete moments the times at which the signal can exist—a process known as *sampling.* Sampling simply measures the amplitude value of the signal at equally spaced moments in time. The most important parameter to define in a sampler is its *sampling frequency* (or equivalently, *sampling rate*)—the number of times per second that the value of the continuous time signal is recorded. The reciprocal of the sampling frequency, which is just the time between samples, is known as the *sampling period.*

Here is an example that we've seen before—a quantized 100-Hz sinusoid being sampled every 0.5 ms, or, equivalently with a sampling frequency of 2 kHz [12 bits were used in the quantization (4096 levels) so you won't see the steps]:

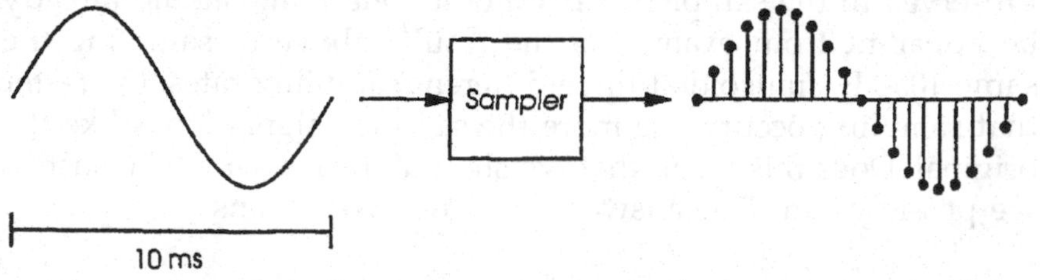

We could sample this waveform at a slower rate, say 1 kHz:

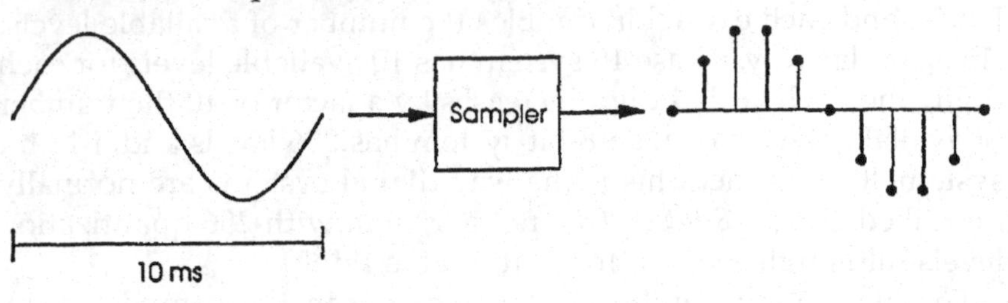

Or even more slowly, at 400 Hz:

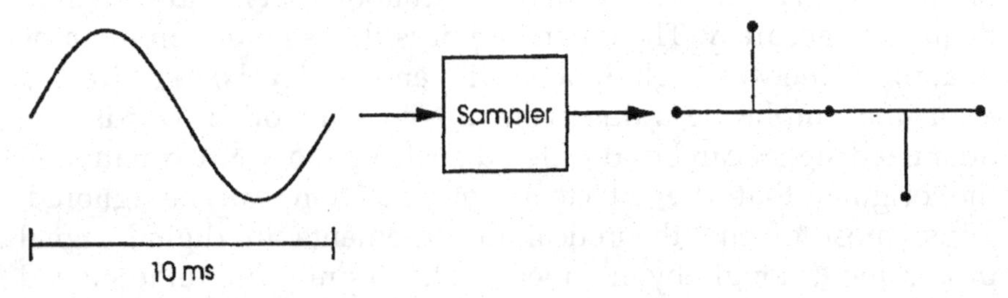

At the end of sampling (which we have assumed would follow quantization), we have our digital signal ready for input to a digital system. Of course, it's possible to sample any signal, not just sinusoids. Here, for example, is a 5-kHz sampled version of a short stretch of the quantized vowel /ɑ/ shown above (p. 318):

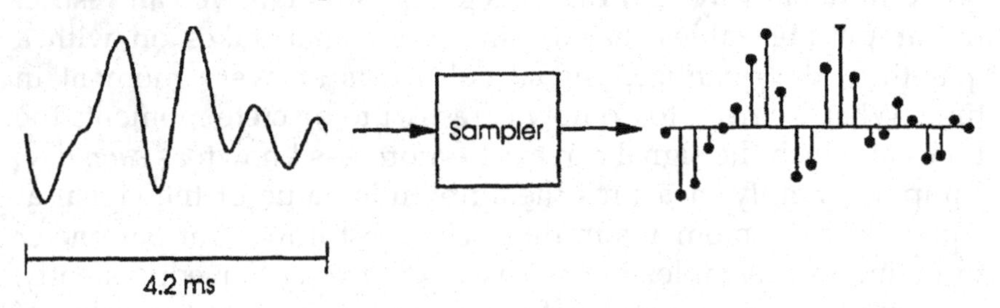

How fast does sampling have to be? The sampling theorem

Although we can now convert a continuous time signal into a discrete time one (via sampling), we still have to address the problem of how well the information in the original signal is preserved in the sampled version of it. One thing should already be apparent from examining the results above of sampling the same 100-Hz sinusoid at three different sampling rates: the faster that sampling occurs, the more the sampled signal looks like the original. Does this mean that we should always sample as fast as we possibly can? The answer is no, for two reasons.

First, there are the practical problems. Fast sampling rates obviously lead to more data points per second than slow sampling rates. For a mode of storage with fixed capacity (like a compact disc or MP3 player), halving the sampling rate would allow twice as much music. Also, there are technical problems in making fast sampling devices and the solutions to them tend to be expensive.

Second and much more importantly, there is a result, known as the *sampling theorem*, which gives the minimum sampling rate which allows the original signal to be reconstructed perfectly. It turns out that going faster than this doesn't help. The limitations on sampling, then, are quite different to those on quantization. The more levels of quantization, the better, because the original signal is better and better approximated (even though the practical advantage may be small beyond a certain point). There is no point at which *all* the information is retained. Sampling a sinusoid at an increasing rate does not ensure better approximation. Below the sampling rate set by the sampling theorem, the information at the input to the sampler is not preserved in the output—above this rate it is.

The rule given by the sampling theorem is simple—as long as the sampling rate is more than twice the frequency of the sinusoid, no information is lost. This rate is known as the *Nyquist rate.* We can get some intuitive understanding of this result by looking at what happens if we sample a sinusoid more slowly than the Nyquist rate.

Let's go back to our 100-Hz sinusoid. The sampling theorem says that a sampling rate greater than 200 Hz (the Nyquist rate) will allow a reconstruction of the continuous time signal from its sampled version. We've already seen the results for rates rather faster than this above, but let's draw the digital waveform obtained by sampling at 1 kHz in a slightly different way, with the sample points as solid circles superimposed on the original waveform:

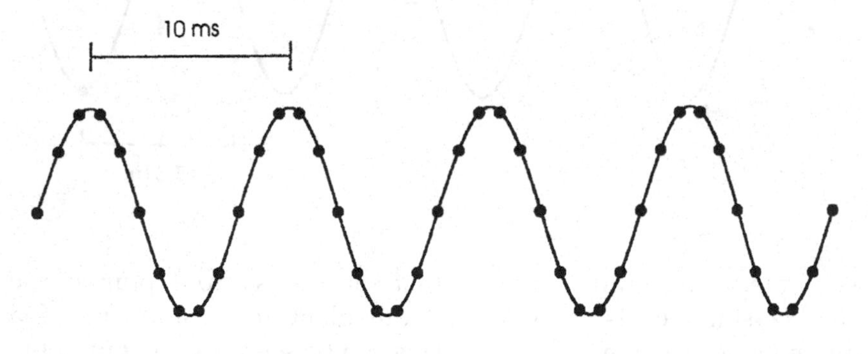

What happens if we now sample at exactly the Nyquist rate, 200 Hz?

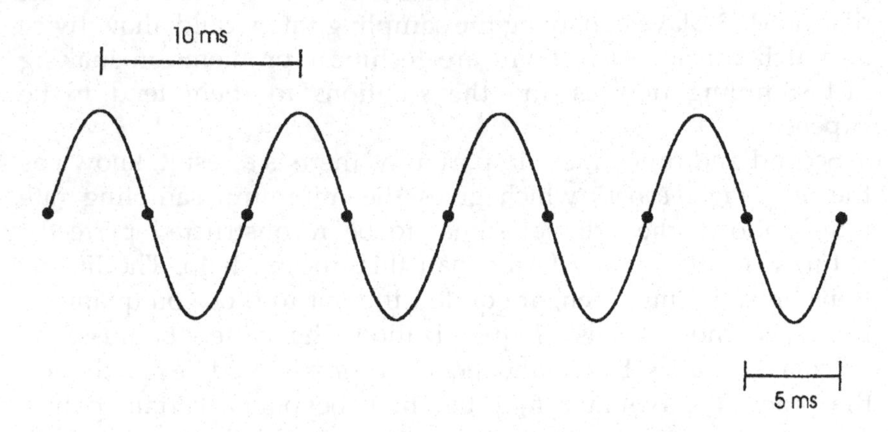

It should be obvious that this sampling rate is not fast enough. All the samples have a value of 0. In other words, the output of a 200-Hz sampler to a 100-Hz sinusoid is indistinguishable from no signal at all! Another way to think of this is as a direct current (DC) level, or a sinusoid of 0 Hz. Therefore, it is as if the 200-Hz sinusoid has been transformed into a sinusoid of 0 Hz. If we sample more slowly than the Nyquist rate of 200 Hz, even stranger things can happen. Here is the same 100-Hz sinusoid sampled now at a rate of about 133.3 Hz (with an exact sampling period of 7.5 ms):

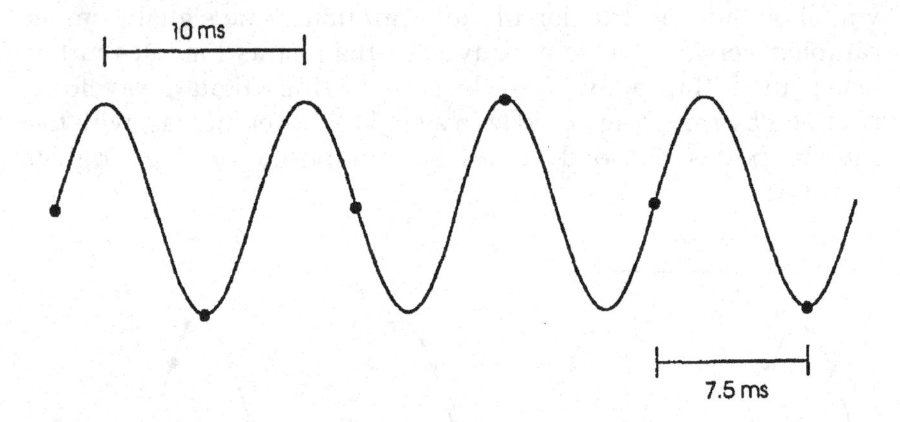

The interesting thing here is that, at the sample points, the 100-Hz sinusoid is completely equivalent to a sinusoid of a much lower frequency, 33.3 Hz. This can be seen if we draw in the

33.3-Hz sinusoid as well, as a dotted line:

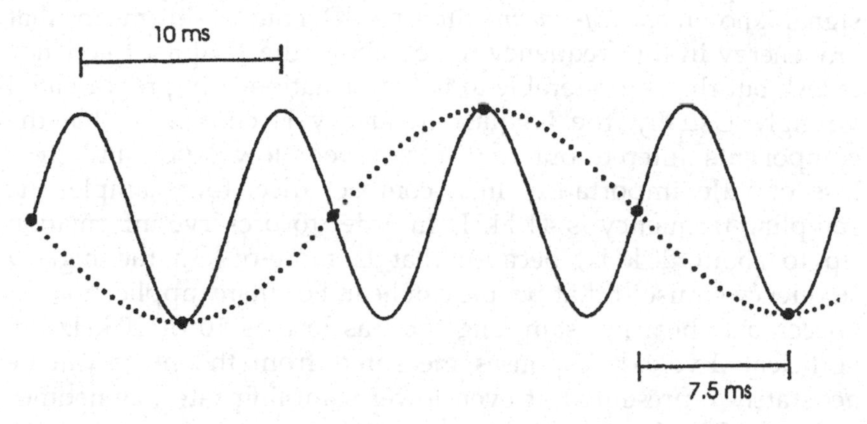

Thus, inputs of 100 Hz and 33.3 Hz are indistinguishable at the output of the sampler. Again, it is as if the 100-Hz sinusoid has been transformed into one of a lower frequency.

This transformation of sinusoids into different frequencies is known as *aliasing*. In everyday parlance, an alias is an identity you take on in place of your genuine identity. Here, a 100-Hz sinusoid has taken on the identity of one at 33.3 Hz.

It can be shown that, for a given sampling rate, all sinusoids will be represented at the output of the sampler as sinusoids of a frequency somewhere between DC (0 Hz) and half the sampling rate (known as the *Nyquist frequency*). So, sinusoids of any frequency at the input of a 200-Hz sampler will be represented at the output as sinusoids of frequencies between 0 and 100 Hz. Only for input sinusoids of frequencies below 100 Hz (that is, less than the Nyquist frequency) will the output match the input. At frequencies higher than this, input sinusoids will be aliased into lower frequencies. This is why the sampling theorem requires a sampling rate at least twice the frequency of the sinusoid to preserve the information in it.

Sampling complex signals

Having now determined the sampling rate necessary for sinusoidal signals, we can easily generalize this result to complex signals. In order to reconstruct any analogue signal from the output of a sampler, the sampling must be done at a rate more than twice as high as the highest frequency component in the signal.

Often, the sampling rate is fixed. It is important then to ensure that there are no frequency components present in the analogue signal above the Nyquist frequency; otherwise they will alias to lower frequencies. This is usually done by low-pass filtering the signal (known as *anti-aliasing* filtering). Of course, this means that any energy in the frequency region above the Nyquist frequency is lost, but this is preferable to the information being represented wrongly. Usually, the Nyquist frequency is chosen so that the components filtered out are of relatively low level, and their loss of little importance. In a compact disc, for example, the sampling frequency is 44.1 kHz in order to preserve information up to about 20 kHz, because that is more-or-less the highest frequency sinusoid that people can hear. For many applications in speech and hearing, sampling rates as low as 10 or 20 kHz are sufficient. Evoked responses measured from the brain can be accurately represented at even lower sampling rates, sometimes as low as 500 Hz.

Processing the digital signal

At this point, if the quantization and sampling have been done properly, we have an appropriate digital signal ready for processing of some kind. We will not, in fact, detail any particular digital system just yet, but you should have some idea of the types of processing that are possible.

One popular use of digital systems is to mimic analogue systems. It's possible to create digital filters of all types (e.g. high- or low-pass), as we will see later in this chapter, and these can be much more flexible than their equivalent analogue counterparts. For example, it's relatively easy to design digital filters with such desirable characteristics as linear phase responses.

There is also much digital processing that would be extremely difficult, if not impossible, to implement using analogue electronics. Two techniques common in speech processing, *cepstral* and *linear-predictive analyses*, are of this sort.

A crucial point is that many of the digital systems in use may be considered to be LTI (although they are normally termed linear *shift*-invariant systems, with identical meaning). Hence, all the concepts that we've developed throughout this book have their counterparts in the digital world, although often in modified form.

Reconverting back to analogue form

Having now processed the signal, we want to get it back into analogue form. The first step is to take the amplitude values and convert them from numbers back into voltages with a digital-to-analogue converter (or DAC, pronounced 'dak'). This is done at a sampling rate identical to the one initially performed, and essentially reverses the operation of the sampler—we are back to the quantized continuous time signal. Here's our digital 100-Hz sinusoid (sampled at 2 kHz), before and after D-to-A conversion:

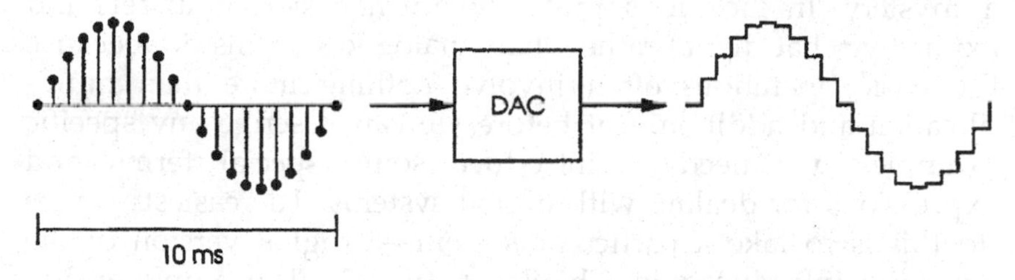

Clearly, this output doesn't look the same as the original analogue sinusoid—nor would it sound the same. Analogue sinusoids don't have steps in them as this 'sinusoid' does. These steps are *not* the result of quantization. Although the signal is indeed still quantized, the quantization is fine enough to be ignored. Even with no quantization, such steps would still occur, as they arise from the operation of sampling. Analogue signals can change continuously since they exist at all moments of time; digital signals reconverted into analogue form must show steps, as changes can only occur at discrete moments of time.

As we've discussed before, fast changes like this in a signal represent high frequencies. But we've already shown that only frequencies up to the Nyquist frequency can be represented. Hence, we can low-pass filter this stepped signal (at half the sampling rate) to remove these meaningless high frequencies, while at the same time not throwing away any correct information. This will, in effect, smooth out the steps and leave us with the original sinusoid:

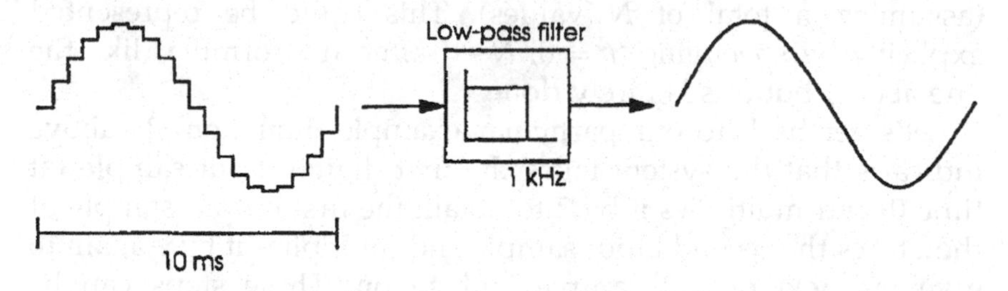

This low-pass filtering step (unlike the anti-aliasing filter discussed previously) is mandatory. If the original analogue signal is already limited in frequency content to less than half the sampling frequency (as in our 100-Hz sinusoid sampled at 2 kHz), low-pass anti-alias filtering prior to sampling would not be necessary (because there is no energy in the signal above 1 kHz). It would still be necessary after D-to-A conversion, though, in order to eliminate the 'steps' in the waveform.

A digital amplifier

Although you should now have a pretty clear idea of what a digital signal is like, digital signal processing may still seem rather a mystery. In fact, it is probably much easier to understand explicitly what digital rather than analogue systems do, because the basic operations often involve nothing more than multiplication and addition. But before we can describe any specific examples, we need to introduce some special terms and expressions for dealing with digital systems. The easiest way to do this is to take a particular system—a digital version of the amplifier introduced in Chapter 4 (p. 47). The amplifier has as its output a signal that is twice its input (roughly a 6 dB gain). This can be represented as a digital system in the following way:

$$y[n] = 2 \cdot x[n] \tag{1}$$

$x[]$ represents the input signal, $y[]$ represents the output signal, and the operator '$\cdot$' represents multiplication. Above, in our tabulation of a digital sinusoid, we used genuine values of time, spaced every ½ s. In fact, it is more typical to simply number the samples of a digital signal, which effectively means that the numbers specifying the time axis do not need to be stored (with the sampling rate noted separately). Therefore, the letter 'n' in the bracket of $x[n]$ indicates that the nth digital input sample is being referred to. Similarly $y[n]$ represents the nth output sample. 'n' would typically start at 0, because when you ask a computer scientist to count to 5, they say '0, 1, 2, 3, 4, 5'! Also, we conventionally think of waveforms as starting at time 0, so calling the first possible sample the 0th sample is not so farfetched.

When processing a digital signal then, n is incremented so that it steps through the entire set of samples from the first to the last (assuming a total of N values). This could be represented explicitly by specifying '$n = 0, N-1$' after any formula (like the one above) but this is rarely done.

Let's get back to our particular example. Equation (1) above indicates that the system takes the first digital input sample (at time 0) and multiplies it by 2 to obtain the first output sample. It then takes the second input sample and multiplies it by 2 again to give the next output sample and so on. These steps can be represented as follows:

$$y[0] = 2 \cdot x[0]$$

$$y[1] = 2 \cdot x[1]$$

$$\ldots$$

$$y[N-1] = 2 \cdot x[N-1]$$

Just as for analogue systems, a digital system can be described by a frequency response, or equivalently, an impulse response.

Let's determine the impulse response first. A digital impulse is very similar to an analogue one, but even simpler to understand. Just like an analogue impulse, a digital impulse only exists at one moment in time. Because digital signals are theoretical notions anyway, you shouldn't have any difficulty imagining such a signal, whereas it's hard to think about an analogue impulse because it is infinitely narrow. A digital impulse also has a finite amplitude of 1, so you need not imagine something with infinite amplitude. Here's what a digital impulse looks like:

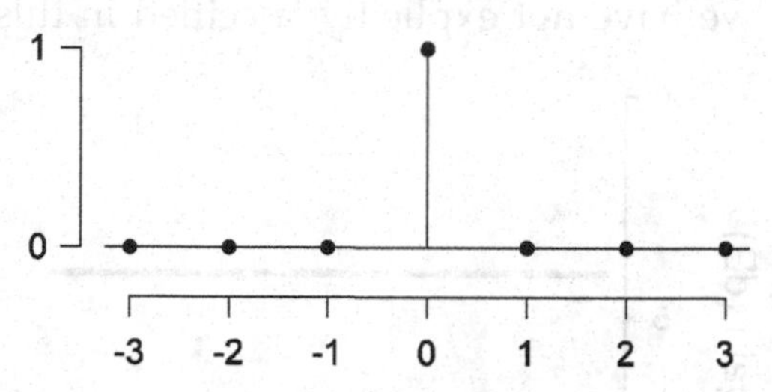

It should be obvious to you what the impulse response of this digital amplifier must be. Remember, all you need do is put an impulse (a single sample value of 1 at time 0) into the digital amplifier. For all values of n not equal to 0, $x[n] = 0$, so the output of the filter will also be 0. Only for $n = 0$ need you do the calculation:

$$y[0] = 2 \cdot x[0] = 2 \cdot 1 = 2$$

Therefore, the impulse response would look like this:

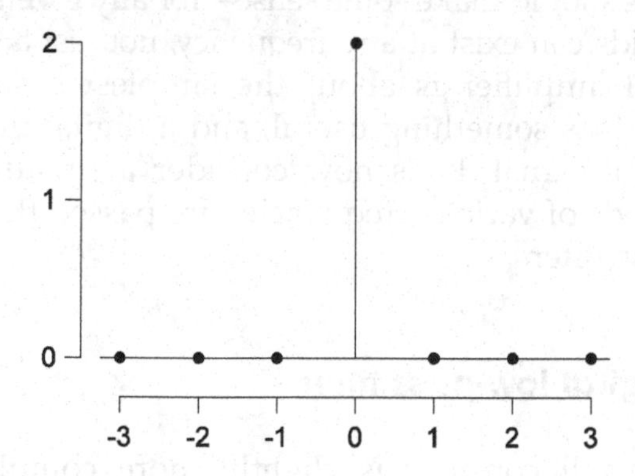

Just as you might expect, the amplitude response of this digital filter could be calculated from the spectrum of the impulse response using a Fourier Transform. But a standard Fourier transform only works on analogue signals, so you would need a modified transform known as the *discrete Fourier transform* (DFT). A DFT takes a digital signal and determines what set of digital sinusoids (frequencies, amplitudes and phases) needs to be added together to synthesize that particular signal. You may also have heard of an FFT—a *fast Fourier transform*. An FFT gives exactly the same result as a DFT, but in a cleverer way, which makes the computations much faster. Therefore, computers always use FFTs to make their calculations.

However, no such calculation is really necessary for such a simple system. Clearly, any sinusoid passed through this system will have its amplitude doubled, so the amplitude response will have a gain of 2 (or 6 dB) for all frequencies. But remember that the range of frequencies will be limited to half the sampling rate (sr/2), which we have not explicitly specified in this example:

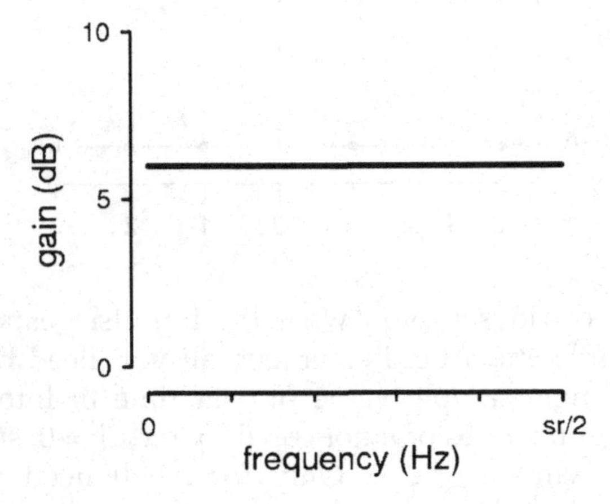

Take special note that the amplitude response of a digital *system* is continuous, existing at all frequencies (at least over a restricted frequency range) even though a digital *signal* only exists at discrete moments. This should make some sense—for any given sample rate, digital sinusoids can exist at any frequency, not just a specified set.

This digital amplifier is about the simplest system one can imagine that does something useful, and a digital impulse is the simplest useful signal. Let's now consider a situation in which digital sinusoids of various frequencies are passed through a very basic low-pass filter.

A simple digital low-pass filter

The system we'll consider is slightly more complex than the amplifier:

$$y[n] = \frac{x[n] + x[n-1]}{2} \tag{2}$$

Unlike the digital amplifier, this system has two x's on the right-hand side. The first x is followed by n, which indicates the current input value, and the second is followed by $n-1$, which means the *preceding* input value. For example, if the current sample is at $n = 2$, then $n-1 = 1$ is the previous sample. Hence this formula indicates the current and the preceding inputs should be added and divided by two to obtain the output—which is to say, averaged.

Equation (2) makes it clear that we are calculating an average, but it will be generally more useful for us to rewrite this equation in a more conventional form as:

$$y[n] = 0.5 \cdot x[n] + 0.5 \cdot x[n-1] \tag{3}$$

The values that are multiplied against each of the input samples (here equal to 0.5) are known as the *coefficients* of the $x[n]$ and $x[n-1]$ terms. In fact, they are not essential for the low-pass characteristic of this filter, but simply change its overall gain. They are included here to introduce you to the idea of filter coefficients which we'll discuss further later.

For an input signal, we'll begin with a 500-Hz sinusoid sampled at 1000 times per second (1 kHz). The sampling theorem indicates that this sampling rate is the lowest rate that will accurately represent the 500-Hz sinusoid. The process of sampling results in each period of the sinusoid being sampled at two equally spaced points in time:

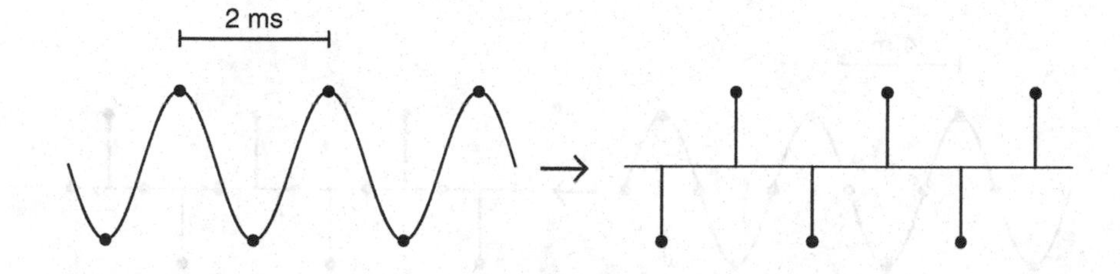

As you can see, the sinusoid varies in peak amplitude between +1 and −1. The first sample is at the trough (−1) and the second point at the peak (+1). If we specified the sequence of values in a list, the values of the sampled signal would be:

$$\ldots -1, +1, -1, +1, -1, +1 \ldots$$

Let's now see what the system of equation (3) does to this signal. We'll start at the second sample in the sequence ($n = 1$) so that we've got a previous sample to average it with ($n = 0$). It is easy to see that the two input samples we need to calculate $y[1]$ are $x[1] = +1$ and $x[0] = -1$. Thus:

$$y[1] = (0.5 \cdot +1) + (0.5 \cdot -1) = +0.5 - 0.5 = 0$$

Stepping to the next (third) sample in the sequence ($n = 2$), now $x[2] = -1$ and $x[1] = +1$, so:

$$y[2] = (0.5 \cdot -1) + (0.5 \cdot +1) = -0.5 + 0.5 = 0$$

The output is zero again. This is true whatever adjacent pair of samples is used. Thus, for this sequence alternating between +1 and −1 (a sampled version of a 500-Hz sinusoid), all output amplitudes are 0. Effectively this particular input signal is removed, or *filtered out* by this simple operation:

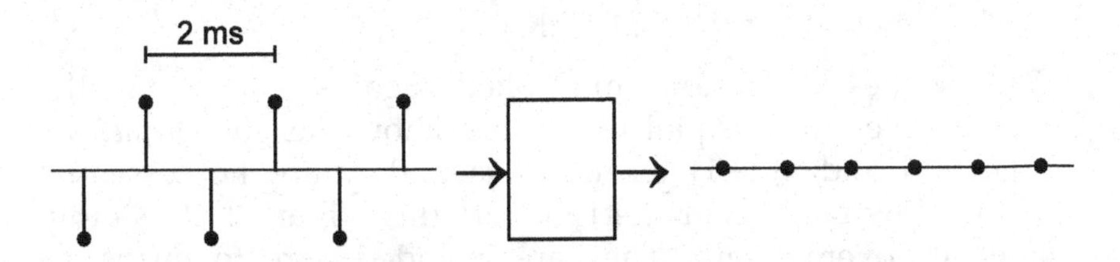

What would this system do with a lower frequency sinusoid, say at 250 Hz? By now you should be able to work out that at the 1000-Hz sampling rate, the signal will repeat after four samples (one period of this sinusoid) and that starting at the same phase as for the 500-Hz sinusoid, the sample values will be:

$$\ldots -1, 0, +1, 0, -1, 0, +1, 0, \ldots$$

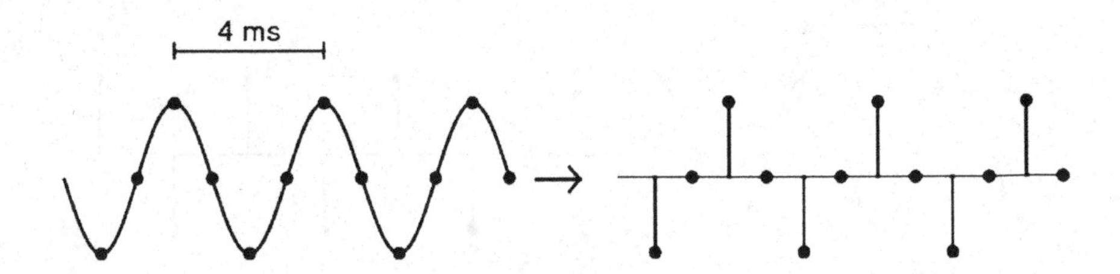

The output values for $y[n] = 0.5 \cdot x[n] + 0.5 \cdot x[n-1]$ need calculating for four pairs of values before they repeat:

For the first pair, when $n = 1$:

$$x[n-1] = -1, x[n] = 0, \text{ hence } y[1] = -0.5$$

For the second pair, when $n = 2$:

$$x[n-1] = 0, x[n] = +1, \text{ hence } y[2] = +0.5$$

For the third pair, when $n = 3$:

$$x[n-1] = +1, x[n] = 0, \text{ hence } y[3] = +0.5$$

For the fourth pair, when $n = 4$:

$$x[n-1] = 0, x[n] = -1, \text{ hence } y[4] = -0.5$$

This set of values then repeats, because the input signal is periodic, leading to:

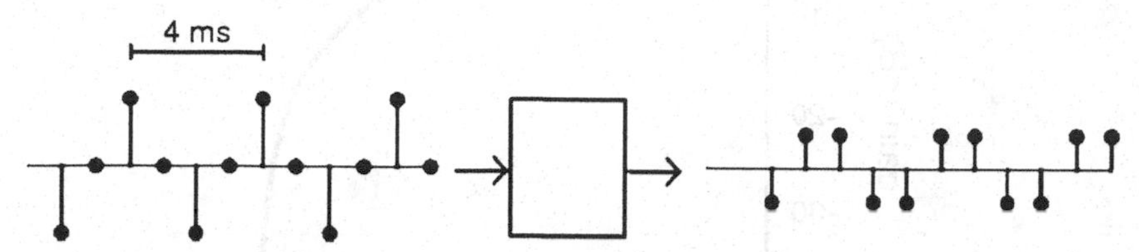

The system produces some output for this signal (that is, it is not filtered out completely) but it is attenuated to some extent. You may also note that a *phase shift* has been introduced, because the digital output wave looks different from the input wave, even though they are both digital sinusoids at the same frequency. This is especially apparent because the relatively low sampling rate leads to a rather sparse (but fully sufficient) representation.

Since the 500-Hz signal is filtered out completely but the 250-Hz signal is not, the system can be described as a crude form of low-pass filter. This can be more readily seen if we use an even lower frequency sinusoid (125 Hz) and examine the input and output (you'll be asked to do the calculations in the exercises):

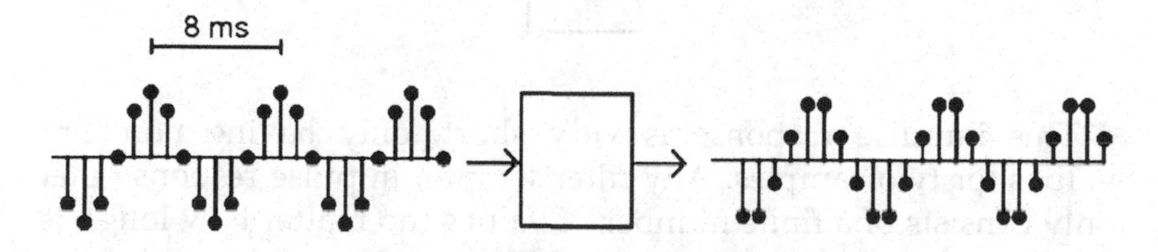

Finally, we will put a signal through our system which has a frequency of 0 Hz (DC). If we let its initial value be +1, then all subsequent values would also be +1, because by definition, this wave cannot vary in amplitude. So the response of $y[n] = 0.5 \cdot x[n] + 0.5 \cdot x[n-1]$ to all inputs is also +1:

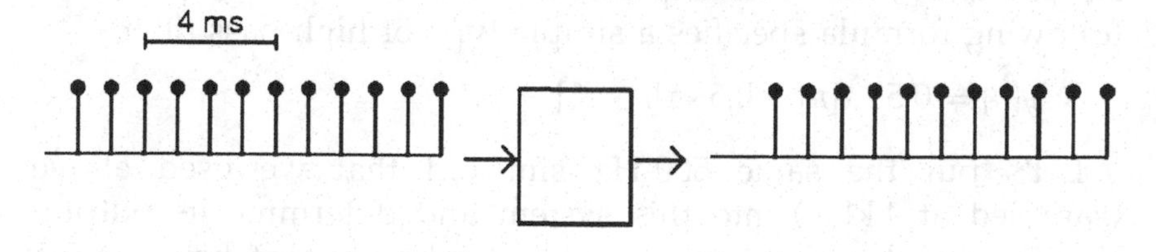

As usual, we want to summarize these measurements in an amplitude response, which shows the extent to which sinusoids of any frequency are passed through the system. As seen on the next page, this is clearly a low-pass filter.

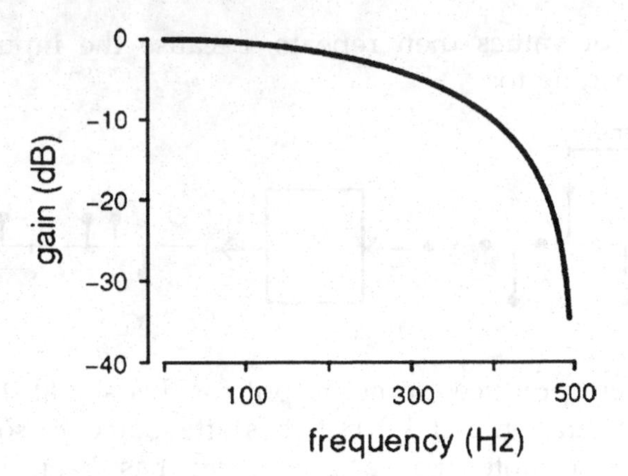

Note that the gain at 500 Hz is not actually drawn on the graph. The system completely filters out this frequency, leading to a gain of 0 on linear scales, and hence one which cannot be expressed in dB.

It is also interesting to examine the response of this system to an impulse:

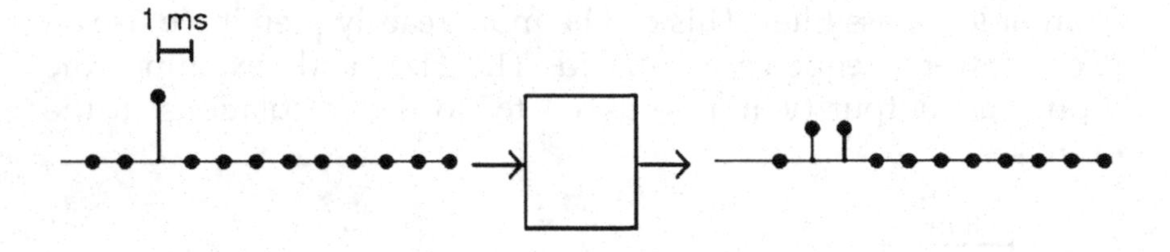

This impulse response is very short, only having non-zero values for two samples. Any filter with an impulse response that only consists of a finite number of values (no matter how long), is known as a *finite impulse response* (FIR) filter. Keep this in mind for later.

A simple digital high-pass filter

As you know, the other important type of filter is high pass. The following formula specifies a simple type of high-pass filter:

$$y[n] = 0.5 \cdot x[n] - 0.5 \cdot x[n-1] \tag{4}$$

Let's put the same 500-Hz sinusoid that we used above (sampled at 1 kHz) into this system and determine its output. You'll remember that the sampled signal alternated between -1 and $+1$.

We start at the second sample in the sequence ($n = 1$) to ensure we've got a previous sample to calculate with ($n = 0$) as we did before. The two input samples we need to obtain $y[1]$ are

$x[1] = +1$ and $x[0] = -1$. Doing the mathematics (remembering that subtracting a negative value means adding that value) gives:

$$y[1] = (0.5 \cdot +1) - (0.5 \cdot -1) = +0.5 - (-0.5) = +1$$

For the next (third) sample in the sequence ($n = 2$), the input samples are now $x[2] = -1$ and $x[1] = +1$, so:

$$y[2] = (0.5 \cdot -1) - (0.5 \cdot +1) = -0.5 - (0.5) = -1$$

The numbers for all subsequent n, $n-1$ pairs are either $[+1, -1]$ or $[-1, +1]$, which give $+1$ and -1 as the result. Consequently, the original input signal is passed by this system unchanged:

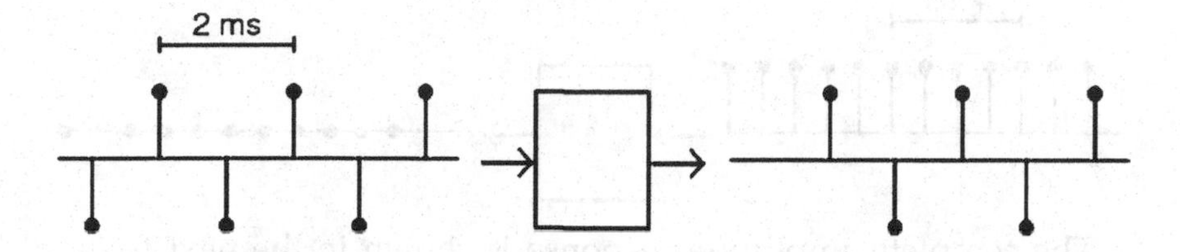

Next we'll put the 250-Hz sinusoid through the system. The values of the 250-Hz signal repeat every fourth term (..., −1, 0, +1, 0, −1, 0, +1, 0, ...). Thus, the output values need calculating for four pairs of input values before they repeat, as shown at the left of the following table. The output values ($y[n]$) obtained by substituting these values into equation (4) are shown in the right column:

$x[n]$	$0.5 \cdot x[n]$	$x[n-1]$	$0.5 \cdot x[n-1]$	$y[n]$
$x[1] = 0$	0	$x[0] = -1$	−0.5	$y[1] = +0.5$
$x[2] = +1$	+0.5	$x[1] = 0$	0	$y[2] = +0.5$
$x[3] = 0$	0	$x[2] = +1$	+0.5	$y[3] = -0.5$
$x[4] = -1$	−0.5	$x[3] = 0$	0	$y[4] = -0.5$

Thus, the 250-Hz input leads to the following output:

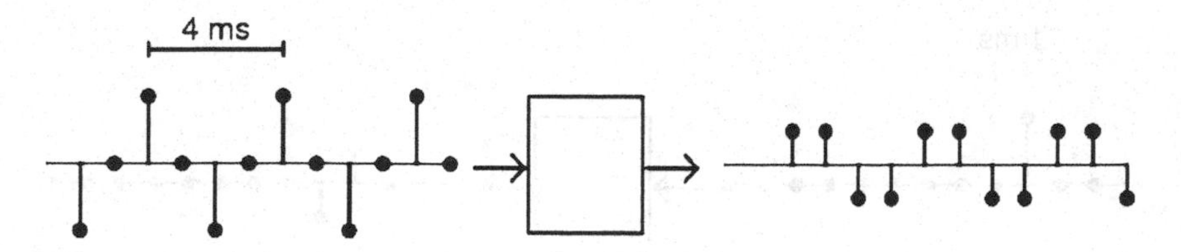

You should be able to do the calculations yourself now, for the 125-Hz sinusoid shown at the top of the next page.

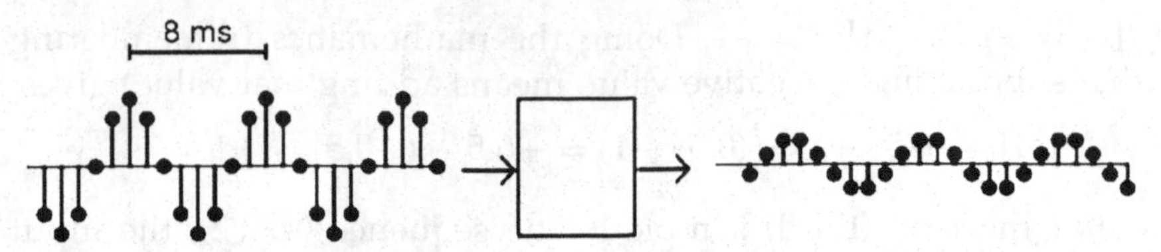

Finally, consider the particularly interesting case of the lowest frequency that can exist—DC. Assuming again that all values of this signal are +1, the response of $y[n] = 0.5 \cdot x[n] - 0.5 \cdot x[n-1]$ will always be zero. This is true whatever particular value the DC signal has. In other words, this system has completely filtered out the lowest frequency signal:

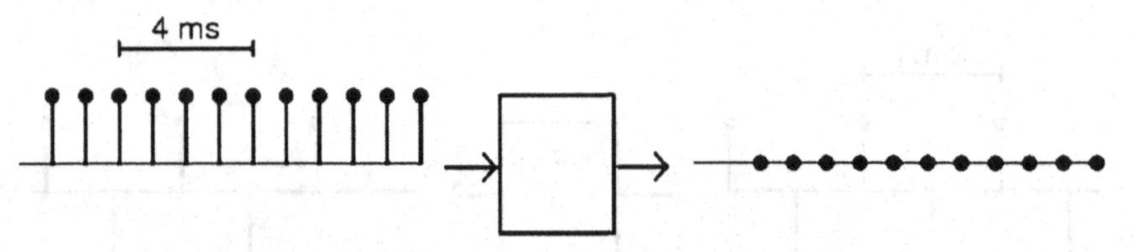

The complete amplitude response is shown in the next figure. This shows that this system has a high-pass characteristic with output levels going down as frequency decreases until it reaches zero at DC. (Again, of course, the gain of zero cannot be represented on dB scales.) You should find this a convincing example of a high-pass filter:

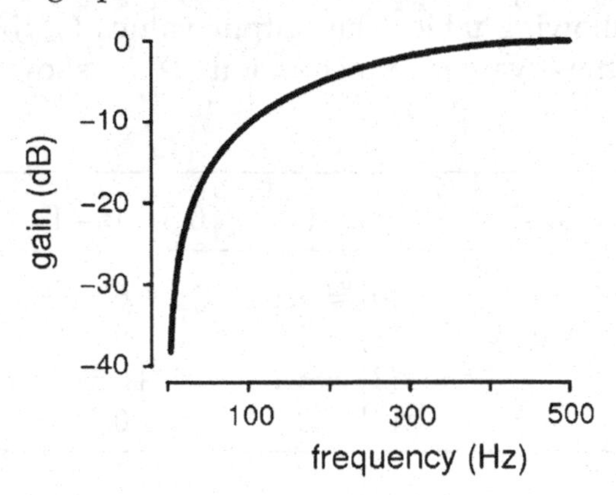

Consider, too, its impulse response, which we know to contain all the information about its properties:

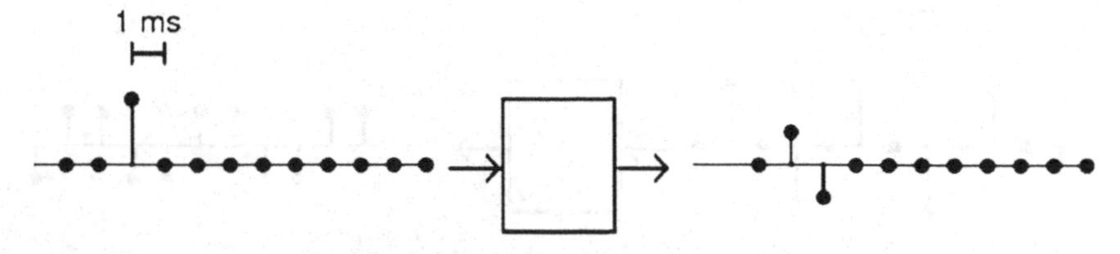

Like the low-pass system, the impulse response is very short, again only having non-zero values for two samples. This is another example of an FIR system.

A simple infinite impulse response system

We have now shown three simple digital systems, all of which had impulse responses which were finite (in fact, quite short). As you might have guessed by our emphasis of this fact, not all digital systems are like this. FIR filters operate on a weighted sum of past inputs. Therefore, if the input becomes zero and stays at that value, then eventually the output will become zero as well. Therefore, the impulse response lasts only a finite time. Contrasted to this are *infinite impulse response* (IIR) filters for which the impulse response extends to infinity. Perhaps surprisingly, it is very easy to make a digital system that is IIR. All the system needs do is to reuse the *output* values of the system in its calculations, in addition to the input values. Here's an example:

$$y(n) = 0.5 \cdot x[n] + 0.5 \cdot y[n-1] \qquad (5)$$

The important difference between this system and the three digital FIR systems we have considered earlier is that y terms appear on the left *and* right side of the equation. In this system, the current output ($y[n]$) is based on half the current input ($0.5 \cdot x[n]$) and half the previous output ($0.5 \cdot y[n-1]$). The ($0.5 \cdot y[n-1]$) term gives the system input from the past and, for this reason, IIR systems are sometimes referred to as systems with *memory*. In addition, because this implies information is being recycled, these are also known as *recursive* systems.

We can examine the response of this system just as we did for the FIR filters. Again we will use the 500-Hz signal as input. We'll start the system with current input $x[1] = +1$ and assume the previous output is zero ($y[0] = 0$). Putting these values into equation (5) gives:

$$y[1] = (0.5 \cdot 1) + (0.5 \cdot 0) = +0.5$$

The next input is -1 ($x[2] = -1$) and we have a past value of output ($y[1] = +0.5$). Using these values in equation (5) gives:

$$y[2] = (0.5 \cdot (-1)) + (0.5 \cdot 0.5) = -0.25$$

The next two values are as follows:

$$y[3] = (0.5 \cdot 1) + (0.5 \cdot -0.25) = 0.375$$

$$y[4] = (0.5 \cdot (-1)) + (0.5 \cdot 0.375) = -0.3125$$

For a 500-Hz sinusoid, the FIR filters discussed earlier repeated their output after four samples. However, this system has a memory so the outputs do not repeat exactly. This is apparent when we calculate the output $y[5]$:

$$y[5] = (0.5 \cdot 1) + (0.5 \cdot -0.3125) = 0.34375$$

For this system, $y[1\,] = 0.5$ and $y[5] = 0.34375$ (that is, the values do not repeat). This is because of what is known as a *start-up transient*, due to the fact that the input wave started at a particular moment in time, and was zero before that. Eventually, the output does settle down to being periodic as you can see from the output wave here:

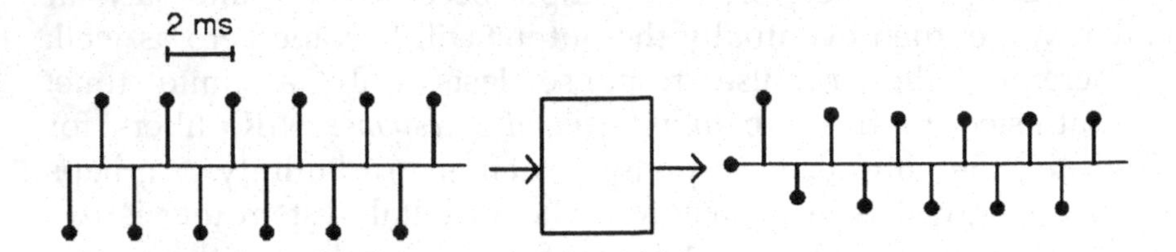

You'll note that more cycles of the input wave have been pictured, in order to allow the start-up transient to die away in the output. Also the y-axis of the output signal has been multiplied by a factor of 2, here and in the following three figures, so as to make the output waveforms easier to see.

Next we put the 250-Hz sinusoid into the system. Again we assume the output value to be zero before the signal started so $y[1] = 0$. The output values ($y[n]$) obtained by substituting the appropriate values into equation (5) are shown in the right column:

$x[n]$	$0.5 \cdot x[n]$	$y[n-1]$	$0.5 \cdot y[n-1]$	$y[n]$
$x[1] = 0$	0	$y[0] = 0$	0	$y[1] = 0$
$x[2] = +1$	+0.5	$y[1] = 0$	0	$y[2] = +0.5$
$x[3] = 0$	0	$y[2] = +0.5$	+0.25	$y[3] = +0.25$
$x[4] = -1$	−0.5	$y[3] = +0.25$	0.125	$y[4] = -0.375$

Here's what the input and output signals look like. Note how the output level (once it has reached equilibrium) has increased a bit from that obtained with an input frequency of 500 Hz:

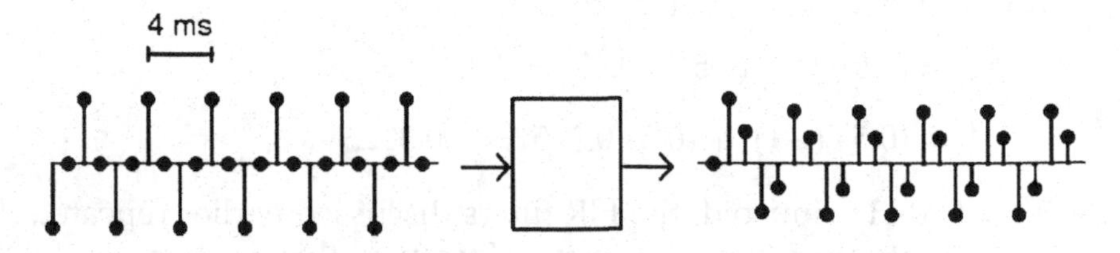

The output amplitude increases again for a 125-Hz sinusoid (you will be asked to do these calculations in the exercises). Only three cycles are shown here because the start-up transient is very short:

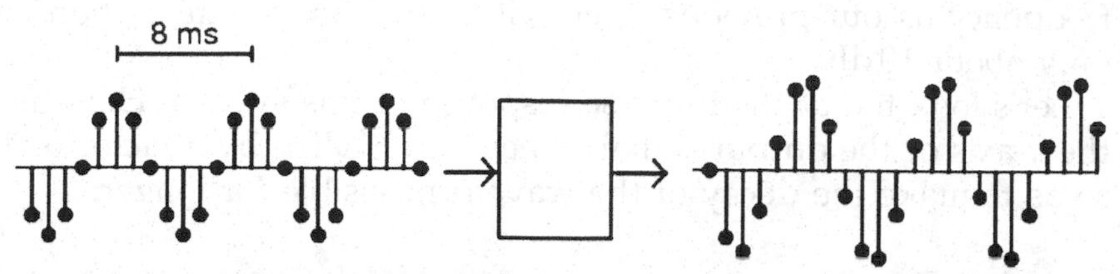

Putting a DC signal through our system and letting the first output value be $y[0] = 0$ (as previously) and all input values be +1, the response can be calculated as follows:

$x[n]$	$0.5 \cdot x[n]$	$y[n-1]$	$0.5 \cdot y[n-1]$	$y[n]$
$x[1] = +1$	+0.5	$y[0] = 0$	0	$y[1] = +0.5$
$x[2] = +1$	+0.5	$y[1] = +0.5$	+0.25	$y[2] = +0.75$
$x[3] = +1$	+0.5	$y[2] = +0.75$	+0.375	$y[3] = +0.875$
$x[4] = +1$	+0.5	$y[3] = +0.875$	+0.4375	$y[4] = +0.9375$

which looks like this:

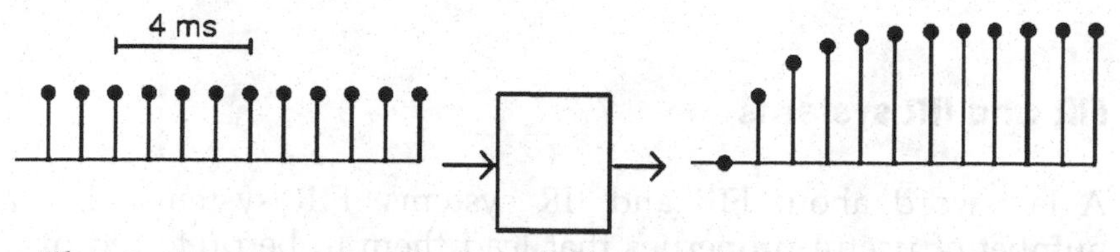

The start-up transient is especially prominent in this case, as the filter rises gradually in the output amplitude over its first five or six values. It then reaches an equilibrium level larger than any other input frequency we have tried.

Given that the output amplitude from this system has been increasing as the input frequency has been progressively lowered, it should not surprise you to find that this is an example of a low-pass filter:

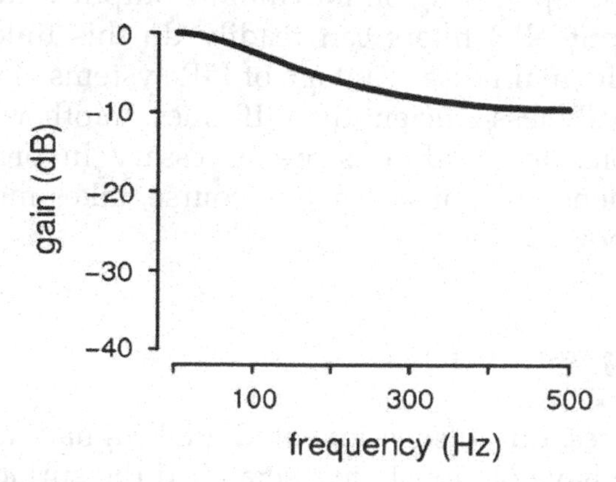

This low-pass filter does not vary so much in its effect over frequency as our previous filter, with a maximum attenuation of only about 10 dB.

Let's look too at the impulse response of this system. Here the the y-axis of the output signal as been multiplied by a factor of 4 so as to make the decay of the waveform visible for longer:

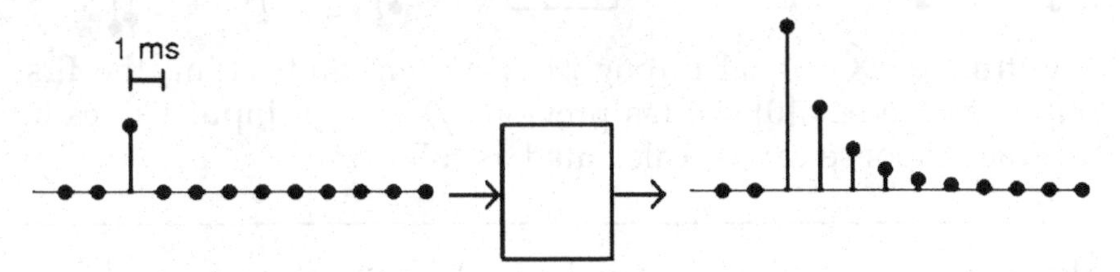

You can see that the impulse response goes on for an extended period of time, in theory, forever. In fact, the quantisation of values represented digitally will mean that the number will become so small, it will round to zero, as it appears to have done here. Theoretical developments typically assume, as we mentioned above, that quantisation effects can be ignored. So we still talk about an IIR even if this will not happen in a case like this in any real digital system.

FIR and IIR systems

A last word about FIR and IIR systems. FIR systems have a number of useful properties that lead them to be preferred over IIR systems for certain applications. One is that they can easily be designed to have a linear phase response by making the impulse response symmetric. Linear phase responses are sometimes desirable as they delay all sinusoidal components by the same amount so they tend to preserve waveform shapes (see p. 103 for further discussion of this). Additionally, FIR systems are simpler to understand and design and are always *stable* (which is to say, they never end up veering off into infinite output values no matter what the input—IIR filters can readily do this unless carefully designed). The main disadvantage of FIR systems is that they are computationally less efficient than IIR ones. In other words, more multiplications and additions are necessary in order to get a similar frequency response. This, of course, takes more time and computing power.

Concluding remarks

This completes our discussion of digital signals and systems, although we have obviously just scratched the surface. Given the ubiquity of digital signal processing, it may not surprise you to learn that we have already been using these techniques throughout the book, even though it was never mentioned! Perhaps the best examples of this can be found in Chapter 11, where all but a few of the spectrograms were made on a computer, hence required the use of digital signals and systems. Even our FM sweep was generated digitally.

Exercises

1. A signal containing no energy above 5 kHz is sampled at 10 kHz and stored on a computer. If this is to be output through a DAC at 10 kHz, sketch the amplitude response of a realistic output filter that would attenuate by at least 12 dB all components in the signal above 4.5 kHz.

2. What sampling rate would you recommend if you were recording digitally (a) a 100-Hz sinusoid, (b) a 200-Hz sinusoid, (c) a 100-Hz square wave, (d) a 200-Hz square wave, (e) a 100-Hz triangle wave and (f) a 200-Hz triangle wave? What factors govern your choice?

3. If you wanted to construct a digital metronome which ticked twice a second, what output rate would you use and how many bits on the DAC would be appropriate? (Assume that it is only important that the tick occur regularly, with the form of the pulse giving the tick being irrelevant.)

4. Design and sketch an input–output function for a 1-bit DAC that would be appropriate for quantizing sinusoids with peak-to-peak amplitudes of 1 V. Sketch a sinusoid before and after quantizing. Do the same for a 2- and 3-bit quantizer. Comment on the three quantized sinusoids.

5. A signal generator was set up to produce a triangular signal which was sampled at the points indicated. Was the sampling performed correctly? If not, say in what way the process went wrong and suggest remedial measures.

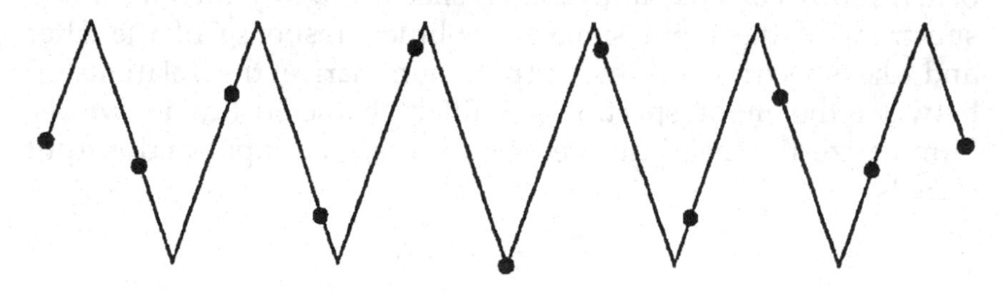

6. Cine film is recorded at the rate of 24 images per second. If you watch a cowboy film, the wheels on the wagons appear

to move slower than we know they would. Write an account of why this occurs. (You may include diagrams if you think this helps.)

7. Here are the values for one period of the 125-Hz digital sinusoid that was used in the examples of simple digital systems:

−0.71 −1.00 −0.71 0.00 0.71 1.00 0.71 0.00

Using these, do the calculations for two complete cycles of the input to check the output waves shown on pages 332, 335 and 338.

8. The following complex periodic signals are to be filtered, sampled, stored in a computer, and then played out again:

(a) 100-Hz fundamental and 19 higher harmonics at constant amplitude.
(b) 100-Hz fundamental and seven higher harmonics at constant amplitude.
(c) 100 Hz fundamental and 19 higher harmonics dropping at 12 dB/octave.
(d) 100-Hz fundamental and seven higher harmonics dropping at 12 dB/octave.
(e) 100-Hz fundamental and 19 higher harmonics dropping at 6 dB/octave.
(f) 100-Hz fundamental and seven higher harmonics dropping at 6 dB/octave.

The sampling rates, filter cutoffs and filter slopes of the filters used prior to sampling these signals (and in reconversion to analogue form) are tabulated below. (i) For each filter specification indicate which signals have been filtered properly so that the output signal accurately represents at least some range of the spectrum of the input signal. (ii) For the first signal, summarize the relationship between the filter cutoffs and the sampling rate in octaves. (iii) For this same signal, sketch roughly the amplitude spectrum of the input signal, amplitude response of the filter and the spectrum of the output. Summarize the relationship between the input spectra and filter characteristics in words. Summarize the major differences between the input and output signals.

Sampling rate	Filter cutoff	Filter slope frequency
1 kHz	250 Hz	Infinitely steep
1 kHz	250 Hz	6 dB/octave
1 kHz	250 Hz	3 dB/octave
1 kHz	–	No filtering
1 kHz	500 Hz	Infinitely steep
1 kHz	500 Hz	6 dB/octave
1 kHz	500 Hz	3 dB/octave
1 kHz	–	No filtering
1 kHz	1 kHz	Infinitely steep
1 kHz	1 kHz	6 dB/octave
1 kHz	1 kHz	3 dB/octave
1 kHz	–	No filtering
1 kHz	2 kHz	Infinitely steep
1 kHz	2 kHz	6 dB/octave
1 kHz	2 kHz	3 dB/octave
1 kHz	–	No filtering
2 kHz	1 kHz	Infinitely steep
2 kHz	1 kHz	6 dB/octave
2 kHz	1 kHz	3 dB/octave
2 kHz	–	No filtering
2 kHz	2 kHz	Infinitely steep
2 kHz	2 kHz	6 dB/octave
2 kHz	2 kHz	3 dB/octave
2 kHz	–	No filtering
2 kHz	4 kHz	Infinitely steep
2 kHz	4 kHz	6 dB/octave
2 kHz	4 kHz	3 dB/octave
2 kHz	–	No filtering
4 kHz	1 kHz	Infinitely steep
4 kHz	1 kHz	6 dB/octave
4 kHz	1 kHz	3 dB/octave
4 kHz	–	No filtering
4 kHz	4 kHz	Infinitely steep
4 kHz	4 kHz	6 dB/octave
4 kHz	4 kHz	3 dB/octave
4 kHz	–	No filtering
4 kHz	8 kHz	Infinitely steep
4 kHz	8 kHz	6 dB/octave
4 kHz	8 kHz	3 dB/octave
4 kHz	–	No filtering
4 kHz	16 kHz	Infinitely steep
4 kHz	16 kHz	6 dB/octave
4 kHz	16 kHz	3 dB/octave
4 kHz	–	No filtering

Appendix

Answers to selected exercises

Chapter 3

3. 20 Hz.

5–6. All three rectangular waves have the same RMS amplitude of 0.7071 V. The duty cycles of the second two pulse trains are 10% and 70%.

8. RMS amplitude $= V/10 \times \sqrt{d}$.

11. 94.8 dB SPL.

12. Two sinusoids of identical frequency added together in phase that are 11.74 dB different will have a level that is 2 dB greater than the larger of the two components being added.

Chapter 5

2. Such a system is additive and homogeneous, and hence linear, but is time-varying. For the latter property, note that the output to a single pulse depends upon the time of its presentation to the system (relative to the switching on and off). Similarly, the value of the sinusoid at onset of each "on" interval will differ according to the initial phase of the sinusoid.

Chapter 6

1. The values of the output sinusoids of System D, a band-stop filter, are given by:

Frequency (Hz)	Output voltage (V)
100	2
200	1.42
300	0.03
400	0.17
500	1.42
600	1.98
700	2
800	2
900	2

Frequency (Hz)	Output voltage (V)
1000	2
1100	2
1200	2
1300	2
1400	2
1500	2
1600	2
1700	2
1800	2
1900	2
2000	2

4. The corrected amplitude response of the system is:

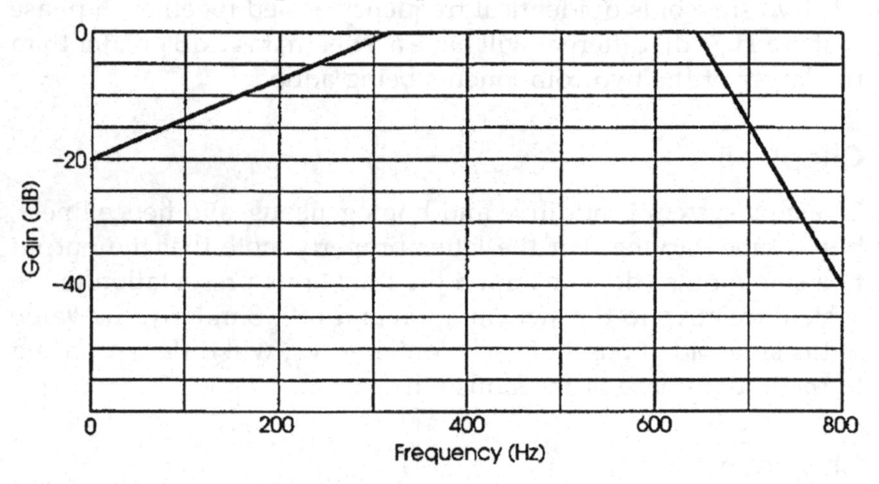

while the unwrapped phase looks like:

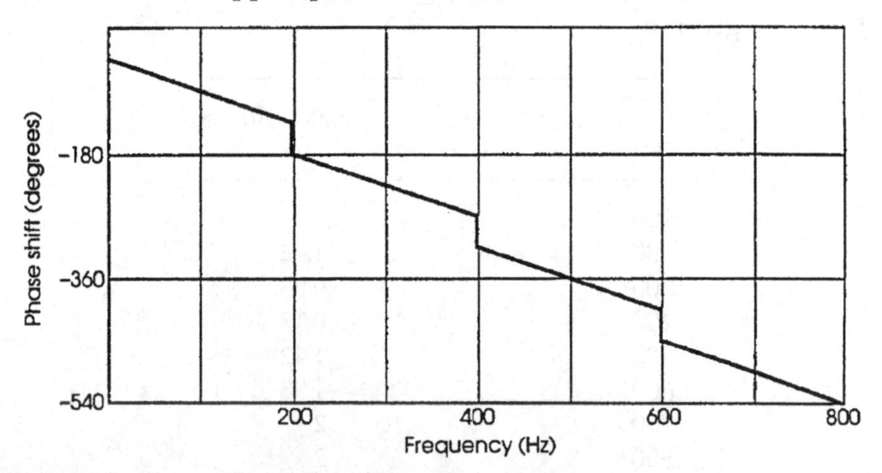

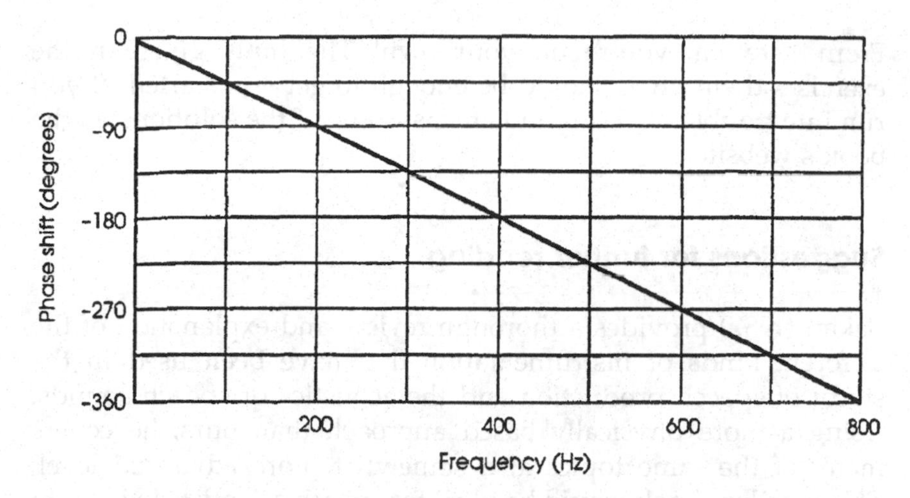

7. The amplitude response of the cascaded system is:

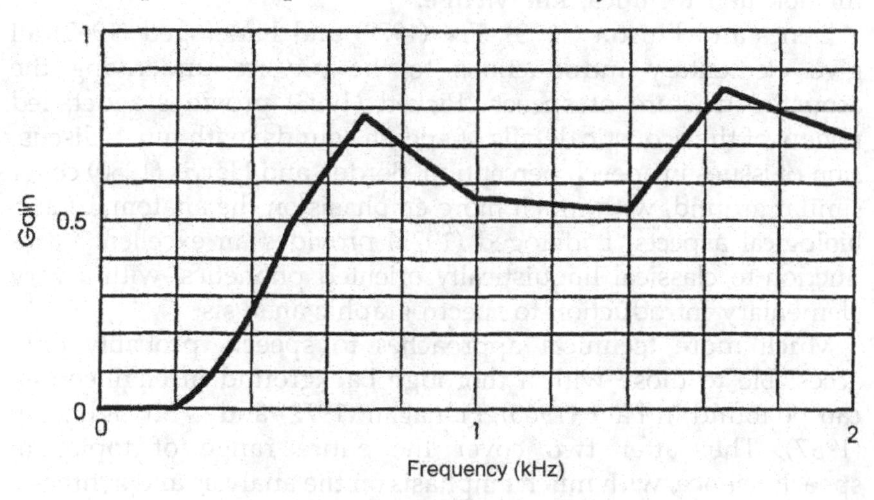

while the unwrapped phase looks like this:

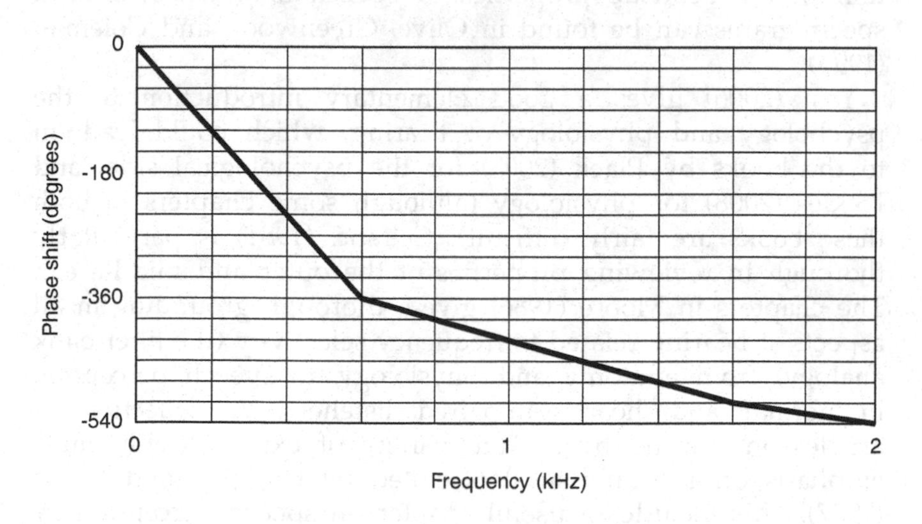

From here on, you're on your own! The hints given in the exercises themselves should be enough to get you started. If you run into trouble, you can find at least some of the solutions on the book's website.

Suggestions for further reading

Baken (1987) provides a thorough review and explanation of the different kinds of instrumentation that have been used in the study of speech production and the acoustics of speech sounds. Using a more physically based approach than ours, he covers many of the same topics at a somewhat more advanced level. This excellent book would be ideal for speech scientists looking to further their technical knowledge.

Denes and Pinson (1973), Fry (1979) and Ladefoged (1962) all give elementary introductions to the physics underlying the acoustic structure of speech. Pickett (1998) provides a detailed review of the acoustic details of speech sounds, with much discussion of issues in speech perception. Borden and Harris (1980) cover similar ground, with much more emphasis on the anatomical and biological aspects. Ladefoged (1975) provides an excellent introduction to classical linguistically oriented phonetics, with a very elementary introduction to spectrographic analysis.

Much more technical approaches to speech, probably only accessible to those with a thorough background in engineering, can be found in Fant (1960), Flanagan (1972) and O'Shaughnessy (1987). The latter two cover the entire range of topics in speech science, with much emphasis on the analysis and synthesis of speech. A particularly wide-ranging and thorough exploration of the acoustic properties of speech sounds as seen in spectrograms can be found in Olive, Greenwood and Coleman (1993).

Yost (2006) gives a good elementary introduction to the psychology and physiology of hearing, which could lead on to the books by Plack (2005) for the psychological side, and Pickles (2008) for physiology (although some chapters in both these books are fairly difficult). Gelfand (1981) is particularly thorough in reviewing properties of the outer and middle ear. The chapters in Moore (1986) give a thorough grounding in all aspects of hearing related to frequency selectivity (the filter-bank analogy) from anatomy and physiology, to speech perception in normal and hearing-impaired listeners. A collection of articles on hearing by a wide variety of experts (with much emphasis on clinical issues) is edited by Haggard and Evans (1987). This includes a useful chapter on speech perception by Summerfield.

For the student wishing to learn more about linear systems analysis *per se*, a knowledge of calculus is mandatory. Students with sufficient mathematical training (not necessarily trained as engineers) should be able to use Bracewell (1986), Lathi (1965) and McGillem and Cooper (1984). The book by Siebert (1986) is thorough and learned, but perhaps only appropriate for the electrical engineering students it is aimed at. Hartmann's (1998) book covers almost all aspects of signals and systems set specifically in the context of their relevance for hearing research. All four of these books also cover topics related to digital signal processing to some extent. Lynn, Fuerst and Thomas (1997) provide a good introduction to digital signal processing that is more accessible and practical than other texts.

Notes on the figures and text

Chapter 4

The data on the middle ear (page 53) are extracted from Figure 6 of Guinan and Peake (1967). The basilar membrane data (figure on page 54) are extracted from the top part of Figure 7 of Rhode (1971). The tape-recorder used in constructing the figures on pages 55–58 was a Technics M250.

Chapter 6

Figures on pages 72, 75, 78 and 81 and the left-hand sides of the figures on pages 79 and 80 represent the amplitude response of a fourth-order Butterworth low-pass filter with a cutoff of 1 kHz given by:

$$\frac{1}{\sqrt{1+\left(\frac{f}{1000}\right)^{8}}}$$

where f is frequency.

The right-hand sides of the figures on pages 79 and 80 represent a second-order Butterworth filter with the same cutoff frequency:

$$\frac{1}{\sqrt{1+\left(\frac{f}{1000}\right)^{4}}}$$

The figure on page 73 is a smoothed curve fit by eye to the top half of Figure 12 of Guinan and Peake (1967).

The figure on page 82 (lower) represents a high-pass Butterworth filter of order 3 with a cutoff of 1 kHz:

$$\frac{1}{\sqrt{1+\left(\frac{1000}{f}\right)^6}}$$

The figure on page 89 is a simple resonance whose amplitude response is generally given by:

$$\frac{C^2+\left(\frac{B}{2}\right)^2}{\sqrt{(f-C)^2+\left(\frac{B}{2}\right)^2}\sqrt{(f+C)^2+\left(\frac{B}{2}\right)^2}}$$

where B is bandwidth, f is frequency as before and C is the centre frequency of the resonance (Fant, 1960).

The technique for determining vocal tract amplitude responses described in the figures on page 95 is based on a study by Fujimura and Lindqvist (1971). The data for the figures on pages 97 and 98 were extracted from Figure 2 of that study.

The figure on page 106 shows the amplitude and phase response of a system with an impulse response $h(t)$ that is exponentially decaying with time t, that is:

$$h(t) = \alpha e^{-\alpha t}$$

The amplitude response can be shown to be given by:

$$\frac{\alpha}{\sqrt{\alpha^2+(2\pi f)^2}}$$

and the phase response by:

$$-\tan^{-1}\left(\frac{2\pi f}{\alpha}\right)$$

with f indicating frequency in both cases. For this figure, $\alpha = 2700$.

The figure on page 107 shows the amplitude and phase response of a system with an impulse response $h(t)$ that consists of a scaled exponentially decaying sinusoid, that is:

$$h(t) = \left(\frac{1}{2\pi f_0}\right)e^{-\alpha t}\sin(2\pi f_0 t)$$

The amplitude response can be shown to be given by:

$$\frac{1}{\sqrt{(\alpha^2+\omega^2)+2\omega_0^2(\alpha^2-\omega^2)+\omega_0^4}}$$

and the phase response by:

$$-\tan^{-1}\left(\frac{2\alpha\omega}{\alpha^2 - \omega^2 + \omega_0^2}\right)$$

where $\omega = 2\pi f$, with f indicating frequency. For this figure, $\alpha = 2700$ while $f_0 = 1000\,\text{Hz}$ (the centre frequency of the band-pass filter). In fact, this is the same transfer function as the simple resonance mentioned above, only parameterized in a different way.

Chapter 7

The oscillations in the approximations to the sawtooth, square wave and pulse waveforms at the points of discontinuity in the original waveforms (illustrated on pages 126 (lower), 132, 133, 135 (upper), 137 (upper) and 141) reflect what is known as the *Gibbs' phenomenon*. The overshoots and undershoots are to the same extent no matter what number of finite number of terms is used in the Fourier series, although the area they enclose shrinks as more terms are included. The oscillations arise from the effectively rectangular windowing of the true frequency spectrum which results from taking a finite number of terms. The oscillations can be reduced by appropriate windowing of the spectrum, that is, by reducing the amplitude of the terms included for higher and higher harmonics. See McGillem and Cooper (1984, pp. 99–100, 146–150) and Siebert (1986, pp. 481–482) for further information.

In the figures on pages 140, 142 (lower), 144 (lower) and 145, the spectral envelopes are drawn based on a decomposition of the pulse train into a set of complex exponentials which can be considered to exist at positive and negative frequencies. These figures show only the non-negative half of the frequency axis. This is different from the convention used previously, where the waveforms have been decomposed into real-valued sinusoids. The complex exponential representation has been used here because the shape of the spectral envelope is simpler, giving a smooth curve from 0 Hz to the fundamental component. If a representation in terms of sinusoids had been used, the component at 0 Hz would have to be half its size relative to the other components. Alternatively, one could consider the spectra given to represent the spectrum of the given pulse train, but which had its DC level adjusted. None of the substantive results are affected by this minor change.

The figure on page 147 (lower) was made using a real-time spectrum analyzer (Ono Sokki CF-910) operating over a 5-kHz bandwidth. The incoming signal was Hanning-windowed over 80 ms for analysis.

Chapter 8

The low-pass filter used in the figures on pages 160–162 is a third-order Butterworth with a cutoff frequency of 250 Hz. Its amplitude response is given by:

$$\frac{1}{\sqrt{1+\left(\frac{f}{250}\right)^{6}}}$$

whereas its phase response is given by:

$$-\tan^{-1}\left\{\frac{x^{3}-2x}{-2x^{2}+1}\right\}$$

with:

$$x=\frac{f}{250}$$

Chapter 9

System Z represents a system with a decaying exponential impulse response given by:

$$10.5e^{-10500t}$$

which is a simple low-pass filter with a cutoff frequency of about 1.7 kHz (see notes for Chapter 6 above). Note the general smoothing effect that this filter has on input waveforms, as, for example, on the tip of the triangle (page 175).

The figure on page 181 was made with a Connevans CE8 circumaural headphone on a B & K Artificial Ear Type 4153. This artificial ear is meant to mimic the impedance of the normal human ear, but not its geometry. Signals were analyzed using a real-time spectrum analyzer (Ono Sokki CF-910) operating over a 2-kHz bandwidth. The amplitude response of these headphones is rather poor for mid-frequencies, but relatively good at low frequencies. These headphones can present very high levels in the frequency range 100–500 Hz with relatively little distortion, important properties for their primary use in testing profoundly hearing-impaired listeners with residual low-frequency hearing.

Chapter 10

As described for Chapter 7, the spectra for the pulse stimuli of the figures on pages 184, 185 and 186 (upper) are based on a decomposition of the signal into complex exponentials, again for simplicity of representation. The level of the 0-Hz (DC) component has been set correctly, so, for a decomposition into sinusoids, every value except at 0 Hz should be doubled. These comments do not apply to the figures on pages 188 and 189, the spectra of which could represent a decomposition into either sinusoids or complex exponentials with a simple change in y-axis scale.

The windowing described starting on page 191 is typically known as a $\cos^2$ ramp. You may recall from the study of trigonometry that:

$$\cos^2\theta = ½ + (½)\cos 2\theta$$

The second term on the right-hand side (½ cos 2θ) is, of course, just a cosine wave with twice the frequency of the original, and half the amplitude. It therefore has values running from –½ to ½. The first term on the right-hand side (½) simply shifts the cosine wave upwards to now run from 0 to 1. Therefore, by picking an appropriate ½ cycle of this wave, you get either a smooth curve ramping up from 0 to 1, or a smooth curve ramping down from 1 to 0.

The figure on page 193 is the result of applying rectangular time windows to the signal, which accounts for the shapes of the spectra. Other window shapes could make for smoother spectra.

The figures on pages 195–200 are based on the two so-called gammatone filters of order 2, whose impulse responses are given by:

$$te^{-2\pi bt}\cos(2\pi f_0 t)$$

where f_0 is the centre frequency of the band-pass filter, and b is proportional to filter bandwidth.

Chapter 11

The linear frequency-modulated tone actually had a duration of 440 ms, over which it varied from 485 Hz to 815 Hz with linear onsets and offsets each of 20-ms duration. Thus, at full amplitude, the frequency sweep was 500–800 Hz over 400 ms. The figure on page 208 was made by playing the FM tone to a real-time spectrum analyzer. The analysis used a rectangular window which completely contained the tone. The spectrum shown in this figure has been limited to about 50 dB down from the peak spectral value. The filter-bank analyses (figures on pages 210, 211, 217–219) used the ILS system for filter design and actual filtering of a digital version of the FM tone. All the older spectrograms were made using a Kay Sona-Graph 7029A.

The figure on page 221 is reproduced from the Sona-Graph 6061A maintenance (1966) manual courtesy of KayPENTAX.

The graphic equalizer shown on page 223 is manufactured by AudioSource, who kindly provided the illustration. See http://www.audiosource.net/equalizers.php

Chapter 12

The figure on page 260 (middle) (and the same data on page 264) is redrawn from Figure 7 of Wiener and Ross (1946). The data for the figure on page 265 are redrawn from Figure 9 of Shaw (1976).

The figure on page 267 is based on data (selected to give a smoother curve) from Figure 11 (top) of Puria, Peake, and Rosowski (1997).

The figure on page 269 is based on a figure in Møller (1972) which originally appeared in Békésy (1960). The figure on page 271 is based on Figure 1 of Békésy (1949), while the lower figure on page 272 is based on Figure 7 of the same paper. The quote on page 273 can be found in Békésy (1949, p. 247). Figures on pages 273–276 are based on data found in Rhode (1971).

The figure at the bottom of page 278 is modified from Figure 16A of Evans (1975), with permission from Springer-Verlag. The sinusoid has been flipped over to make the relationship between the nerve firings and a particular phase of the sinusoid clearer.

The figure on page 283 used the cochlear spectrogram facility of SFS (Speech Filing System) which can be found at http://www.phon.ucl.ac.uk/resource/sfs/.

Chapter 13

The figure on page 286 is reproduced with permission from Rosen and Fourcin (1986). The figure on page 288 (lower) is modelled on part of a figure found in Hirano (1981). The recording of a soprano used to construct the figure on page 297 was kindly provided by Thomas Millhouse and Dianna Kenny (Millhouse, 2009; Millhouse and Kenny, 2008). The two vocal tract shapes in the figure on page 299 are based on those found in Figure 1.9 of Ladefoged (1975). The discussion of voiceless fricatives was much informed by Shadle (1985). The figure on page 303 was, in part, based on her Figure 4.1.

The synthetic sounds analyzed in the figures on pages 295 and 296 were created by passing digital sawtooths (with a sampling rate of 40 kHz) through a cascade of 12 digitally simulated simple resonators. Ten of these were at the appropriate frequencies (500, 1500, 2500, 3500 Hz and so on) with a bandwidth of 100 Hz. Two resonances were used to correct for the lack of higher resonances (the so-called higher-pole correction). These syntheses did not include the effects of radiation (accounting for the fact that speech sounds emanate from a mouth contained in a head—modelled as an opening in a sphere). In fact, as radiation has an effect that can be roughly approximated by a 6 dB/octave spectral lift, and the glottal source is often assumed to have a −12 dB/octave slope, the resulting −6 dB/octave spectral slope is equivalent to that used here (due to the −6 dB/octave slope of the sawtooth). For further details, see Fant (1960), Flanagan (1972) and O'Shaughnessy (1987).

Chapter 14

The discussion of aliasing is slightly misleading in considering the sampling of a 100-Hz sinusoid at the Nyquist rate of 200 Hz. In fact, the sampled waveform represents a sinusoid of 100 Hz (not 0 Hz—although it is indistinguishable from it), but the sampling points happen to lie at times where the sinusoid is 0. A shift of 2.5 ms in the start of sampling would lead to an output that was clearly related to a sinusoid of 100 Hz, and not one at 0 Hz. A sampling rate of 100 Hz would always lead to an aliased DC component.

Useful links

Many newer figures (for example, those illustrating specific digital systems in Chapter 14) were constructed using the graphical facilities of the freely available statistical package R: http://cran.r-project.org/

Most of the new spectrograms and spectral cross-sections used the freely available speech-processing package Praat (written by Paul Boersma and David Weenink), which has now become the standard in the field: http://www.fon.hum.uva.nl/praat/

Extremely useful software for illustrating many aspects of the topics explored in this book has been written by Dr Mark Huckvale, of the Department of Speech, Hearing and Phonetic Sciences at UCL: http://www.phon.ucl.ac.uk/resource/software.html

Particularly useful for this book are ESECTION, ESYNTH (relevant to Chapter 7) and ESYSTEM (relevant to Chapters 6–10 and 14). ESYSTEM has extensive (and highly recommended) tutorial material under its 'Help' menu, concerning both analogue and digital systems. Also available is a web tutorial about logarithms: http://www.phon.ucl.ac.uk/cgi-bin/wtutor?tutorial=t-log

CochSim is a cochlear simulation useful for some of the topics in Chapter 12.

An example of a one-term course using the first edition of this book (which needs little revision for this newer edition) can be found at: http://www.phon.ucl.ac.uk/courses/spsci/sigsys/

References

Baken RJ (1987) *Clinical Measurement of Speech and Voice* (Taylor & Francis Ltd: London).

Békésy G von (1949) On the resonance curve and the decay period at various points on the cochlear partition. *Journal of the Acoustical Society of America* **21**: 245–254. Also reprinted in G von Békésy (1960) *Experiments in Hearing*, pp. 446–460 (McGraw-Hill).

Békésy G von (1960) Experimental models of the cochlea with and without nerve supply. In *Neural Mechanisms of the Auditory and Vestibular Systems* GL Rasmussen and WF Windle (eds) pp. 3–20 (Thomas: Springfield, Illinois).

Borden GJ, Harris KS (1980) *Speech Science Primer: Physiology, Acoustics and Perception of Speech* (Williams & Wilkins: Baltimore).

Bracewell RN (1986) *The Fourier Transform and its Applications*, 2nd edition, revised (McGraw-Hill: New York).

Denes PB, Pinson EN (1973) *The Speech Chain: The Physics and Biology of Spoken Language* (Anchor Books: Garden City, New York).

Evans EF (1975) Cochlear Nerve and Cochlear Nucleus. In *Handbook of Sensory Physiology: Auditory System* WD Keidel and WD Neff (eds) pp. 1–108 (Springer Verlag: Berlin).

Fant GM (1960) *Acoustic Theory of Speech Production* (Mouton: 's-Gravenhage).

Flanagan JL (1972) *Speech Analysis, Synthesis and Perception* (Springer-Verlag: Berlin).

Fry DB (1979) *The Physics of Speech* (Cambridge University Press: Cambridge).

Fujimura O, Lindqvist J (1971) Sweep-tone measurements of vocal-tract characteristics. *Journal of the Acoustical Society of America* **49**: 541–558.

Gelfand SA (1981) *Hearing: An Introduction to Psychological and Physiological Acoustics* (Marcel Dekker: New York).

Guinan JJ, Peake WT (1967) Middle ear characteristics of anesthetized cats. *Journal of the Acoustical Society of America* **41**: 1237–1261.

Haggard MP, Evans EF (eds) (1987) Hearing. *British Medical Bulletin* 43 (Churchill Livingstone: Edinburgh).

Hartmann WM (1998) *Signals, Sound, and Sensation* (American Inst. of Physics).

Hirano M (1981) *Clinical Examination of Voice* (Springer-Verlag: New York).

Ladefoged P (1962) *Elements of Acoustic Phonetics* (University of Chicago: Chicago).

Ladefoged P (1975) *A Course in Phonetics* (Harcourt Brace Jovanovich: New York).

Lathi BP (1965) *Signals, Systems and Communication* (Wiley: New York).

Lynn PA, Fuerst W and Thomas B (1997) *Introductory Digital Signal Processing with Computer Applications* (Wiley: Chichester, England).

McGillem CD, Cooper GR (1984) *Continuous and Discrete Signal and System Analysis*, 2nd edition (CBS Publishing Japan Ltd: New York).

Millhouse TJ (2009). *Auditory-acoustic characterisation of variability in the operatic vowel*, PhD thesis, University of Sydney, Australia.
Millhouse TJ, Kenny DT (2008). Vowel placement during operatic singing: 'Come si parla' or 'aggiustamento'? *Proceedings of the International Conference of Spoken Language Processing* (ICSLP 2008), Brisbane, Australia.
Møller AR (1972) *Horselns fysiologi* (Fysiologiska institutionen II, Karolinska Institutet: Stockholm, Sweden).
Moore, BCJM (ed.) (1986) *Frequency Selectivity in Hearing* (Academic Press: London).
Olive JP, Greenwood A and Coleman J (1993) *Acoustics of American English Speech: A Dynamic Approach* (Springer: New York).
O'Shaughnessy D (1987) *Speech Communication: Human and Machine* (Addison-Wesley: Reading, Massachusetts).
Pickett JM (1998) *The Acoustics of Speech Communication: Fundamentals, Speech Perception Theory, and Technology* (Allyn & Bacon).
Pickles JO (2008) *An Introduction to the Physiology of Hearing*, 3rd edition (Emerald: Bingley).
Plack C (2005) *The Sense of Hearing* (Erlbaum).
Puria S, Peake WT and Rosowski JJ (1997) Sound-pressure measurements in the cochlear vestibule of human-cadaver ears. *Journal of the Acoustical Society of America* **101**: 2754–2770.
Rhode WS (1971) Observations of the vibration of the basilar membrane in squirrel monkeys using the Mössbauer technique. *Journal of the Acoustical Society of America* **49**: 1218–1231.
Rosen S, Fourcin AJ (1986) Frequency selectivity and the perception of speech. In *Frequency Selectivity in Hearing*. BCJ Moore (ed.), pp. 373–487 (Academic Press: London).
Shadle CH (1985) The acoustics of fricative consonants. *Research Laboratory of Electronics Technical Report 506* (MIT: Cambridge, Massachusetts).
Shaw EAG (1976) Transformation of sound pressure level from the free field to the eardrum in the horizontal plane. *Journal of the Acoustical Society of America* **56**: 1848–1861.
Siebert WMcC (1986) *Circuits, Signals, and Systems* (MIT Press: Cambridge, Massachusetts).
Wiener FM, Ross DA (1946) The pressure distribution in the auditory canal in a progressive sound field. *Journal of the Acoustical Society of America* **18**: 401–408. Also in: ED Schubert (ed.) (1979) *Psychological Acoustics*, pp. 32–39 (Dowden, Hutchinson & Ross: Stroudsberg, Pennsylvania).
Yost WA (2006) *Fundamentals of Hearing: An Introduction*, 5th edition (Emerald: Bingley).

Index

ted in the United States
3ookmasters